THE CHEMISTRY
OF THE ATMOSPHERE
AND OCEANS

THE CHEMISTRY OF THE ATMOSPHERE AND OCEANS

HEINRICH D. HOLLAND

**Department of Geological Sciences
Harvard University**

A WILEY-INTERSCIENCE PUBLICATION

JOHN WILEY & SONS, New York • Chichester • Brisbane • Toronto

Library of Congress Cataloging in Publication Data

Holland, Heinrich D.
    The chemistry of the atmosphere and oceans.

    1. Atmospheric chemistry—Collected works.
2. Chemical oceanography—Collected works.
I. Title.
QC879.6.H64   551.5'11'08   77-28176

    ISBN 0-471-03509-2

Printed in the United States of America

10 9 8 7 6 5 4 3 2 1

To the memory of my father

OTTO HOLLAND

1900–1969

Do not go gentle into that good night...

PREFACE

The atmosphere and the oceans are parts of an extremely complicated chemical system. A considerable effort has been made, particularly during the past 30 years, to understand the operation of this system and to establish a set of differential equations describing the transfer of major and minor components among the system's various parts. The results of this effort are still incomplete; the present volume is therefore in the nature of a progress report.

Beginning with a discussion of the earth as a chemical system, Chapter I discusses the notion of reservoirs, of transfers between reservoirs, characteristic times, and some rather simple equations that relate the quantity of components in reservoirs to the time dependence of their inputs and outputs in systems that do, and in those that do not, include feedback mechanisms. Examples are chosen to illustrate the operation of rather simplified aspects of the real world. Chapter II breaks into the exogenic geochemical cycle with a discussion of chemical weathering, the major process by which CO_2 and O_2 have been removed from the atmosphere. The rate of weathering depends on the operation of the hydrologic cycle. Chapter III describes the present relationships among evaporation, rainfall, and runoff from the continents, and shows that continental runoff has probably varied severalfold during geologic time. Potential effects of the probable variations in runoff are examined in Chapter IV. The concentration of the major dissolved constituents of river water is shown to vary inversely with respect to runoff, so that the variability in the quantity of total dissolved solids annually delivered to the oceans is much smaller than the variability in runoff itself. The relationship between runoff and concentration is not the same for all the major ions in river water; the proportion of the major constituents of world average river water is, therefore, somewhat dependent on the value of the average annual world river runoff.

The response of the oceans to the world river flux and to other inputs is explored in Chapter V. The variety of mechanisms by which the oceans dispose of river inputs is impressive. Near-equilibrium seems to prevail in some processes, for example those involving cation exchange and the precipitation and dissolution of $CaCO_3$. The system is wildly out of equilibrium for components such as the N_2–O_2 content of the atmosphere and the NO_3^-–H^+ content of seawater. Kinetics play a major role in the removal of numerous components, sulfur and phosphorus, for example, from the oceans; seawater cycling through the oceanic crust, mainly in the vicinity of mid-ocean ridges, is probably important for the marine budget of Mg^{2+}. The quantitative functional relationships between the concentration of ions in seawater and their rate of removal from the oceans are still somewhat uncertain. It seems virtually certain, however, that the composition of seawater has changed rather little during the past 600 million years and that the ocean–atmosphere system has been reasonably stable in response to the internal driving forces of the earth and to the external, largely solar, driving forces.

The atmosphere is similar to the oceans in that nearly all of its components are cycled through in geologically short periods of time. The variety of processes that control the partial pressure of many of the atmospheric constituents are examined in Chapter VI. P_{CO_2} has been roughly controlled by mineralogical equilibria; P_{O_2} has been determined largely by the balance between the rate of photosynthetic oxygen production and the rate of oxygen loss via oxidation reactions in the oceans and on land.

The history of the ocean–atmosphere system will be discussed in a companion volume to this book. Until recently a single volume had been planned to cover both the present state of the system and its history. A first draft of much of this volume was written during a sabbatical leave at the University of Hawaii during 1968–1969. The completion of the draft was delayed by the usual press of other work on my return to Princeton, and little progress was made after my move to Harvard until my sabbatical leave in 1975–1976. During that year the first draft was nearly completely rewritten, in large part because advances during the intervening years had made much of the earlier version obsolete. The decision to rewrite extensively was surely correct, but the end of the sabbatical year again left me with an incomplete text. A second rewriting several years hence seemed unattractive, and, since John Wiley & Sons was agreeable to the publication of two volumes, the first, dealing with the present state of the system, became a separate entity.

This volume owes a great deal to many people. Its origins can be traced to my doctoral dissertation under J. L. Kulp at Columbia University on the uranium, ionium, and radium content of the oceans and marine sediments. It received considerable stimulus from H. C. Urey while we were both at Oxford University during 1956–1957. Continuing discussions with R. M. Garrels over a period of 20 years have been extremely instructive and stimulating. Many of my colleagues at Princeton, particularly A. G. Fischer, D. J. J. Kinsman, K. S. Deffeyes, S. Judson, S. Manabe, and K. Bryan, as well as K. Chave and R. Moberly at the University of Hawaii and M. B. McElroy at Harvard, have had an important impact on the development of the book. L. G. Sillén has left his mark on my work as he has on the work of so many geochemists interested in the ocean–atmosphere system, and W. S. Broecker's cheerful skepticism has served as an important antidote for much premature optimism.

Several postdoctoral fellows and a number of former students have contributed a great deal by their data and ideas. Among the latter J. I. Drever, K. S. Russell, A. C. Lasaga, and M. J. Mottl have been particularly important. R. F. Quirk's talent for careful organization, and P. Solomon's patience through the final versions of the manuscript were essential for the completion of the book.

I am particularly indebted to J. I. Drever, F. T. Mackenzie, S. Manabe, A. C. Lasaga, J. Thrailkill, K. S. Deffeyes, F. L. Sayles, P. B. Hostetler, R. Siever, M. B. McElroy, S. C. Wofsy, S. Judson, D. Langmuir, K. Bryan, A. G. Fisher, and R. M. Garrels for critical reviews of the manuscript.

The volume could not have been written without a great deal of moral support from my wife, and the generous financial support of the National Science Foundation, the University of Hawaii, the John Simon Guggenheim Memorial Foundation, and the Committee on Experimental Geology at Harvard University. To all of them I extend my sincere thanks and appreciation.

HEINRICH D. HOLLAND

Cambridge, Massachusetts
February 1978

CONTENTS

THE CHEMISTRY
OF THE ATMOSPHERE
AND OCEANS

CHAPTER I

THE EARTH AS A CHEMICAL SYSTEM

The history of the earth records a series of unique events, unidirectional processes, and cyclic processes. The accretion of the earth and the formation of the core are examples of unique events. The loss of helium from the atmosphere into interplanetary space and the decay of the long-lived radio-elements are examples of unidirectional processes. The geochemical cycle of Figure 1-1 and the hydrologic cycle, which is such an important contributor to near-surface chemical processes, are major examples of cyclic processes. Many of the important unique events occurred very early in earth history. Since most unidirectional processes are slow when compared to cyclic processes, the geologic record tends to be dominated by the consequences of the operation of geochemical cycles, and is still adequately described by James Huttons' famous observation: "We find no vestige of a beginning—no prospect of an end."

Much effort has therefore been spent on understanding the geochemical cycles of the elements. During the first half of this century V. M. Gold-schmidt and his students contributed a great deal to this undertaking. Their results are summarized in numerous diagrams of varying complexity, partic- ularly in *Geochemistry*, by Rankama and Sahama (1950). Since midcentury the emphasis has gradually shifted toward the application of physical chemistry to the study of geochemical processes, and to the quantification of cycling processes. Barth (1952) first used the concept of the characteristic time in the analysis of the chemistry of the oceans. Since then box models have been developed and applied with considerable success, particularly to the oceans and to the ocean–atmosphere system (see, for instance, Broecker, 1974). Ultimately, one would, of course, like to have at one's disposal a set of partial differential equations that describe the transfer of all of the elements

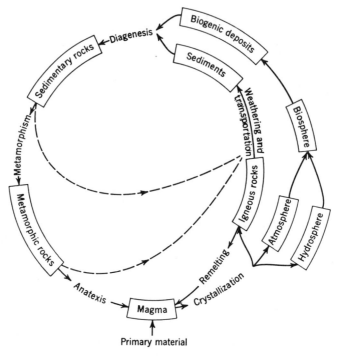

Figure 1-1. The geochemical cycle (Mason, 1966).

among the several parts of the earth and between the earth and interplanetary space, and that are capable of predicting the response of the entire system to changes in a variety of environmental parameters. Progress toward this end has been substantial. The size of reservoirs and their composition are reasonably well known, and most of the important sinks and sources that feed and drain the major reservoirs have been identified. Functional relationships among the state of the reservoirs, their inputs, and their outputs are, however, poorly understood, and a complete working model of the earth is still a rather distant goal. This chapter summarizes some of the concepts that have proved useful in describing reservoir behaviors and lays the ground work for the application of these concepts in later chapters of the book.

RESERVOIRS AND TRANSFER BETWEEN RESERVOIRS

Traditionally, the earth has been subdivided into "spheres": the atmosphere, biosphere, hydrosphere, and lithosphere. Each of these can be considered as a single reservoir, but for many purposes a subdivision into smaller spheres is useful. The exosphere is frequently separated from the remainder of the atmosphere when problems of planetary escape are discussed, and sea water above the thermocline is frequently separated from the deeper parts of the oceans when mixing of the oceans is considered. Separation of the lithosphere into geographic and time-related units has always been an important part of geology.

The definition of reservoirs depends on the problem to be solved and on the availability of data. If the chosen reservoirs are very large, they are usually heterogeneous. Treatments based on assumptions of reservoir homogeneity are therefore apt to be rather imprecise. If the chosen reservoirs are small, they are probably numerous, and the amount of data necessary to treat many-reservoir systems is difficult to obtain and frequently unavailable. Most geochemical earth models are therefore rather imprecise; nevertheless, many of the first-order models that have been developed are useful both in their own right and as starting points for second-generation models.

Reservoirs are either accumulative or nonaccumulative with respect to their constituent components.* If a component i moves from reservoir 1 to reservoir 2, as in Figure 1-2a, its quantity $_2n_i$ in reservoir 2 is given by the expression

$$_2n_i = {}_2^0n_i + \int_0^t \frac{d_{1,2}n_i}{dt}\,dt \qquad (1\text{-}1)$$

where $_2^0n_i =$ the quantity of component i in reservoir 2 at $t=0$, and $d_{1,2}n_i/dt =$ the rate of flow of component i from reservoir 1 to reservoir 2. If reservoir 2 contained no component i initially, then

$$_2n_i = \left(\overline{\frac{d_{1,2}n_i}{dt}} \right)\Delta t \qquad (1\text{-}2)$$

where $\overline{d_{1,2}n_i/dt}$ is the mean rate of accumulation of component i in reservoir 2 during the time interval Δt. It follows that

$$\frac{_2n_i}{\overline{d_{1,2}n_i/dt}} = \Delta t \qquad (1\text{-}3)$$

i.e., that the time Δt during which reservoir 2 has been accumulating component i equals the total amount of component i in the reservoir at time t divided by the mean input rate. This ratio is called the characteristic time (see Barth, 1952). If it is known that a system has been accumulating a component throughout the history of the earth, then the characteristic time for the component has a value near 4.5×10^9 yr. On the other hand, if the

---

*A component in the sense used here is a chemical element or a compound.

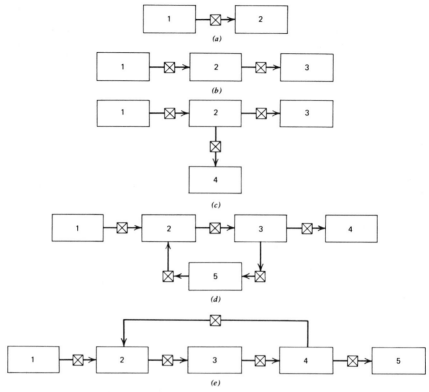

Figure 1-2. Some simple arrangements of reservoirs and material transfer among them (for discussion see text).

characteristic time turns out to have a much smaller value, then the values chosen for $_2n_i$, $\overline{d_{1,2}n_i/dt}$, or both are wrong, or the reservoir is nonaccumulative with respect to component i.

The atmosphere seems to be nearly accumulative for all of the rare gases except helium and xenon. On the other hand, the characteristic times of the major cations in the oceans are rather different and much smaller than the age of the earth (see Chapter V); the oceans are therefore nonaccumulative reservoirs for all of the major cations (Barth, 1952).

Reservoir 2 in Figure 1-2b is nonaccumulative. The amount of component i in the reservoir therefore depends on its rate of influx and on its rate of outflow. Since

$$\frac{d_2n_i}{dt} = \frac{d_{1,2}n_i}{dt} - \frac{d_{2,3}n_i}{dt} \tag{1-4}$$

it follows that

$$_2n_i = {_2^0}n_i + \int_0^t \left(\frac{d_{1,2}n_i}{dt} - \frac{d_{2,3}n_i}{dt} \right) dt \tag{1-5}$$

If the system is in dynamic equilibrium, the rate of influx equals the rate of outflow, and the quantity $_2n_i$ is constant with time. In this case the characteristic time $_2\tau_i$, defined by the expressions

$$\frac{_2n_i}{d_{1,2}n_i/dt} = \frac{_2n_i}{d_{2,3}n_i/dt} = {_2\tau_i} \tag{1-6}$$

is the time required for $_2n_i$ mol of component i to pass through reservoir 2. Therefore, $_2\tau_i$ is the mean residence time of component i in reservoir 2 when the reservoir is in steady state with respect to component i. The existence of a steady state always requires proof, and the validity of residence time calculations depends as much on the quality of this proof as on the parameters used in the calculation of τ.

In general, earth reservoirs are not at steady state. Climatic variations during the past 10^6 yr have undoubtedly affected the concentration of components of the atmosphere and oceans whose characteristic time in these reservoirs is $\leqslant 10^6$ yr. Tectonic cycles span longer periods of time and must influence the concentration of even the longest-lived components of the atmosphere and oceans. Functions that describe the response of the quantities $_jn_i$ to changes in the input rates and to the effects of environmental parameters on the relationship among the $_jn_i$'s and the output rates $(d_{j,(j+1)}n_i)/dt$ are therefore vital ingredients of earth models.

NONSTEADY STATE RESERVOIRS WITHOUT FEEDBACK

The variation of the quantity of a component in a given reservoir depends on the time variation of the input rate and of the output rate. Two cases should be distinguished: the simpler, where the inputs to a given reservoir are independent of the state of the reservoir, and the more complex, where the inputs are not independent of the state of the reservoir. In the first case

$$\frac{d_jn_i}{dt} = {_jf_i}(t) - {_jk_i}g({_jn_i}) \tag{1-7}$$

where $_jk_i$ may be a complicated function of the quantity of the other components in reservoir j. A particularly simple example is given by the case in which the input rate is constant for a period of time long enough for the

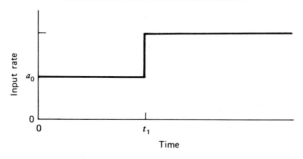

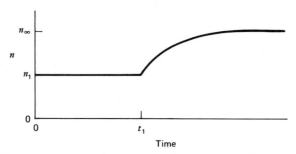

Figure 1-3. The response of the quantity n of a component in a reservoir whose behavior is governed by Equations 1-9a and 1-9b.

system to reach steady state and then suddenly changes to a new, constant level as shown in Figure 1-3, where

$$_jf_i(t) = a_0 \quad 0 < t \leqslant t_1 \tag{1-8a}$$

and

$$_jf_i(t) = a_1 \quad t_1 < t. \tag{1-8b}$$

If the output rate is proportional to the quantity of component i in the reservoir, then, dropping subscripts yields

$$\frac{dn}{dt} = a_0 - kn; \quad 0 < t \leqslant t_1 \tag{1-9a}$$

and

$$\frac{dn}{dt} = a_1 - kn; \quad t_1 < t \tag{1-9b}$$

Since the interval between 0 and t_1 has been taken to be long enough for steady state to be attained, it follows that

$$n = n_1 = \frac{a_0}{k}; \qquad t = t_1 \tag{1-10}$$

and that

$$n = n_\infty - (n_\infty - n_1)e^{-k(t-t_1)}; \qquad t > t_1 \tag{1-11}$$

where n_∞ is the value of n at its new steady state value

$$n_\infty = \frac{a_1}{k} \tag{1-12}$$

The rate of approach to the new equilibrium state is determined by the value of the constant k. At the new steady state

$$\frac{n_\infty}{a_1} = \frac{1}{k} = \tau \tag{1-13}$$

as defined by equation 1-6. In this particular instance there is, therefore, a simple, direct relationship between the characteristic time τ at steady state and the response time of the system to sudden changes in the input rate.

The input rate can, of course, be a more complicated function of time, and the output rate is usually a more complicated function of n. If, for example, the input has a sinusoidal form

$$f(t) = a + b \sin \omega t \tag{1-14}$$

so that

$$\frac{dn}{dt} = a + b \sin \omega t - kn \tag{1-15}$$

then

$$n = \left(n_0 - \frac{a}{k} + \frac{\omega}{k^2 + \omega^2} \right) e^{-kt} + \frac{a}{k} + \frac{1}{k^2 + \omega^2} (k \sin \omega t - \omega \cos \omega t) \tag{1-16}$$

where n_0 is the value of n when $t = 0$. The first term on the right side of equation 1-16 represents the approach to a steady state, in which the value of n fluctuates about the mean value a/k with a frequency ω and a phase which is different from that of the driving function $f(t)$.

The ^{14}C content of the atmosphere and the reservoirs that exchange carbon readily with the atmosphere follow equations such as 1-11 and 1-16. If the rate of ^{14}C production is constant for a period of time that is long when compared to the half-life of ^{14}C (5600 yr), the ^{14}C content of the exchange reservoirs is essentially constant. Sudden changes in the production rate produce a gradual approach to the new steady state as in equation 1-11. The onset of a sinusoidally varying production rate produces a ^{14}C inventory that varies with time as defined by equation 1-16.

Several other concentration parameters in the atmosphere–ocean system respond in a similar fashion. An increase in the NaCl content of the oceans probably produces a nearly proportional increase in the rate of removal of NaCl via evaporite formation, given the same amount of evaporation to dryness in evaporite basins (see Chapter V). Similarly, the output of SO_4^{2-} from the oceans via reduction accompanied by pyrite formation seems to be nearly proportional, within limits, to the SO_4^{2-} content of sea water, provided all other parameters remain constant (see Chapter V). The rate of Mg^{2+} loss from the oceans via sea water cycling through mid-ocean ridges is probably proportional, again within limits, to the Mg^{2+} content of seawater, all other parameters remaining constant (see Chapter V). The assumption that outputs are proportional to the concentration of a component in a reservoir can therefore be justified in quite a few instances, and the behavior of the concentration of such a component in response to a variety of inputs can be computed either in closed form or by stepwise integration.

The output of most components of most reservoirs is, however, a more complicated function of their concentration. In some cases output is proportional to the concentration taken to a power greater than or smaller than 1. The removal rate of CO_2 via photosynthesis becomes positive at values of P_{CO_2} in excess of the compensation point, approaches a nearly constant value at higher values of P_{CO_2}, and decreases at even higher values of P_{CO_2} (see Chapter II). The rate of oxygen use during weathering is also a nonlinear function of P_{O_2}. With increasing atmospheric P_{O_2} the removal rate of oxygen via surface oxidation reactions approaches a nearly constant value, which is determined largely by the total rate of rock weathering and by the content of elemental carbon, Fe^{2+}, and S^{2-} in the rocks undergoing weathering (see Chapter VI).

Closed solutions exist for a number of the differential equations in which the exponent of the n_i's is not unity. If, for example, the output rate is proportional to the square of the quantity of a component in a reservoir, if the input rate is constant, and the initial value of n is zero, then

$$\frac{dn}{dt} = a - bn^2 \tag{1-17}$$

and

$$n = \left(\frac{a}{b}\right)^{1/2} \left[\frac{\exp\left(2(ab)^{1/2}t\right) - 1}{\exp\left(2(ab)^{1/2}t\right) + 1} \right] \tag{1-18}$$

Gradually, n approaches the steady state value $(a/b)^{1/2}$. The rate of approach is, however, no longer quite so simply related to the function $\tau = n_\infty/a = 1/(ab)^{1/2}$. This is true also if the exponent of n is not 2, or if the output rate varies exponentially with respect to n or with respect to the departure of n from a constant value. Thus, if

$$\frac{dn}{dt} = a - be^{kn} \tag{1-19}$$

then

$$n = \frac{1}{k} \ln \frac{a}{b + ce^{-akt}} \tag{1-20}$$

where

$$c = ae^{-kn_0} - b \tag{1-21}$$

As t approaches infinity, n approaches the steady state value $(1/k)\ln a/b$. Again, the rate of approach to steady state is not simply related to the characteristic time, τ.

In many instances the output rate of a particular component from a reservoir is determined by the presence of another component. The inorganic precipitation of $CaCO_3$ and the dissolution of $CaCO_3$ in the oceans below the lysocline depend both on the activity of calcium and on the activity of carbonate (see Chapter V). The photosynthetic production of organic matter in the oceans is limited not by the availability of CO_2 or H_2O but by the availability of nitrate and/or phosphate (see Chapter V). The removal of sulfur from the oceans as a constituent of pyrite is frequently limited by the availability of digestible organic matter and sometimes by the availability of iron in marine sediments (see Chapter V). Any set of differential equations which seeks to describe the operation of the real ocean–atmosphere system must therefore contain numerous cross terms. Few of these are well known. The linking of geochemical cycles via these cross terms apparently tends to stabilize the system, and this may account, in part, for the relatively slow and minor excursions in the chemistry of the oceans during the Phanerozoic era (Holland, 1972).

The transfer of elements between reservoirs is rarely represented adequately by diagrams such as those of Figures 1-2a and 1-2b. Normally, more than a single source feeds into and drains a particular reservoir. In Figure 1-2c reservoir 2 is fed from a single source and drains into two sinks. At steady state

$$\frac{d_2 n_i}{dt} = 0 = \frac{d_{1,2} n_i}{dt} - \frac{d_{2,3} n_i}{dt} - \frac{d_{2,4} n_i}{dt} \tag{1-22}$$

The ratios $_2 n_i / (d_{1,2} n_i / dt)$, $_2 n_i / (d_{2,3} n_i / dt)$, and $_2 n_i / (d_{2,4} n_i / dt)$ are normally all different and have different meanings. The first of these ratios describes the overall residence time of element i in reservoir 2. The second and third ratios represent the residence time with respect to removal into reservoirs 3 and 4, respectively, and are somewhat analogous to the individual half-lives of radioisotopes, such as ^{40}K, with branching decay schemes. In the general case of dynamic equilibrium between j inputs into and k outputs from a reservoir h

$$\frac{_h n_i}{\Sigma_j (d_{j,h} n_i / dt)} = \frac{_h n_i}{\Sigma_k (d_{h,k} n_i / dt)} = {_h \tau_i} \tag{1-23}$$

The characteristic times $_{j,h} \tau_i$ and $_{h,k} \tau_i$ for the individual inputs and outputs will, of course, be larger than $_h \tau_i$.

The removal of most elements from the oceans involves more than one sink (see Chapter V, Table 14). Mg^{2+}, for example, is removed as a constituent of carbonates and evaporite minerals, on cation exchange sites, and probably by the formation of smectites and chlorite during the cycling of seawater through the oceanic crust. Sulfur is removed as a constituent of gypsum and anhydrite, and as a constituent of pyrite. Carbon is removed as a constituent of organic matter and of carbonates. Fluctuations in the relative importance of the various removal mechanisms of sulfur and carbon leave a record in the isotopic composition of these elements in marine evaporites and carbonates, and this record can be used to define the dynamics of their behavior in the oceans (see, for instance, Holland, 1973: Junge et al., 1975). The various output rates of the elements are normally different functions of their quantity in each reservoir; the real differential equations governing the operation of the entire system are therefore considerably more complicated than those discussed above.

NONSTEADY STATE RESERVOIRS WITH FEEDBACK

The geochemical cycle of a number of elements and compounds involves feedback loops similar to that shown in Figure 1-2d. Consider, for instance, the cycle of sodium (see Chapters II, IV, and V). Sodium entering the oceans is the sum of the sodium released by weathering and the sodium returned through the atmosphere, largely in the form of sea salt. If reservoir 1 represents the crustal reservoir, reservoir 2 the eroding rocks, reservoir 3 the oceans, reservoir 5 the atmosphere, and reservoir 4 new sediments, then at steady state

$$\frac{d_{2,3}n_{Na}}{dt} = \frac{d_{3,5}n_{Na}}{dt} + \frac{d_{3,4}n_{Na}}{dt} \tag{1-24}$$

and

$$\frac{d_{2,3}n_{Na}}{dt} = \frac{d_{1,2}n_{Na}}{dt} + \frac{d_{3,5}n_{Na}}{dt} \tag{1-25}$$

thus

$$\frac{d_{1,2}n_{Na}}{dt} = \frac{d_{3,4}n_{Na}}{dt} \tag{1-26}$$

i.e., the flux of sodium from the oceans into new sediments at steady state is equal to the rate of sodium released by weathering.

The CO_2 cycle involves a similar return through the atmosphere. Let reservoir 1 in Figure 1-2e represent volcanic reservoirs of CO_2, reservoir 2 the atmosphere, reservoir 3 the rivers, reservoir 4 the ocean, and reservoir 5 the ocean sediments. The major part of CO_2 released from volcanoes enters the atmosphere, reacts with surface rocks, and goes into solution in ground waters, largely as HCO_3^-. From there it passes down rivers into the oceans. A good deal of CO_2 is released there during the precipitation of $CaCO_3$, and returns to the atmosphere to participate in another weathering cycle (see Chapters II, V, and VI).

Feedback processes also play a decisive role in controlling the oxygen content of the atmosphere (see Chapter VI). The major oxygen loss from the atmosphere appears to be the oxidation of elemental carbon, S^{2-}, and Fe^{2+} in rocks exposed to weathering. Most oxygen is produced as a by-product of the burial of organic matter at sea. The rate of burial of organic matter depends on the quantity of organic matter produced annually and on the fraction of the organic matter preserved in marine sediments. This fraction depends on the oxygen content of the atmosphere. The higher the atmospheric P_{O_2}, the larger the oxygen content of seawater, and hence the smaller the fraction of organic matter that is preserved in marine sediments. The exact functional relationship between P_{O_2} and the fraction of organic matter preserved is not known, but Figure 1-4 is probably a fair representation of the system (see Chapter VI). The oxygen input as well as the oxygen output depend on the oxygen content of the atmosphere. The difference

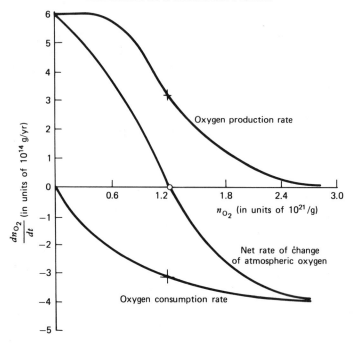

Figure 1-4. Semischematic diagram illustrating the relationship between the oxygen content of the atmosphere, the rate of oxygen production, the rate of oxygen consumption, and the net rate of change of atmospheric oxygen (see Chapter VI).

between the rate of production and the rate of consumption is positive at low values of P_{O_2} and negative at high values of P_{O_2}. The crossover point is a stable control point, since any perturbation away from steady state is compensated so as to bring P_{O_2} back to the value at the crossover point. Both the production and consumption curves depend on a variety of biologic and environmental parameters, and have probably shifted through geologic time. The crossover point has therefore almost certainly shifted as well, and the oxygen pressure has tended to track the movement of the steady state value.

Near the crossover point the response curve can be represented approximately by a straight line, so that

$$\frac{dn_{O_2}}{dt} \cong a - bn_{O_2} \qquad (1\text{-}27)$$

The response of n_{O_2} to changes in the crossover point due to an increase or decrease in the value of a is determined, as in systems following equation

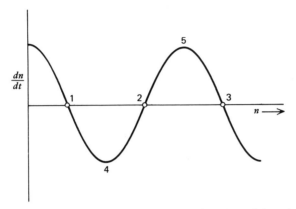

Figure 1-5. A feedback system with three crossover points; two of these (points 1 and 3) represent stable control points, the third (point 2) does not.

1-9, by the term e^{-bt}. The steeper the slope of the oxygen response curve, the more rapid is the adjustment of the oxygen content of the atmosphere to fluctuations in the parameter a.

If a system exhibits more than one crossover point, not all of them represent stable control points. In Figure 1-5, for example, crossover points 1 and 3 are stable control points but point 2 is not. Any perturbation of the system away from point 2 moves the system either to point 1 or to point 3. Small perturbations about points 1 and 3 are compensated by the system. Large perturbations from point 1 in the direction of point 2 may move the system to point 3, and large perturbations from point 3 in the direction of point 1 may move the system to point 1. The ease and frequency with which such a system oscillates between two stable control points depend on the frequency of the disturbances which are large enough to place the system beyond point 2 in Figure 1-5.

THE MAJOR DRIVING FORCES

The chemistry of the atmosphere and oceans is driven by two major sources of energy: radioactive decay within the earth, and the consequences of nuclear reactions in the sun. The influence of tidal forces, meteorite infall, and the flux of non-solar cosmic rays is minor today. Solar energy largely drives the exogenic cycle, terrestrial energy largely drives the endogenic cycle, and the history of the earth has been determined in large part by these cycles and by their interaction.

Life apparently began to complicate the chemistry of the earth more than 3 billion years ago. Many living organisms play and have played the role of catalysts for sluggish chemical reactions in surface and near-surface environments. However, the invention of the production of free oxygen as a by-product of photosynthesis generated additional chemical disequilibria among the atmosphere, the earth's crust, and many sediments. The need to eliminate these disequilibria opened a large number of new biologic niches, and has contributed a great deal toward the present complexity of the earth as a chemical system. This complexity is explored in the succeeding chapters of the book; its history will be treated in a subsequent volume.

REFERENCES

Barth, T. F. W., 1952, *Theoretical Petrology*, Wiley, New York, 387 pp.

Broecker, W. S., 1974, *Chemical Oceanography*, Harcourt, Brace, Jovanovich, New York, 214 pp.

Holland, H. D., 1972, The geologic history of sea water—An attempt to solve the problem *Geochim. Cosmochim. Acta* **36**, 637–651.

Holland, H. D., 1973, Systematics of the isotopic composition of sulfur in the oceans during the Phanerozoic and its implications for atmospheric oxygen, *Geochim. Cosmochim. Acta* **37**, 2605–2616.

Junge, C. E., Schidlowski, M., Eichmann, R., and Pietrek, H., 1975, Model calculations for the terrestrial carbon cycle: Carbon isotope geochemistry and evolution of photosynthetic oxygen, *J. Geophys. Res.* **80**, 4542–4552.

Mason, B., 1966, *Principles of Geochemistry*, 3rd ed., Wiley, New York, 329 pp.

Rankama, K. and Sahama, Th. G., 1950, *Geochemistry*, University of Chicago Press, Chicago, Ill., 911 pp.

CHAPTER II

WEATHERING AND OTHER NEAR-SURFACE REACTIONS

The interface between the atmosphere and the surface of the continents is chemically very reactive. Carbon, nitrogen, and water are fixed, and oxygen is released photosynthetically. Respiration and decay reverse these processes. Rainwater dissolves certain minerals and catalyzes the oxidation and carbonation of others. The remains of biological processes, solutes derived from the dissolution and decomposition of minerals, and solid weathering residues are ultimately transported to the oceans. As a consequence of these events, net quantities of water, CO_2, N_2, and probably O_2 are removed from the atmosphere. These removal rates are complicated functions of numerous biological, physical, and chemical parameters. Although the parameters themselves are now reasonably well understood, many of the functional relationships are not. Any system of equations that attempts to describe the interaction of the atmosphere with continental surfaces is, therefore, still rather tentative. This chapter is devoted to a description of the most important of the near-surface processes and to a discussion of the status of their quantification.

SIMPLE DISSOLUTION REACTIONS

Processes that simply involve the solution of minerals in water without disproportionation or the participation of other compounds are relatively rare. Quartz is probably the most important mineral to dissolve in this fashion. Its solubility at $25\,^{\circ}C$ is in the vicinity of 6 ppm. Most ground waters contain more dissolved silica than this, because most of the silica in ground waters, and hence in rivers, is derived not from the solution of quartz but

15

from the weathering of silicate minerals, such as the feldspars, into clays. The rate of silica release during such reactions is discussed below.

The important evaporite minerals—halite (NaCl), anhydrite ($CaSO_4$), and gypsum ($CaSO_4 \cdot 2H_2O$)—dissolve simply. The solubility of NaCl is 357 and 391 g/1000 ml of H_2O at 0° and 100°C, respectively. In an area where the runoff is equal to the world average of 30 cm/yr and is saturated with respect to NaCl, the rate of erosion of an outcrop of NaCl is about 11 g/cm^2 yr, that is, some three orders of magnitude more rapid than the erosion rate of silicate rocks. Preferential erosion of halite in such an area would, therefore, continue until the access of surface waters to salt becomes restricted, or until the entire body of salt is removed by subsurface leaching. The solubility of gypsum and anhydrite in surface and ground waters is between 2 and 3 g/1000 of H_2O (Blount and Dickson, 1973). In an area with a runoff of 30 cm/yr, dissolution rates of outcrops of these minerals exceed normal erosion rates by about one order of magnitude. Both minerals would, therefore, behave much like halite but to a less dramatic extent. It is not surprising, then, that gypsum, anhydrite, and particularly halite crop out only in arid areas where their rate of solution is not much greater than the rate of erosion of the rock units surrounding the evaporites.

CARBONATION REACTIONS

The Weathering of Carbonate Rocks

The weathering of carbonate rocks is now well understood. The chemistry of carbonate equilibria has been known for a number of decades, and its application to geologic processes at earth surface temperatures is discussed extensively by Garrels and Christ (1965), Berner (1971), and Stumm and Morgan (1970). Data pertaining to carbonate equilibria at higher temperatures have been summarized by Holland and Malinin (1978).

The concentration, m_{CO_2}, of CO_2 dissolved in rain water is proportional to P_{CO_2}, the partial pressure of CO_2 in the atmosphere, for all geologically reasonable values of that pressure.

$$m_{CO_2} = BP_{CO_2}. \qquad (2\text{-}1)$$

The value of B decreases with increasing temperature up to about 175°C (Ellis, 1959). Values of B in the range 0–50°C are listed in Table 2-1. Although CO_2 dissolved in water is largely unhydrated (see Garrels and Christ, 1965, p. 76), the symbol $m_{H_2CO_3}$ is retained to express the sum of the concentration of hydrated and unhydrated CO_2 in solution. The first ioniza-

Table 2-1

The Value of Some Constants of Importance
in Carbonate Equilibria; $pK_i = -\log K_i$

Temperature (°C)	$pK_B{}^a$	$pK_1{}^b$	$pK_2{}^b$	$pK_C{}^c$	$pK_D{}^c$
0	1.11	6.58	10.62	8.340	16.6
5	1.19	6.52	10.56	8.345	16.6
10	1.27	6.47	10.49	8.355	16.7
15	1.34	6.42	10.43	8.370	16.8
20	1.41	6.38	10.38	8.385	16.9
25	1.46	6.35	10.33	8.40±0.02	17.0±0.2
30	1.52	6.33	10.29	—	—
40	1.64	6.30	10.22	—	—
50	1.72	6.29	10.17	—	—

[a] Data from Harned and Davis (1943).
[b] Data from compilation of Garrels and Christ (1965), p. 89.
[c] Data from Langmuir (1971); see also Thrailkill (1972).

tion constant of carbonic acid, K_1,

$$K_1 = \frac{a_{H^+} \cdot a_{HCO_3^-}}{a_{H_2CO_3}}$$ (2-2)

and the second ionization constant, K_2,

$$K_2 = \frac{a_{H^+} \cdot a_{CO_3^{2-}}}{a_{HCO_3^-}}$$ (2-3)

both increase in the interval 0–50°C as shown in Table 2-1.

In solutions saturated with respect to calcite, the product of the activity of calcium and carbonate ions equals K_C, the solubility product of calcite.

$$K_C = a_{Ca^{2+}} \cdot a_{CO_3^{2-}}$$ (2-4)

Values of pK_C are assembled in Table 2-1. In solutions saturated with respect to stoichiometric dolomite

$$K_D = a_{Ca^{2+}} \cdot a_{Mg^{2+}} \cdot a_{CO_3^{2-}}^2$$ (2-5)

where K_D is the solubility product of dolomite.

If rain water falls into a depression in pure limestone and becomes saturated with respect to calcite, relations 2-1, -2, -3, and -4 are satisfied. Furthermore, the condition of electrical neutrality of the solution demands that

$$2m_{Ca^{2+}} + m_{H^+} = m_{HCO_3^-} + 2m_{CO_3^{2-}} + m_{OH^-} \tag{2-6}$$

Even at CO_2 pressures well below the present partial pressure of CO_2 in the atmosphere, equation 2-6 can be simplified without serious loss of accuracy (Garrels and Christ, 1965, Chap. III) by neglecting m_{H^+}, $m_{CO_3^{2-}}$, and m_{OH^-}, so that

$$2m_{Ca^{2+}} \cong m_{HCO_3^-} \tag{2-7}$$

If equation 2-2 is divided by equation 2-3, activities are replaced by concentrations*, BP_{CO_2} is substituted for $m_{H_2CO_3}$, $2m_{Ca^{2+}}$ for $m_{HCO_3^-}$, and $K_C/m_{Ca^{2+}}$ for $m_{CO_3^{2-}}$, it follows that

$$m_{Ca^{2+}}^3 \cong \frac{K_1 K_C B}{4K_2} P_{CO_2} \tag{2-8}$$

At 25°C and at the present value of atmospheric P_{CO_2} ($10^{-3.50}$ atm) $m_{Ca^{2+}}^3$ is, therefore, equal to

$$m_{Ca^{2+}}^3 \cong \frac{10^{-6.35} 10^{-8.40} 10^{-1.46}}{10^{0.60} 10^{-10.33}} 10^{-3.50}$$

$$\cong 10^{-9.98} \, (mol/kg)^3$$

Thus

$$m_{Ca^{2+}} \cong 10^{-3.33} \, mol/kg \cong 19 \, mg/kg$$

and

$$m_{HCO_3^-} \cong 2 \times 10^{-3.33} \, mol/kg \cong 57 \, mg/kg$$

As shown in Figure 2-1 the solubility of calcite increases toward lower temperatures and higher atmospheric CO_2 pressures. At a CO_2 pressure of

---

*This is permitted because the activity coefficients of the species in these solutions is very close to unity.

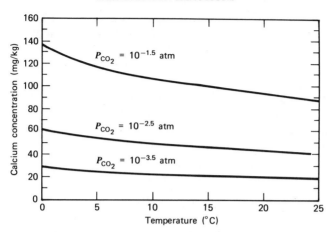

Figure 2-1. The concentration of calcium in pure water saturated with respect to calcite as a function of temperature and atmospheric CO_2 pressure; calculations based on data in Table 2-1.

$10^{-1.5}$ atm, that is, 100 times the present value, the calcium concentration is 87 mg/kg.

In an area where the runoff is equal to the world average of 30 cm/yr, the rate of dissolution of limestone at 25°C and at the present CO_2 pressure is approximately equal to

$$19 \ \mu g \, Ca/ml \times \frac{100 \ g/mol \ CaCO_3}{40 \ g/mol \ Ca} \times 30 \ ml/cm^2 \ yr = 1.4 \ mg/cm^2 \ yr$$

or 1.4 g/cm^2 1000 yr. This is only one-tenth of the mean rate of continental erosion today (see Chapter IV). Contrary to gypsum and halite, limestone should therefore crop out and be preferentially dissolved only in areas where rainfall is abnormally high or where the mean erosion rate is much less than the world average. This does not seem to be the case, largely because the effective CO_2 pressure during limestone dissolution is usually increased by the decay of vegetation in soils to values between 10 and 100 times the atmospheric CO_2 pressure.

Limestones are rarely pure $CaCO_3$. They frequently contain dolomite and a variety of silicate minerals. In rain water that is saturated with respect to dolomite alone

$$m^3_{Ca^{2+}} = m^3_{Mg^{2+}} \cong \frac{K_1 \sqrt{K_D} \ B}{16 K_2} P_{CO_2} \qquad (2\text{-}9)$$

At 25°C and in equilibrium with present-day atmospheric CO_2, concentrations of Ca^{2+} and Mg^{2+} calculated with the data of Table 2-1 are

$$m_{Ca^{2+}} = m_{Mg^{2+}} \cong 10^{-3.56} \text{ mol/kg}$$

which corresponds to 11 mg/kg of calcium and 6.7 mg/kg of magnesium.

Rain water which equilibrates with a carbonate rock consisting of both calcite and dolomite becomes saturated with respect to both phases. At saturation

$$\frac{K_C^2}{K_D} = \frac{a_{Ca^{2+}}^2 \cdot a_{CO_3^{2-}}^2}{a_{Ca^{2+}} \cdot a_{Mg^{2+}} \cdot a_{CO_3^{2-}}^2} = \frac{a_{Ca^{2+}}}{a_{Mg^{2+}}} \qquad (2\text{-}10)$$

Since most of such waters are very dilute, the activity ratio, $a_{Ca^{2+}}/a_{Mg^{2+}}$, is nearly equal to the concentration ratio, $m_{Ca^{2+}}/m_{Mg^{2+}}$. Thus in solutions saturated with respect to calcite and dolomite at 25°C

$$\frac{m_{Ca^{2+}}}{m_{Mg^{2+}}} \cong \frac{10^{-16.80 \pm 0.04}}{10^{-17.0 \pm 0.2}} = 10^{0.20 \pm 0.24} = 1.6^{+1.1}_{-0.6}$$

The kinetics of equilibration with respect to dolomite are quite slow, and this accounts for a good deal of the uncertainty of the value of K_D and in the calculated Ca/Mg ratio.

The solubility product of dolomite increases toward lower temperatures. Dolomite, like calcite, is, therefore, more soluble in cold than in warm climates. The solubility of both minerals is proportional to the cube root of the CO_2 pressure, so that an increase in P_{CO_2} from $10^{-3.50}$ to $10^{-0.50}$ atm increases the concentration of Ca^{2+} and Mg^{2+} in solutions saturated by dolomite by a factor of 10.

Weathering of silicate minerals in limestones or intercalated shales normally supplies cations in addition to Ca^{2+} and Mg^{2+}; these affect the concentration of calcium in solutions in equilibrium with calcite (Holland and Borcsik, 1965) and of calcium and magnesium in solutions saturated with dolomite. The electrical neutrality equation 2-6 can be written in the form

$$2m_{Ca^{2+}} + \sum_i z_i m_i = m_{HCO_3^-} + \sum_j z_j m_j \qquad (2\text{-}11)$$

where $\sum_i z_i m_i$ is the sum of the charge z_i times the concentration m_i of all cations except calcium, and where $\sum_j z_j m_j$ is the sum of the charge z_j times the concentration m_j of all anions except bicarbonate. Sodium, potassium, and magnesium are normally the major contributors to the term $\Sigma_i z_i m_i$; chloride and sulfate are normally the major contributors to the term $\Sigma_j z_j m_j$.

By following the steps outlined for the derivation of equation 2-8, it can be shown that in solutions saturated with calcite

$$m_{Ca^{2+}}\left[m_{Ca^{2+}} + \frac{\left(\sum\limits_i z_i m_i - \sum\limits_j z_j m_j\right)}{2}\right]^2 \cong \frac{K_1 K_C B}{4 K_2} P_{CO_2} \qquad (2\text{-}12)$$

This equation reduces to equation 2-8 when

$$\sum_i z_i m_i = \sum_j z_j m_j$$

From this point of view there are three kinds of aqueous solutions:

type 1: $\qquad \sum\limits_i z_i m_i > \sum\limits_j z_j m_j$

type 2: $\qquad \sum\limits_i z_i m_i = \sum\limits_j z_j m_j$

type 3: $\qquad \sum\limits_i z_i m_i < \sum\limits_j z_j m_j$

In solutions of type 1 the charge contribution of cations other than calcium exceeds that of anions other than bicarbonate. As shown in Chapter IV, the mean composition of river waters of the world belongs to this type. In such solutions the presence of the additional cations i and anions j reduces the concentration of calcium in solutions saturated with calcite. In solutions of type 2 the presence of the additional cations and anions affects the concentration of calcium only via their effect on the activity coefficient of Ca^{2+} and of HCO_3^- and via the formation of complex ions in solutions. In solutions of type 3 the charge contribution of the additional cations is less than that of the additional anions, and the calcium concentration is greater than in the absence of the additional ions. Ocean water is a type 3 solution, but the solubility relationships of $CaCO_3$ and dolomite in this medium are complicated by complexing and by the very considerable deviation of single ion activity coefficients from unity (see Chapter V).

The application of the equations developed above to the chemistry of surface and ground waters in carbonate terrains has shown that the concentration of CO_2 dissolved in many such waters is 10 to 100 times greater than that in solutions equilibrated with atmospheric CO_2. The data in Table 2-2 illustrate this point for ground water from carbonate terrains in central Pennsylvania. Similar results have been obtained in many other areas, including Virginia (Holland et al., 1964), Florida (Hanshaw, Back, and Rubin, 1965), and Kentucky (Thrailkill, 1970, 1972). Clearly, these solutions have absorbed a great deal of CO_2 between their arrival on the ground as rain water and their appearance in caves, springs, and wells. Most of the excess CO_2 is derived from soil air. The data of Tables 2-3 and 2-4 (Russell, 1961) show that the CO_2 content of the air in the surface layers of arable soil is usually between 0.15 and 0.65%, and that under tropical conditions the CO_2 content of soil air may be much greater (Lundegardh, 1927). The highest reported CO_2 content in Tables 2-3 and 2-4 is 11.5%. As the atmosphere contains a near-constant 0.03% of CO_2, soil air is normally enriched in CO_2 by a factor of 5 to 22, and may be enriched by a factor of as

Table 2-2

Chemical Data for Some Carbonate Ground Waters in Central Pennsylvania[a]

	Spring	Wells				
	1	59	66	332	SC-11	303
T(°C)	10.0	14.4	10.5	13.0	10.6	11.8
Ca^{2+}	47 mg/kg	91 mg/kg	55 mg/kg	56 mg/kg	41 mg/kg	75 mg/kg
Mg^{2+}	24	35	15	29	26	35
Na^+	3.8	3.1	2.5	1.0	1.4	5.7
K^+	—	0.3	1.8	0.8	1.3	1.6
HCO_3^-	234	321	195	263	228	352
SO_4^{2-}	14	36	12	17	12	26
Cl^-	10	37	15	4.7	5.1	32
NO_3^-	14	33	24	27	16	23
pH	7.63	7.19	7.45	7.55	7.41	7.16
Dissolved O_2 (% saturation)	—	73	85	102	77	30
$\log P_{CO_2}$ (atm)	−2.35	−1.76	−2.25	−2.21	−2.13	−1.70
$SI_C{}^d$	−0.04	−0.04	−0.21	+0.04	−0.31	−0.14
$SI_D{}^d$	−0.09	+0.12	−0.39	+0.02	−0.31	−0.19
Country rock	ls.[b]	ls.	ls.	dol.[c]	dol.	dol.

[a]Data from Langmuir (1971).
[b]ls. = limestone.
[c]dol. = dolomite.
[d]SI = saturation index.

Table 2-3

Composition of the Air in Soils[a]

Soil	Usual Composition		Extreme Limits Observed		Analyst
	Oxygen	Carbon Dioxide	Oxygen	Carbon Dioxide	
Arable, no dung for 12 months	19–20	0.9	—	—	J. B. Boussingault and Lévy
Pasture land	18–20	0.5–1.5	10–20	0.5–11.5	Th. Schloesing *fils*
Arable, uncropped, no manure:					E. Lau, mean of determinations made frequently during a period of 12 months. Values at depths of 15, 30, and 60 cm, not widely different. (30-cm values given here.)
sandy soil	20.6	0.16	20.4–20.8	0.05–0.30	
loam soil	20.6	0.23	20.0–20.9	0.07–0.55	
moor soil	20.0	0.65	19.2–20.5	0.28–1.40	
Sandy soil, dunged and cropped:					
potatoes, 15 cm	20.3	0.61	19.8–21.0	0.09–0.94	
serradella, 15 cm	20.7	0.18	20.4–20.9	0.12–0.38	
Arable land, fallow	20.7	0.1	20.4–21.1	0.02–0.38	
unmanured	20.4	0.2	18.0–22.3	0.01–1.4	E. J. Russell and A. Appleyard
dunged	20.3	0.4	15.7–21.2	0.03–3.2	
Grassland	18.4	1.6	16.7–20.5	0.3–3.3	

[a]Percent by Volume (Russell, 1961).

Table 2-4

The Oxygen and Carbon Dioxide Content of Soil Air[a]

Depth of Sampling (cm)	Oxygen Content		Carbon Dioxide Content			Carbon Dioxide Gradient (%/cm)	
	Wet Oct.–Jan.	Dry Feb.–May	Wet Oct.–Jan.	Early Dry Feb.	Late Dry April–May	Wet Oct.–Jan.	Dry April–May
10	13.7	20.6	6.5	1.0	0.5	0.65	0.05
25	12.7	19.8	8.5	2.1	1.2	0.13	0.06
45	12.2	18.8	9.7	4.3	2.1	0.04	0.07
90	7.6	17.3	10.0	6.7	3.7	0.01	0.06
120	7.8	16.4	9.6	8.5	5.1	−0.01	0.06
Observed CO_2 diffusion rate from the soil (liters/m^2 day)			6.8	7.5	17.7		

[a]In percent by volume, under Cacao (Rivers Estate, Trinidad) (Russell, 1961).

much as 400 (Makarov, 1960). Althouth the oxidation of ancient organic matter in rocks undergoing weathering produces some CO_2, most of the observed excess is generated during root respiration and by the microbial decay of recent plant matter in soils. During most of geologic time the present land cover of vascular plants was absent. It is therefore likely that the CO_2 content of soil air prior to the Silurian was considerably smaller, unless the CO_2 pressure in the atmosphere was very much greater than it is at present.

Cave waters are generally supersaturated with calcite and not infre-quently with dolomite (Holland et al., 1964). Supersaturation is generally due to CO_2 escape from cave waters (Plummer et al., 1976), although in caves such as Carlsbad Caverns evaporation of water is also of importance (Thrailkill, 1964). Spring and well waters in carbonate terrains are usually just saturated or undersaturated with calcite and dolomite. This is illustrated by the data in Table 2-2. The saturation indexes, SI, of calcite and dolomite, respectively, are defined by the expressions

$$SI_C = \log \frac{a_{Ca^{2+}} \cdot a_{CO_3^{2-}}}{K_C} \qquad (2\text{-}13)$$

and

$$SI_D = \log \frac{a_{Ca^{2+}} \cdot a_{Mg^{2+}} \cdot a_{CO_3^{2-}}^2}{K_D} \qquad (2\text{-}14)$$

Spring 1 water is just slightly undersaturated with respect to both minerals. Well C-11 water is considerably undersaturated with respect to both miner-als. Downward movement of CO_2 from the soil zone is a likely cause of the observed undersaturation (Thrailkill, 1972), although some of the observed undersaturations are not easily explained by such a mechanism.

The Weathering of Silicate Rocks

The weathering of silicate rocks is enormously more complicated than that of carbonates. Few of the silicate minerals dissolve completely during weathering, and many of the new minerals formed during the attack of silicates by oxidizing and carbonated ground waters have a complicated chemistry that depends both on thermodynamics and on the kinetics of weathering processes. This is true even for the behavior of the relatively simple minerals of ultramafic rocks. The system MgO–SiO_2–H_2O is now reasonably well known. Figure 2-2 shows the field of solutions under-saturated with respect to the stable phases in this system and the boundaries along which solutions are saturated with respect to one or more mineral phases. If a ground water containing an initial H_2CO_3 concentration,

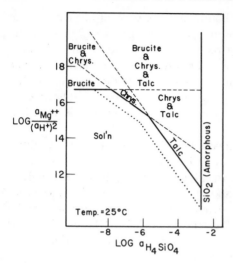

Figure 2-2. Stability relations of some phases in the system $MgO-SiO_2-H_2O$ at $25°C$ and 1 atm total pressure as a function of $\log a_{Mg^{2+}}/a_{H^+}^2$ vs. $\log a_{H_4SiO_4}$. Dotted lines represent the stability field of talc and chrysotile calculated using data from Robie and Waldbaum (1968) and Hostetler and Christ (1968). Mineral fields bounded by the continuous and dashed lines are fields in which the solution is supersaturated with one or more of the minerals; Bricker, Nesbitt, and Gunter (1973). Copyright by the Mineralogical Society of America.

$m_{H_2CO_3}^0$, is brought into contact with one of the stable mineral phases, this phase should dissolve, and the composition of the solution should follow a path defined by the initial composition, the stoichiometry of the dissolving phase, and the availability or lack of availability of additional H_2CO_3 to replenish the H_2CO_3 used in the dissolution reaction. If the dissolving phase is chrysotile, $Mg_3Si_2O_5(OH)_4$, and if H_2CO_3 is not added to the solution during the dissolution process, then

$$m_{H_2CO_3}^0 = m_{H_2CO_3} + m_{HCO_3^-} + m_{CO_3^{2-}} \tag{2-15}$$

Generally, the pH of water in soil is sufficiently low, so that $m_{CO_3^{2-}} \ll m_{HCO_3^-}$; thus

$$m_{H_2CO_3} \cong m_{H_2CO_3}^0 - m_{HCO_3^-} \tag{2-16}$$

In such solutions magnesium complexes are present in only negligible concentrations (Christ et at., 1973). so that

$$2m_{Mg^{2+}} \cong m_{HCO_3^-} \tag{2-17}$$

From equations 2-16 and 2-17 it follows that

$$K_1 \cong \frac{m_{H^+} \cdot m_{HCO_3^-}}{m_{H_2CO_3}} \cong \frac{2m_{H^+} \cdot m_{Mg^{2+}}}{\left(m_{H_2CO_3}^0 - 2m_{Mg^{2+}}\right)}$$

Thus

$$\frac{m_{Mg^{2+}}}{m_{H^+}^2} \cong \frac{4m_{Mg^{2+}}^3}{K_1^2 \left(m_{H_2CO_3}^0 - 2m_{Mg^{2+}} \right)^2} \qquad (2\text{-}18)$$

From the stoichiometry of the dissolution process of chrysotile,

$$m_{Mg^{2+}} = \tfrac{3}{2} m_{H_4SiO_4} \qquad (2\text{-}19)$$

Equations 2-17 and 2-18 together define a variety of paths through the solution field in Figure 2-2, in which the concentration of magnesium in solution can be taken as the progress variable (see, for instance, Helgeson, 1971). Different paths correspond to different initial H_2CO_3 concentrations, and to the replenishment, if any, of H_2CO_3 during the dissolution process (Hemley et al., 1977).

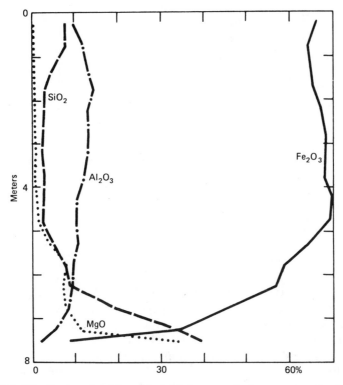

Figure 2-3. Distribution of oxides with depth in a weathering sequence developed on serpentine at Kalimantan, Borneo; Schellmann (1964) as shown in Loughnan (1969).

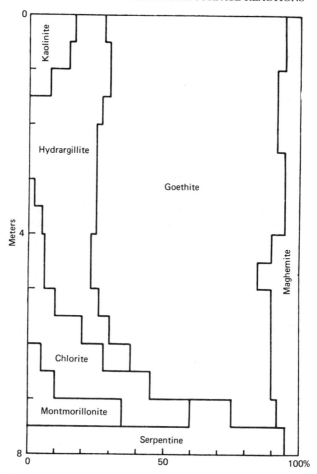

Figure 2-4. Quantitative depth variation of the mineralogy of the laterite profile on Figure 2-3; Schellmann (1964) as shown in Loughnan (1969).

In actual dissolution experiments the observed reaction paths are more complicated than those predicted by simple theory (Luce et al., 1972). There is, first, a rapid exchange of surface magnesium ions with hydrogen ions followed by a longer period of hydrogen exchange and extraction of internal magnesium and silicon. The observed kinetics are consistent both with non-steady state diffusion of ions within the mineral and with quasi-steady state diffusion of ions through a leached shell surrounding the mineral. Although dissolution is therefore usually incongruent during the early phase of acid attack, serpentine and the other magnesium silicates do dissolve congruently during the later stages of equilibration with aqueous solutions (Hostetler and Christ, 1968, and Christ et al., 1973).

The weathering of real ultramafic rocks is much more complicated than that of rocks within the system $MgO–SiO_2–H_2O$. Most of the complications are traceable to the presence of FeO and Al_2O_3 as components of ultramafic rocks. After an initial period during which nontronite-montmorillonite forms, iron released above the water table is usually oxidized and precipitated as a constituent of goethite or hematite. Aluminum released during attack by ground water is largely precipitated as a constituent of goethite and aluminosilicates (Marval, 1968, Zeissink, 1969). During intense weathering, the aluminosilicates are decomposed, and gibbsite is the normal residue. The resilicification of this mineral near the top of the weathered layers tends to produce kaolinite. This sequence is well illustrated in Figures 2-3 and 2-4, taken from Schellmann's (1964) description of a complete weathering sequence developed on a serpentine-rich rock in Kalimantan, Borneo. During the early stages of weathering, MgO is removed more rapidly than SiO_2. In later stages both MgO and SiO_2 are present in only minor amounts. Al_2O_3 and Fe_2O_3 are strongly concentrated in the soil zone, although the increase upward in the TiO_2/Fe_2O_3 and in the TiO_2/Al_2O_3 ratios of the soil indicates that aluminum and iron are both lost during weathering.

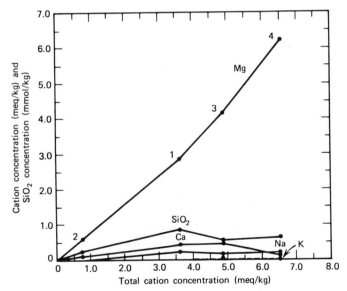

Figure 2-5. The concentration of the major cations and silica in ground water from ultramafic rock types; data from Table 2 in White, Hem, and Waring (1963).

Table 2-5

Analyses of Ground Water from Ultramafic Terrains[a]

	Analysis							
	1		2		3		4	
	Bushveld ultra-mafics, Pretoria district, Union of South Africa		*Peridotite, Webster, N.C.*		*Serpentine, Lake Roland, Md.*		*Serpentine, Nottingham, Pa.*	
	Date of collection							
	Dec. 17, 1940		——		*Mar. 19, 1954*		*Sept. 21, 1925*	
	ppm	epm	ppm	epm	ppm	epm	ppm	epm
SiO_2	50		16		31		40	
Al					0.2			
Fe			0.08		0.06		0.28	
Mn					0.00			
Cu					0.00			
Zn					0.05			
Ca	9.6	0.48	2.5	0.12	9.5	0.47	2.1	0.10
Mg	35	2.88	7.7	0.63	51	4.19	76	6.25
Na	} 5.8	} 0.25	{ 0.2	0.01	4.0	0.17	4.2	0.18
K			{ 0.0	0.00	2.2	0.06	1.0	0.03
Total cations		3.61		0.76		4.89		6.56
HCO_3	168	2.75	44	0.72	276	4.52	329	5.39
CO_3	0			0.0	0		0	
SO_4	0.5	0.01	0.0	0.00	2.6	0.05	8.5	0.18
Cl	14	0.40	0.7	0.02	12	0.34	30	0.85
F			0.3	0.02	0.0	0.00		
NO_3	28	0.45			6.8	0.11	2.7	0.04
PO_4					0.0			
Total anions		3.61		0.76		5.02		6.46
Total, as reported	311		71		395		494	
Specific conductance, (micromhos at 25°C)						427		
pH		7.6		8.5		8.3		
Temperature (°C)								

[a]Data from White et al. (1963); ppm = parts per million; epm = equivalents per million.

In an area such as Kalimantan, where removal of the major cations and of SiO_2 from the soil cover is virtually complete, the relative proportions of the major cations and SiO_2 in ground water should be the same as in the parent rock. This is not true for the major portions of the continents where weathering is incomplete or where the depth and/or composition of the weathering zone has not attained steady state. Figure 2-5 and Table 2-5 reproduce four analyses of ground waters from ultramafic terrains (White et al., 1963). Since Mg^{2+} is the dominant cation in ultramafic rocks, its dominance in these ground waters is hardly surprising. On the other hand, the high Mg^{2+}/SiO_2 ratio, especially in analysis 4, indicates that Mg^{2+} and SiO_2 are not removed in the ratio in which they are present in the parent rock. SiO_2 is almost certainly retained in the weathering zone, probably as a constituent of one or more aluminosilicates.

The presence of Al_2O_3 affects the weathering of basaltic and granitic rocks even more profoundly than the weathering of ultramafics. Feldspars undergo a complex series of reactions during the first days or weeks of contact with CO_2-rich water. On the longer, geologically more important, time scale feldspar weathering seems to be controlled largely by diffusion through and between the grains of an aluminosilicate surface layer (Wollast, 1967; Pǎces, 1973; Busenberg and Clemency, 1976). Simple dissolution paths such as those proposed for rocks in the system $MgO–SiO_2–H_2O$ are therefore ruled out except for dissolution during the first few minutes or hours of ground water attack on fresh feldspar surfaces.

The chemical patterns observed in the composition of ground waters in basaltic-gabbroic and in granitic terrains are surprisingly similar. Figure 2-6

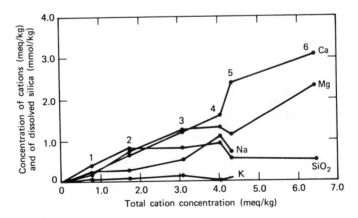

Figure 2-6. The concentration of the major cations and dissolved silica in ground water from gabbros and basalts; data from Table 1 in White, Hem, and Waring (1963).

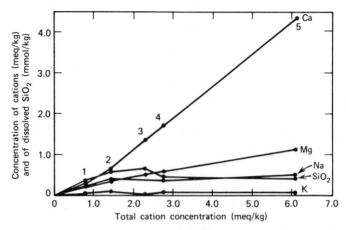

Figure 2-7. The concentration of the major cations and dissolved silica in ground water from granites; data from Table 1 in White, Hem, and Waring (1963).

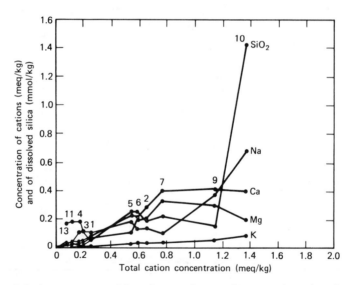

Figure 2-8. Average composition of ground water from granites; data from Tardy (1971).

32

and Table 2-6 reproduce ground water analyses from gabbros and basalts, and Figure 2-7 and Table 2-7 reproduce ground water analyses from a number of American granites (White, Hem, and Waring, 1963). Only analyses in which the HCO_3^- ion concentration is equivalent to more than two-thirds of the total cation charge have been included; ground water whose composition is strongly influenced by sources of sulfate and chloride have been excluded. Figure 2-8 and Table 2-8 contain averages of the composition of ground water in granites from regions representing a wide range of latitudes and rainfall (Tardy, 1971). Analyses 6 and 7, the average of ground waters from Brittany and Corsica, have been adjusted before plotting in Figure 2-8 by eliminating an amount of sodium equivalent to the concentration of Cl^-, since sea spray is almost certainly responsible for the high Cl^- content of these waters. In ground water both from basaltic and gabbroic terrains calcium generally is the dominant cation. Potassium is a minor constituent in all of the ground waters. With increasing total cation concentration, magnesium becomes progressively less important than calcium, and the concentration of SiO_2 first rises more or less linearly and then, with few exceptions, stays approximately constant.

The relative mobility of cations during the weathering of basic rocks is, therefore, quite different from their relative mobility during the weathering of granites. In Table 2-9 the percentage of the major cations in ground water analysis 5 of Figure 2-7 has been divided by the cation percentage in average granite. There is obviously an enormous range of relative mobility. The relative mobility of calcium is one order of magnitude larger than that of sodium; sodium, in turn, is five times more mobile than potassium, which is more than twice as mobile as SiO_2. These figures are not particularly precise, because the analyzed rocks from which the ground water was drawn probably did not have the composition of average granite in Table 2-9. However, the differences in relative mobility are so large that errors in the choice of parent rock composition are unlikely to affect the order of relative cation mobilities. Clearly, however, the values of relative mobilities are somewhat different for the more dilute ground waters shown in Figure 2-7, since the dominance of calcium in these ground waters is much less pronounced. In none of the waters, however, is the ratio of the cation concentration in the ground water equal to the ratio of the cation concentration in the parent rock.

The difference in the relative mobility of calcium, magnesium, sodium, and potassium is much less pronounced in ground water from basaltic and gabbroic terrains than in ground water from granitic terrains. Table 2-10 shows that the difference in the relative mobility of these four cations in ground water 5 from Figure 2-6 is only a factor of 2. The calculated order of the relative cation mobilities in ground water from basalts and gabbros is, therefore, not as meaningful a parameter as in ground water from granites.

Table 2-6

Analyses of Ground Waters from Gabbroic and Basaltic Terrains[a]

	Analysis						
	1	2	3	4		5	6
	Gabbro, Waterloo, Md.	Gabbro, Harrisburg, N.C.	Basalt, Camas, Wash.	Columbia River Basalt, Farmington, Oreg.	Basalt, Moses Lake, Wash.	Basalt, Shoshone, Idaho	Deccan Basalt, Purna, Hyderabad, India
	Date of Collection						
	Dec. 23, 1952	Feb. 22, 1955	May 17, 1949	May 15, 1951	May 1, 1950	Oct. 30, 1956	—
	ppm	epm	ppm	epm	ppm	epm	ppm	epm	ppm	epm	ppm	epm	ppm	epm
SiO_2	39		56		49		50		55		33		30	
Al	0.0										0.0			
Fe	5.1		0.06		0.04		0.43		0.03		0.00			
Mn	0.19		0.00								0.00			
Cu	0.21													
Zn	2.7													

34

	S1 ppm	S1 epm	S2 ppm	S2 epm	S3 ppm	S3 epm	S4 ppm	S4 epm	S5 ppm	S5 epm	S6 ppm	S6 epm	S7 ppm	S7 epm
Ca	5.1	0.25	32	1.60	13	0.65	24	1.20	29	1.45	48	2.40	62	3.09
Mg	2.3	0.19	16	1.32	9.0	0.74	15	1.23	19	1.56	14	1.15	28	2.30
Na	6.2	0.27	25	1.09	6.6	0.29	12	0.52	12	0.52	16	0.70	24	1.04
K	3.2	0.08	1.1	0.03	2.8	0.07	5.3	0.14	3.5	0.09	3.2	0.08		
Total cations		0.79		4.04		1.75		3.09		3.62		4.33		6.43
HCO_3	37	0.61	203.	3.33	88.	1.44	156	2.56	177	2.90	220	3.61	294	4.82
CO_3	0		0		0		0		0		0		0	
SO_4	9.2	0.19	10	0.21	4.9	0.10	1.6	0.03	15	0.31	19	0.40	30	0.62
Cl	1.0	0.03	13	0.37	6.9	0.20	15	0.42	6.9	0.20	8.0	0.23	37	1.04
F	0.0	0.00	0.2	0.01	0.1	0.01	0.2	0.01	0.4	0.02	0.3	0.02		
NO_3	0.3	0.01	2.7	0.04	8.4	0.14	0.1	0.00	9.7	0.16	5.2	0.08		
PO_4	0.0										0.15			
Total anions		0.84		3.96		1.89		3.02		3.59		4.34		6.48
Total, as reported	112		359		189		280		328		367		505	
Specific conductance (micromhos at 25°C)	77		340		388		181		427		404		340	
pH	6.7		6.7		6.8		7.7		7.7		7.9		8.0	
Temperature (°C)					16.7								13.1	

[a]Data from White (1963); ppm = parts per million; epm = equivalents per million.

35

Table 2-7

Analyses of Ground Waters from Granitic Terrains in the United States[a]

	Analysis									
	1		2		3		4		5	
	Granite, West Warwick, R.I.		Granite, McCormick Co., S.C.		Granite, Ellicott City, Md.		Quartz Monzonite, W. of Clayton, Idaho		Granite, Spokane, Wash.	
	Date of Collection									
	May 26, 1955		Nov. 24, 1954		Mar. 21, 1951		Sept. 8, 1954		June 6, 1951	
	ppm	epm	ppm	epm	ppm	epm	ppm	epm	ppm	epm
SiO_2	20		35		39		27		25	
Al	0.0		0.1		0.9		0.1			
Fe	0.19		0.18		1.6		0.05		0.22	
Mn	0.0		0.13		0.0		0.00		0.0	
Cu	0.0		0.0				0.00			
Zn	0.07		0.09							
Ca	6.5	0.32	13	0.65	27	1.35	34	1.70	87	4.34
Mg	2.6	0.21	4.3	0.35	6.2	0.51	7.3	0.60	14	1.15

	ppm	epm	ppm	epm	ppm	epm	ppm	epm	ppm	epm
Na	5.9	0.26	8.4	0.36	9.5	0.41	8.5	0.37	12	0.52
K	0.8	0.02	3.5	0.09	1.4	0.04	3.3	0.08	3.2	0.08
Total cations		0.81		1.45		2.31		2.75		6.09
HCO_3	38	0.62	72	1.18	93	1.52	136	2.23	268	4.39
CO_3	0		0		0		0		0	
SO_4	0.9	0.02	6.9	0.14	32	0.67	20	0.42	33	0.69
Cl	5.0	0.14	3.8	0.11	5.2	0.15	1.2	0.03	14.	0.39
F	0.5	0.03	0.2	0.01	0.0	0.00	0.2	0.01	0.2	0.01
NO_3	1.5	0.02	0.4	0.01	7.5	0.12	0.2	0.00	32	0.52
PO_4	0.0		0.1							
Total anions		0.83		1.45		2.46		2.69		6.00
Total, as reported	82		148		223		238		489	
Specific conductance (micromhos at 25° C)	76		150		258		255		569	
pH	7.6		7.0		6.6		7.5		7.3	
Temperature (°C)	11.1		18.1		6		7.8		10.6	

[a]Data from White et al. (1963); ppm = parts per million; epm = equivalents per million.

37

Table 2-8

Analyses of Ground Waters from Granites;[a]

Location	No.		pH	HCO₃	Cl	SO₄	SiO₂	Na	K	Ca	Mg
1 Norway	28	M	5.4	4.9	5.0	4.6	3.0	2.6	0.4	1.7	0.6
		σ	0.6	3.6	4.0	1.6	1.6	1.7	0.2	1.6	0.3
2 Vosges	51	M	6.1	15.9	3.4	10.9	11.5	3.3	1.2	5.8	2.4
		σ	0.5	13.0	1.9	8.5	6.2	2.0	0.8	4.7	2.4
3 Alrance Spring F	77	M	5.9	6.9	<3	1.15	5.9	2.3	0.6	1.0	0.4
		σ	0.3	1.3	—	0.45	1.6	0.3	0.3	0.3	0.1
4 Alrance Spring A	47	M	6.0	8.1	<3	1.1	11.5	2.6	0.6	0.7	0.3
		σ	0.4	2.4	—	0.5	1.2	0.2	0.2	0.1	0.05
5 Central Massif	10	M	7.7	12.2	2.6	3.7	15.1	4.2	1.2	4.6	1.3
		σ	0.5	6.0	1.5	2.0	10.2	2.8	0.9	3.2	0.9
6 Brittany	7	M	0.5	13.4	16.2	3.9	15.0	13.3	1.3	4.4	2.6
		σ	0.5	6.2	6.0	2.0	7.2	6.1	0.4	3.0	1.1
7 Corsica	25	M	6.7	40.3	22.0	8.6	13.2	16.5	1.4	8.1	4.0
		σ	0.4	36.0	15.3	6.2	7.4	12.1	0.9	7.1	4.0
8 Sahara	8	M	6.9	30.4	4.0	20	9	30	1.8	40	—
		σ	0.5	5.0	2.0	4	3	2	0.3	10	—
9 Senegal	7	M	7.1	43.9	4.2	0.8	46.2	8.4	2.2	8.3	3.7
		σ	0.9	24	2.5	0.7	31.2	5.4	1.1	5.3	2.5
10 Chad	2	M	7.9	54.4	<3	1.4	85	15.7	3.4	8.0	2.5
		σ	—	—	—	—	—	—	—	—	—
11 Ivory Coast (Korhogo, dry season)	54	M	5.5	6.1	<3	0.4	10.8	0.8	1.0	1.0	0.10
		—	—	—	—	—	—	—	—	—	—
12 Ivory Coast (Korhogo, wet season)	59	M	5.5	6.1	<3	0.5	8.0	0.2	0.6	<1	<0.1
		—	—	—	—	—	—	—	—	—	—
13 Malagasy (High Plateaus)	2	M	5.7	6.1	1	0.7	10.6	0.95	0.62	0.40	0.12
		—	—	—	—	—	—	—	—	—	—

[a]Major element concentrations in milligrams per kilogram; adapted from Tardy, 1971. Values in mg/liter; No. = number of analyses; M = mean; σ = standard deviation.

These inferences regarding relative mobilities are borne out by the mineralogy and chemistry of weathered basalts and granites. Loughnan (1969) has summarized much of the available data that show that pyroxene and olivine normally alter to Mg–Fe montmorillonite and then to gibbsite and goethite or hematite, while plagioclase tends to alter to Al-montmorillonites and then, via halloysite and kaolinite, to the bauxite minerals. The details of the mineral sequences and the total extent of chemical alteration depend largely on climatic factors, particularly temperature, rainfall, and their distribution throughout the year (see, for instance, Hay and Jones, 1972).

Table 2-9

Relative Mobility of Cations During the Weathering of Granite

	Average Granite			*Groundwater Analysis 5 in Figure 2-7*		*Relative*
	wt. %	*Moles of Cation Per Kilogram of Rock*	*Cation Percentage*	*Millimoles Per Liter*	*Cation Percentage*	*Mobility (column 5/column 3)*
SiO_2	70.77	11.78	66.1	0.42	11.2	0.17
TiO_2	0.39	0.05	0.3			
Al_2O_3	14.59	2.86	16.1			
Fe_2O_3	1.58	0.20	1.1	<0.01	<0.1	<0.04
FeO	1.79	0.25	1.4			
MnO	0.12	0.02	0.1	0.00	0.0	
MgO	0.89	0.22	1.2	0.57	15.2	12.7
CaO	2.01	0.36	2.0	2.17	57.5	28.8
Na_2O	3.52	1.14	6.4	0.52	13.8	2.2
K_2O	4.15	0.88	5.0	0.08	2.1	0.4
P_2O_5	0.19	0.03	0.2			

Table 2-10

Relative Mobility of Cations During the Weathering of Basalt

	Average Basalt			*Groundwater Analysis 5 in Figure 2-6*		*Relative*
	wt. %	*Moles of Cation Per Kilogram of Rock*	*Cation Percentage*	*Millimoles Per Liter*	*Cation Percentage*	*Mobility (column 5/column 3)*
SiO_2	51.55	8.55	48.2	0.55	17.7	0.37
TiO_2	1.48	0.19	1.1			
Al_2O_3	14.95	2.94	16.6			
Fe_2O_3	2.55	0.32	1.8			
FeO	9.10	1.26	7.1			
MnO	0.20	0.03	0.2			
MgO	6.63	1.65	9.3	0.57	18.4	2.0
CaO	10.00	1.78	10.1	1.20	38.7	3.8
Na_2O	2.35	0.76	4.3	0.70	22.6	5.2
K_2O	0.89	0.19	1.1	0.08	22.6	2.4
P_2O_5	0.30	0.04	0.2			

The concentration of CaO, MgO, Na_2O, and K_2O decreases rapidly and nearly simultaneously during the weathering of basalt. Craig and Loughnan's data (1964) for a weathered sequence at Bathurst, New South Wales, are reproduced in Table 2-11. Between a depth of 2.13 m and 1.68 m nearly all of the CaO, MgO, Na_2O, and K_2O disappear. At a depth of 1.98 m nearly all of the CaO and MgO are still present whereas less than half of the

Table 2-11

Chemical Data for the Weathered Sequence Developed on Basalt
at Bathurst, New South Wales[a]

		Analysis (%)									
No.	Depth (m)	SiO_2	Al_2O_3	Fe_2O_3	CaO	MgO	Na_2O	K_2O	TiO_2	H_2O	Total
B1	0.15	39.9	20.3	16.4	1.7	0.5	0.3	0.6	2.4	18.6	100.1
B2	0.46	36.1	20.6	21.8	0.5	0.3	0.2	0.2	4.1	16.3	100.1
B3	0.76	27.8	18.5	33.2	0.5	0.2	0.2	0.1	5.2	14.3	100.0
B4	1.07	28.9	19.4	30.5	0.5	0.2	0.2	0.1	4.8	15.8	100.4
B5	1.37	33.4	21.4	24.3	0.5	0.3	0.2	0.1	4.1	15.8	100.1
B6	1.68	37.1	23.2	18.1	0.6	0.3	0.2	0.1	3.8	16.8	100.2
B7	1.98	42.2	15.2	14.0	9.0	8.7	1.1	0.9	2.0	7.2	100.3
B8	2.13	45.0	14.6	13.3	9.8	10.1	2.4	1.5	1.9	1.6	100.2

[a]Craig and Loughnan, 1964.

Table 2-12

Chemical Analyses (%) of Fresh and Weathered Morton Gneiss[a]

	Sample						
	1	2	3	4	5	6	7
SiO_2	71.54	69.89	68.09	61.75	70.30	57.53	55.07
Al_2O_3	14.62	16.54	17.31	18.58	18.34	23.57	26.14
Fe_2O_3	0.69	2.33	3.86	1.69	1.55	3.05	3.72
FeO	1.64	0.34	0.36	4.11	0.22	3.59	2.53
MgO	0.77	0.30	0.46	0.76	0.21	0.41	0.33
CaO	2.08	0.06	0.06	0.16	0.10	0.05	0.16
Na_2O	3.84	0.43	0.12	0.10	0.09	0.06	0.05
K_2O	3.92	5.34	3.48	3.54	2.47	0.35	0.14
H_2O^+	0.30	4.00	5.14	5.56	5.58	8.82	10.75
H_2O^-	0.02	0.35	0.47	0.35	0.30	0.70	0.64
CO_2	0.14	0.21	0.05	1.84	0.20	0.77	0.36
TiO_2	0.26	0.14	0.34	0.92	0.21	0.87	1.03
P_2O_5	0.10	0.04	0.03	0.09	0.04	0.08	0.11
MnO	0.04	0.04	0.06	0.21	0.03	0.07	0.03
BaO	0.09	0.13	0.07	0.09	0.05	Trace	0.01
SO_3	—	Trace	0.00	0.03	0.00	Trace	Trace
S	0.02	0.01	0.01	0.05	0.01	0.01	0.04
Total	100.07	100.15	99.91	99.83	99.70	99.93	100.11

[a]Goldich, 1938. Sample 1: representative composite of the Morton granite gneiss. Samples 2–7: weathered clay derived from the granite gneiss.

Na_2O and K_2O are still present. Na_2O and K_2O are therefore slightly more mobile than CaO and MgO.

During the mild weathering of granites, illite and montmorillonites tend to develop. These give way to kaolin group minerals during more intense weathering. Ultimately, these break down to minerals of the bauxite group. Among the major minerals in granites plagioclase is least stable, biotite moderately so, and the potash feldspars are among the most stable. This sequence is reflected in the chemical analyses of fresh and altered Morton gneiss in Table 2-12 from the classic study of Goldich (1938). During weathering CaO is lost first, Na_2O remains slightly longer, MgO loss is erratic and variable, and K_2O is lost only in the most altered samples of the gneiss. With the exception of Mg, the mobility sequence $Ca > Na > Mg > K$ agrees well with that in Table 2-9 derived from ground water analyses.

Garrels (1967) and Garrels and Mackenzie (1971) have shown that the composition of spring waters from granitic rocks in the Sierra Nevada can be explained quantitatively by assigning all of the Ca^{2+} and Na^+ and sufficient SiO_2 to reconstitute kaolinite to plagioclase, all of the Mg^{2+} and a commensurate quantity of K^+ and SiO_2 to reconstitute kaolinite to biotite, and the remaining K^+ with sufficient SiO_2 to reconstitute kaolinite to K-feldspar. The choice of kaolinite as the only important product of chemical weathering of Sierra Nevada granites is suggested by the observed mineralogy of the weathered granites. The proof that the calculation is justified lies in the nearly perfect balance between the amount of SiO_2 in spring water and the quantity needed to accomplish the reconstitution of plagioclase, biotite, and K-feldspar demanded by the Ca^{2+}, Na^+, Mg^{2+}, and K^+ content of spring water. The rate of attack of plagioclase in these granites is more than an order of magnitude greater than that of the K-feldspars. This difference is in reasonable agreement with the difference in mobility inferred from the ground water analyses in Table 2-9.

The behavior of metamorphic rocks during weathering is similar to the behavior of equivalent igneous rocks. Sedimentary rocks behave during weathering much as partially weathered igneous rocks do. The outlines of silicate rock weathering are, therefore, contained, albeit rather sketchily, in the above discussion of the weathering of igneous rocks.

The concentration of solutes in the ground waters of silicate terrains is controlled largely by a very few parameters. Among these, the most important is probably the quantity of CO_2 and organic acids (Silverman and Munoz, 1970) generated in soil zones by root respiration and by the bacterial decay of organic matter. Virtually all of the CO_2 that dissolves in ground water is neutralized to HCO_3^- by reaction with silicate minerals. The total dissolved load of ground waters in silicate rocks therefore depends in the main on the H_2CO_3 content of soil water as it leaves the zone rich in decaying organic matter, and on the extent to which H_2CO_3 used in silicate weathering below this zone is replenished by the downward transport of CO_2. The CO_2 content of soil water in the humus-rich layer depends largely on the quantity and rate of decay of organic matter, and on the rate of diffusion of CO_2 upward out of the soil. The rate of the downward movement of CO_2 dissolved in ground waters is normally less than 10% of the rate of CO_2 loss back to the atmosphere (see Table 6-7). Thus downward CO_2 removal rarely, if ever, exerts a strong effect on the CO_2 content of soil air. The quantity and rate of decay of organic matter in soil is largely a function of temperature, rainfall, and the time variation of these parameters. It is likely that at least some of the extremely rapid weathering in the tropics is due to the reinforcement of the effect of the rapid rates of photosynthesis by the effects of the rapid rates of decay of organic matter in tropical soils. However the effect of climate on ground water chemistry in silicate terrains is not as extreme as it might be, because downward replenishment of CO_2 in highly leached, lateritic soil becomes progressively less efficient as the depth to unweathered rock increases. The relative concentration of the major cations in ground water within silicate terrains depends largely on the relative abundance of the various rock types and their degree of alteration. In well-drained areas, where the degree of weathering is modest, the abundance of major cations in ground water depends largely on their relative mobilities. In well-drained, deeply weathered areas, their relative abundance in ground water approaches their proportions in average fresh rock. In arid areas weathering is not as rapid as in well-drained areas of equivalent temperature; cation concentrations and pH are both increased by evaporative loss of water, and the cation ratios are frequently altered by the precipitation of $CaCO_3$ in the soil zone. At steady state the relative proportion and composition of both dissolved and particulate fluxes from an area are determined largely by the bulk composition of the parent rock and by the addition of CO_2, H_2O, and O_2 from the atmosphere during weathering.

OXIDATION REACTIONS

Many elements change valence state in weathering horizons; however, only four are of quantitative importance for the oxidation state of the atmosphere–ocean–crust system: carbon, sulfur, iron, and oxygen itself. The others are present in such small concentrations that their valence changes are unimportant. They can, however, serve as indicators of the oxidation state of the atmosphere; the behavior of uranium, for instance, has played an important role in discussions of the history of atmospheric oxygen. Others, such as nitrogen, are critical as plant nutrients.

The Oxidation of Carbon Compounds

Photosynthesis is extremely rapid both on land and at sea. Precise numbers for gross rates of photosynthesis are understandably hard to obtain; fortunately, the fairly rough numbers that have been compiled by various authors turn out to be quite adequate. Table 2-13 is an inventory of terrestrial carbon and of the rates of production of organic compounds in a variety of terrestrial settings. The data, which are largely from Olson (1970), agree well with those of Woodwell et al. (1978). The estimated total terrestrial rate of production of carbon in carbon compounds is 48×10^{15} g/yr. Since the mass of carbon in atmospheric CO_2 is 690×10^{15} g, approximately 7% of the atmospheric CO_2 is converted annually to organic matter on land. Photosynthesis at sea accounts for the fixation of some 25×10^{15} g of C/yr (see Table 6-7). Thus approximately 11% of atmospheric CO_2 is "fixed" annually. If none of this carbon were returned, the entire atmospheric carbon reservoir would be exhausted in 9 yr at the present rate of consumption. Fortunately, such an event is impossible because the rate of photosynthesis decreases rapidly with decreasing concentrations of atmospheric CO_2, and because CO_2 would be in part replenished from the oceans. Nevertheless, the calculation demonstrates that the rate of transfer of carbon from the atmosphere to the biosphere is very rapid, and that the system must always be operating close to steady state, so that the rate of fixation of carbon is very nearly equal to that of the return of CO_2 to the atmosphere by organic matter. In detail the system is more complicated, and those details which are critical for maintaining the overall oxidation state of the atmosphere are discussed in Chapter VI.

In Table 2-13 organic matter has been subdivided into material of short and long lifetimes. Incipient fossil fuels, resistant humus, and peat decaying with a lifetime in excess of ca. 1000 yr have been excluded. The decay of terrestrial organic matter clearly involves processes with a spectrum of lifetimes extending from 1 yr or less to more than 1000 yr. If we let W_i be the number of grams per square centimeter of carbon in organic matter with a half-life of t_i yr for oxidation then

$$dW_i / dt \cong R_i - \lambda_i W_i - \phi_i Q \qquad (2\text{-}20)$$

where

R_i = rate of addition of carbon in organic matter i in g/cm^2 yr^{-1}.

$\lambda_i = 0.693 / t_i$ = decay constant of organic matter i.

Q = rate of soil removal by weathering (g/cm^2 yr^{-1}).

ϕ_i = weight fraction of carbon in organic matter i in soil removed by weathering.

Table 2-13

Subdivision of the World Inventory of Terrestial Organic Carbon into Material with Short and Long Lifetimes[a]

Reservoir	Carbon Inventory (10^{15} g)					Net Primary Production (10^{15} g / yr)			Turnover Time (yr)	
	Short Lifetime	Long Lifetime	Total	Living[b]	Dead	Short Lifetime	Long Lifetime	Total[b]	Short Lifetime	Long Lifetime
(1) Woodland or forest										
(a) Temperate and boreal										
deciduous	12.0	160.0	172.0	80.0	92.0	4.0	4.0	8.0	3	40
coniferous	15.0	240.0	255.0	120.0	135.0	3.0	6.0	9.0	5	40
rain forest	1.8	24.0	25.8	12.0	13.8	0.6	0.6	1.2	3	40
dry woodland	4.2	140.0	144.2	70.0	74.2	1.4	1.4	2.8	3	100
(b) Tropical and subtropical										
rain forest	20.0	400.0	420.0	200.0	220.0	10.0	5.0	15.0	2	80
Total	53.0	964.0	1017.0	482.0	535.0	19.0	17.0	36.0		
(2) Nonforest										
tundra-like	3.6	60.0	63.6	7.2	56.4	0.6	0.6	1.2	6	100
grassland	7.8	234.0	241.8	18.2	223.6	3.9	3.9	7.8	2	60
agricultural	4.0	160.0	164.0	15.0	149.0	4.0	2.0	6.0	1	80
desert and semidesert	4.8	64.0	68.8	19.2	49.6	1.6	1.6	3.2	3	40
wetlands	1.0	40.0	41.0	4.0[c]	37.0	1.0	1.0	2.0[c]	1	40
Total	21.2	558.0	579.2	63.6	515.6	11.1	9.1	20.2		
Grand total	74.2	1522.0	1596.2	545.6	1050.6	30.1	26.1	56.2	mean 2.5	mean 58
Rounded	75	1560	1635	550	1085	30	26	56	2.5	60

[a]Keeling, 1973.
[b]All values except for wetlands from Olson (1970, p. 234).
[c]Living carbon assumed to be 10% of long lifetime material.

If soil removal takes place in gullies, so that h cm is the mean depth to which soil is removed, and if all of the organic matter of type i is present in the soil between the surface and depth h then

$$\phi_i = \frac{W_i}{\rho h}$$

where ρ is the density of the soil horizon between the surface and depth h.

At steady state W_i is constant and equation 2-20 becomes

$$0 \simeq R_i - \lambda_i W_i - \frac{W_i Q}{\rho h} \tag{2-21}$$

Thus

$$W_i \approx \frac{R_i}{\lambda_i + (Q/\rho h)} \tag{2-22}$$

The total steady state organic carbon content of the soil, W, is then

$$W = \sum_i W_i \cong \sum_i \frac{R_i}{\lambda_i + (Q/\rho h)} \tag{2-23}$$

Most terrestrial organic matter has a half-life of between 1 and 1000 yr, and hence a λ_i value of between 0.7 and 0.7×10^{-3} yr^{-1}. The mean value of Q today (see Chapter IV) is approximately 12×10^{-3} g/cm^2 yr; ρ is approximately 2 g/cc. If h is taken to be 100 cm, $Q/\rho h$ is about

$$\frac{Q}{\rho h} \approx \frac{12 \times 10^{-3} \text{ g/cm}^2 \text{ yr}}{2 \text{ g/cm}^3 \cdot 100 \text{ cm}}$$

$$\approx 6 \times 10^{-5} \text{ yr}^{-1}$$

This value is negligible compared to the decay constant of most terrestrial organic matter. Virtually all of the terrestrial organic matter therefore decays in place, and, with the exception of organic matter with decay times in excess of 1000 yr,

$$W_i \approx R_i / \lambda_i \tag{2-24}$$

It follows that the total amount of organic matter in soil zones tends to adjust itself with a time constant depending on the spectrum of λ_i's until the rate of production of organic matter is nearly equal to the rate of decay.

Virtually all of the organic matter in soil, including old organic carbon in sedimentary rocks, is decomposed by a large cast of soil microorganisms including aerobic and anaerobic bacteria, actinomycetes, filamentous fungi, and higher fungi (see for instance Alexander, 1961). It seems likely that this cast has changed with time, and that the decomposition efficiency has not remained constant. Such changes have probably been responsible for gradual changes in the quantity of organic matter in the soils of particular climatic zones. The effects of such changes on the carbon cycle have probably been negligible unless the changes took place on a time scale shorter than a century or involved order of magnitude increases in the rate of photosynthesis; neither of these extremes is at all likely.

The Oxidation of Sulfides

Sulfur in the earth's crust is divided nearly evenly between sulfates and sulfides. Gypsum and anhydrite are by far the most important sulfates; pyrite is by far the most abundant, although by no means the most valuable, sulfide. During weathering gypsum and anhydrite are dissolved as described earlier in this chapter. Pyrite and the other iron sulfides are almost always oxidized—the Fe^{2+} to Fe^{3+} and the S_2^{2-} to SO_4^{2-}. The Fe^{3+} follows the same path as the other, generally more important, Fe^{3+} produced by the destruction of Fe^{2+}-containing silicates. SO_4^{2-} is generally flushed out of the weathering zone as a constituent of ground waters, although gypsum is sometimes precipitated in the alteration zones of sulfide ore bodies, most commonly in arid and semiarid areas, and where limestone is abundant.

The oxidation of pyrite has been studied rather extensively, because the scourge of highly acidic mine drainage is due largely to sulfuric acid produced during the oxidation of pyrite. The major steps in the oxidation of pyrite are probably the following (Stumm and Morgan, 1970. p. 541).

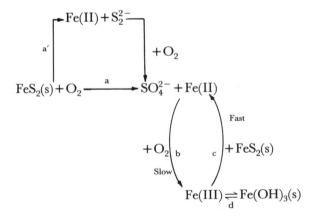

Pyrite dissolves, and the sulfur is oxidized to SO_4^{2-} via the reaction

$$FeS_2 + \tfrac{7}{2}O_2 + H_2O \rightarrow Fe^{2+} + 2SO_4^{2-} + 2H^+ \qquad (2\text{-}25)$$

Subsequently Fe^{2+} is oxidized to Fe^{3+}

$$Fe^{2+} + \tfrac{1}{4}O_2 + H^+ \rightarrow Fe^{3+} + \tfrac{1}{2}H_2O \qquad (2\text{-}26)$$

Generally the solubility product of ferric hydroxide and/or goethite is exceeded, and one or more of these phases precipitate

$$Fe^{3+} + 3H_2O \rightarrow Fe(OH)_3 + 3H^+ \qquad (2\text{-}27)$$

However, some Fe^{3+} remains in solution and catalyzes the oxidation of additional sulfur

$$FeS_2 + 14Fe^{3+} + 8H_2O \rightarrow 15Fe^{2+} + 2SO_4^{2-} + 16H^+ \qquad (2\text{-}28)$$

The hydrogen ions released during this process attack a variety of minerals in the soil zone. If carbonates are present, these tend to neutralize the sulfuric acid. In limestone terrains gypsum tends to form via the reaction

$$CaCO_3 + 2H^+ + SO_4^{2-} + H_2O \rightarrow CaSO_4 \cdot 2H_2O + CO_2 \qquad (2\text{-}29)$$

In silicate terrains H_2SO_4 acts like H_2CO_3 to release Na^+, K^+, Ca^{2+}, and Mg^{2+}, especially from feldspars and ferromagnesian minerals.

Bacteria have been known for many years to be present in acid mine drainage waters; however, their major contribution to acid formation in nature has only recently been demonstrated. Rates of Fe^{2+} oxidation in acid media have been shown to increase by a factor of 10^6 by microbial mediation (Lundgren et al., 1972). The dominant oxidizing bacteria in acid mine drainage are apparently the iron-oxidizing bacteria *Thiobacillus ferrooxidans* and the sulfur-oxidizing bacteria *Thiobacillus thiooxidans*. The former grows poorly above pH 4, and it seems likely that in the pH range 3.5–5.0 the filamentous iron bacterium *Metallogenium* is more important than *T. ferrooxidans* as a catalyst for the oxidation of Fe^{2+} to Fe^{3+} (Walsh and Mitchell, 1972a–c). Above pH 5 the inorganic oxidation of Fe^{2+} to Fe^{3+} is so rapid, that bacteria play only a minor role in mediating the process. Singer and Stumm's (1970) data for the pH-dependence of the rate of inorganic oxidation of Fe^{2+} to Fe^{3+} are shown in Fig. 2-9. Above pH 5 the rate of oxidation of Fe^{2+} is first order in $m_{Fe^{2+}}$ and P_{O_2} and second order in

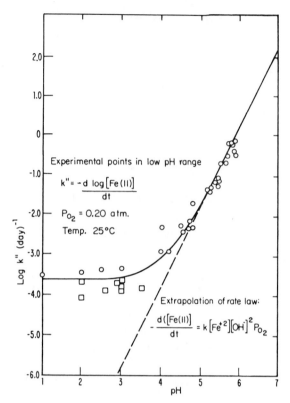

Figure 2-9. Oxygenation rate of ferrous iron as a function of pH (Stumm and Morgan, 1970).

a_{OH^-}, so that

$$-dm_{Fe^{2+}}/dt = km_{Fe^{2+}} \cdot a_{OH^-}^2 P_{O_2} \qquad (2\text{-}30)$$

where $k = (8.0 \pm 2.5) \times 10^{13}$ min^{-1} atm^{-1} mol^{-2}.

At a given pH the rate of oxidation increases about tenfold for a 15°C increase in temperature.

Equation 2-30 can be recast in the form

$$-\frac{d\ln m_{Fe^{2+}}}{dt} = \frac{k_H m_{O_2}(\text{aq})}{a_{H^+}^2} \qquad (2\text{-}31)$$

where k_H at 20°C is 3×10^{-12} min^{-1} mol liter^{-1}. At the temperature and pH of most normal ground water the inorganic oxidation of Fe^{2+} to Fe^{3+} in solution is apparently complete in a matter of hours or days, at any rate in a period of time less than the residence time of most ground waters. Bacterial oxidation is, therefore, probably of minor importance compared to the inorganic oxidation of Fe^{2+}.

The sulfur oxidizing bacteria *T. thiooxidans* oxidizes sulfur most effficiently near pH 3, and only slowly at pH values of 7 (Zajic, 1969, Chap. 4). However, two bacteria, *T. thioparus* and *T. novellus*, apparently operate well in the near-neutral pH range characteristic of most ground waters. Little is known about the importance of these or other sulfur-oxidizing bacteria in the oxidation of pyrite in normal weathering zones. Generally, pyrite is completely destroyed during weathering, and appears only rarely as a constituent of the heavy mineral fraction of stream sediments.

The Oxidation of Ferrous Iron

Most of the ferrous iron in minerals in soil zones that is released as a result of attack by H_2CO_3 and H_2SO_4 is oxidized to ferric ion and is precipitated largely as a constituent of ferric hydroxide, goethite, hematite, gibbsite, or montmorillonite. Even in highly oxidized soil zones a small fraction of the Fe^{3+} is removed in solution; where reducing conditions prevail, Fe^{2+} may not be oxidized to Fe^{3+}, and rather large quantities of iron may be removed with ground waters. Concentrations of Fe^{2+} up to 50 mg/kg have been reported, for instance from siderite-bearing aquifers in New Jersey (Whittemore and Langmuir, 1975; Langmuir, 1969; Coonley et al., 1971), but <1 mg/kg iron is present in most ground waters, and the mobility of Fe^{3+} is generally of the same order as that of Al^{3+}.

Some iron minerals are extremely resistant to oxidative weathering. Magnetite, in particular, stubbornly refuses to be oxidized, even though it is far out of equilibrium with the atmosphere and the oxidation state of most ground waters. Since magnetite contains only a small fraction of the iron in most rock types, its resistance to weathering does not have an important effect on the rate of oxygen use during chemical weathering. The mineral is, however, a very poor indicator of the oxidation state of its environment, and its survival can tell us little about the history of atmospheric oxygen.

SUMMARY AND IMPLICATIONS FOR WEATHERING IN THE PAST

Chemical weathering today is most strongly influenced by rainfall, atmospheric composition, and the land biota. The present climate is abnormally cold, and chemical weathering during the more common, warmer periods was probably more intense, all other parameters being equal. The development of soil zones during the Pleistocene epoch has been strongly affected by glacial scouring and deposition, particularly in the higher latitudes of the northern hemisphere. Chemical weathering in these areas prior to the onset of widespread glaciation must have been similar to weathering in non-glaciated areas of comparable climate today. Very thick soil zones certainly developed in areas where soil formation since the Wisconsin ice age has been minor (e.g., Hunt, 1972, Chap. 9; Basham, 1974).

Total rainfall on the continents has surely varied with time; the probable extent of this variation and its influence on total runoff is assessed in Chapter III. It seems likely that there have always been extremes of rainfall, and that in areas of intense rainfall chemical weathering has been more pronounced than in arid regions. The probable effects of this heterogeneity on the total acid consumption of the continents are described in Chapter IV.

In the weathering of all but the evaporite minerals, the effect of the land biota has turned out to be of major importance. The dissolution of the carbonate minerals and the destruction of the silicates are very dependent on the concentration of H_2CO_3 in soil atmospheres. The oxidation of Fe^{2+} in these minerals is preceded by their dissolution; therefore, oxidation also depends on CO_2 in weathering zones. The high value of P_{CO_2} in soil atmospheres and the considerable concentration of dissolved CO_2 in ground waters is largely due to root respiration and plant decay in soils. In most areas the rate of plant decay averaged over a number of years is nearly equal to the rate of photosynthesis. The rate of chemical weathering is, therefore, strongly influenced by the intensity of photosynthesis. Since this intensity depends on climate (see for instance Cooper, 1975), temperature, rainfall, and light intensity influence the rate of chemical weathering both directly and indirectly.

The rate of photosynthesis depends on other variables as well. Chief among these are the CO_2 and O_2 content of the atmosphere. The rate of

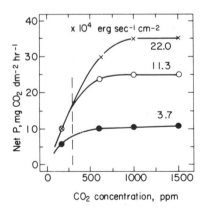

Figure 2-10. Net photosynthesis of intact young spinach plants in relation to CO_2 concentration at different light intensities; light intensities of 3.7, 11.3, and 22.0 × 10^4 erg sec^{-1} cm^{-2}, as shown. The normal CO_2 concentration of air is represented by the dashed line (after Zelitch, 1971).

photosynthesis by most plants rises steeply with increasing CO_2 pressure, and reaches a plateau at CO_2 concentrations which are generally greater than that of the present atmosphere (Gaffron, 1960; Kramer, 1958; Thomas and Hill, 1949; Thomas, 1965; Zelitch, 1971). Practically all the carbon dioxide response curves are similar to those in Fig. 2-10, which shows that the present carbon dioxide concentration of the air is insufficient to produce optimum rates of photosynthesis. The photosynthesis yield of many plants can be increased by a factor of 2 by simply increasing the CO_2 content of the air. There is, however, a CO_2 pressure, beyond which the rate of photosynthesis begins to decline. For some plants this occurs when the CO_2 concentration rises beyond 0.02 atm, that is roughly 60 times the present CO_2 pressure (Rabinowitch, 1951, pp. 901–903). Different species can withstand different pressures of carbon dioxide. Spoehr (1926, p. 129) has pointed out that mosses and the lower plants are especially resistant to the effects of asphyxiating gases. However, even algal growth is inhibited when the CO_2 pressure reaches 0.3 atm.

Photosynthesis can take place at CO_2 pressures very much lower than the present atmospheric value. However, the rate of photosynthesis decreases rapidly with decreasing CO_2 pressure, and for many plants drops below the CO_2 production rate via respiration at CO_2 pressures between roughly one-third and one-tenth of the present atmospheric CO_2 pressure (Zelitch, 1971). Figure 2-11 shows that the net rate of photosynthesis for tobacco leaves at 2500 ft-c and 25°C is 0 when the CO_2 content of the atmosphere is 50 ppm, that is, one-sixth of the present atmospheric value. However, some land plants, for instance certain types of maize, have CO_2 compensation points as low as a few parts per million of CO_2. Compensation points for

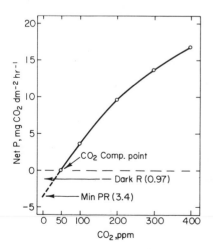

Figure 2-11. Effect of CO_2 concentration on net photosynthesis of tobacco leaves at 2500 ft-c and 25°C. Each upper point was the mean of 45 observations made on attached leaves sealed in a chamber, and the CO_2 compensation point was the mean of 15 determinations. Dark respiration was measured after the other measurements were completed and the leaves had been in darkness for at least 10 min; (Zelitch, 1971).

CO_2 in the same range have also been observed for Chlorella, Scenedesmus, and Fucus (Brown and Tregunna, 1967).

The geologic record of an abundant vascular terrestrial flora since Devonian time indicates that the CO_2 content of the atmosphere has not been below the CO_2 compensation point of today's higher land plants during the past 400 million yr. The conversion of this dictum into usable numbers depends, of course, on a knowledge of the CO_2 compensation point of ancient plants.

REFERENCES

Alexander, M., 1961, *Introduction to Soil Microbiology*, Wiley, New York, 472 pp.

Basham, I. R., 1974, Mineralogical changes associated with deep weathering of gabbro in Aberdeenshire, *Clay Minerals* **10**, 189–202.

Berner, R. A., 1971, *Principles of Chemical Sedimentology*, McGraw-Hill, New York, 240 pp.

Blount, C. W. and Dickson, F. W., 1973, Gypsum–anhydrite equilibria in systems $CaSO_4$–H_2O and $CaSO_4$–$NaCl$–H_2O, *Am. Mineral.* **58**, 323–331.

Bricker, O. P., Nesbitt, N. H., and Gunter, W. D., 1973, The stability of talc, *Am. Mineral.* **58**, 64–72.

Brown, D. L. and Tregunna, E. B., 1967, Inhibition of respiration during photosynthesis by some algae, *Can. J. Bot.* **45**, 1135–1143.

Busenberg, E. and Clemency, C. V., 1976, The dissolution kinetics of feldspars at 25°C and 1 atm. CO_2 partial pressure, *Geochim. Cosmochim. Acta* **40**, 41–49.

Christ, C. L., Hostetler, P. B. and Siebert, R. M., 1973, Studies in the system MgO–SiO_2–CO_2–H_2O (III): The activity-product constant of sepiolite, *Am. J. Sci.* **273**, 65–83.

Coonley, L. S., Jr., Baker, E. B., and Holland, H. D., 1971, Iron in the Mullica River and in Great Bay, New Jersey, *Chem. Geol.* **7**, 51–63.

Cooper, J. P., Ed., 1975, *Photosynthesis and Productivity in Different Environments*, Cambridge Univ. Press, Cambridge, England. 715 pp.

Craig, D. C. and Loughnan, F. C., 1964, Chemical and mineralogical transformations accompanying the weathering of basic volcanic rocks from New South Wales, *Aust. J. Soil Res.* **2**, 218–234.

Ellis, A. J., 1959, The solubility of carbon dioxide in water at high temperatures, *Am. J. Sci.* **257**, 217–234.

Gaffron, H., 1960, Energy storage: Photosynthesis, in *Plant Physiology, A Treatise*, Vol. 1B, Chap. 4, F. C. Steward, Ed., Acedemic Press, New York.

Garrels, R. M., 1967, Genesis of some ground waters from igneous rocks, in *Researches in Geochemistry*, Vol. 2, P. H. Abelson, Ed., Wiley, New York, pp. 405–420.

Garrels, R. M. and Christ, C. L., 1965, *Solutions, Minerals and Equilibria*, Harper & Row, New York, 450 pp.

Garrels, R. M. and Mackenzie, F. T., 1971, in *Evolution of Sedimentary Rocks*, Chap. 6, Norton, New York.

Goldich, S. S., 1938, A study of rock weathering, *J. Geol.* **46**, 17–58.

Hanshaw, B. B., Back, W., and Rubin, M., 1965, Carbonate equilibria and radio-carbon distribution related to ground water flow in the Floridian limestone aquifer, U.S.A., in International Association Scientific Hydrologists, Symposium at Dubrovnik, Paris, pp. 601–604.

Harned, H. S. and Davis, R., Jr., 1943, The ionization constant of carbonic acid in water and the solubility of carbon dioxide in water and aqueous salt solutions from 0° to 50°, *J. Amer. Chem. Soc.* **65**, 2030–2037.

Hay, R. L. and Jones, B. F., 1972, Weathering of basaltic tephra on the Island of Hawaii, *Bull. Geol. Soc. Am.* **83**, 317–332.

Helgeson, H. C., 1971, Kinetics of mass transfer among silicates and aqueous solutions, *Geochim. Cosmochim. Acta* **35**, 421–469.

Hemley, J. J., Montoya, J. W., Christ, C. L., and Hostetler, P. B., 1977, Mineral equilibria in the MgO–SiO_2–H_2O system: I talc-chrysotile-forsterite-brucite stability relations, *Am. J. Sci.*, **227**, 322–351.

Holland, H. D., Kirsipu, T. V., Huebner, J. S., and Oxburgh, U. M., 1964, On some aspects of the chemical evolution of cave waters, *J. Geol.* **72**, 36–67.

Holland, H. D. and Borcsik, M., 1965, On the solution and deposition of calcite in hydrothermal systems, pp. 364–374 in *Symposium Problems of Postmagmatic Ore Deposition*, Vol. II, M. Štemprok, Ed., The Geological Survey of Czechoslovakia, Prague.

Holland, H. D. and Malinin, S. D., 1978, The solubility and occurrence of non-ore minerals, in *Geochemistry of Hydrothermal Ore Deposits*, 2nd ed., H. L. Barnes, Ed., Holt, Rinehart, and Winston, New York.

Hostetler, P. B. and Christ, C. L., 1968, Studies in the system MgO–SiO_2–CO_2–H_2O (I): The activity-product constant of chrysotile, *Geochim. Cosmochim. Acta* **32**, 485–497.

Hunt, C. B., 1972, *Geology of Soils*, W. H. Freeman, San Francisco, 344 pp.

Keeling, C. D., 1973, The carbon dioxide cycle: reservoir models to depict the exchange of atmospheric carbon dioxide with the oceans and land plants, in *Chemistry of the Lower Atmosphere*, S. I. Rasool, Ed., Chap. 6, Plenum Press, New York.

Kramer, P. J., 1958, Photosynthesis of trees as affected by their environment, in *The Physiology of Forest Trees*, K. V. Thimann, Ed., Chap. 8, Ronald Press, New York.

Langmuir, D., 1969, Geochemistry of iron in a coastal-plain groundwater of the Camden, New Jersey area, U.S. Geological Survey Professional Paper 650-C, 224–235.

Langmuir, D., 1971, The geochemistry of some carbonate ground waters in central Pennsylvania, *Geochim. Cosmochim. Acta* **35**, 1023–1045.

Loughnan, F. C., 1969 *Chemical Weathering of the Silicate Minerals*, American Elsevier, New York, 154 pp.

Luce, R. W., Bartlett, R. W., and Parks, G. A., 1972, Dissolution kinetics of magnesium silicates, *Geochim. Cosmochim. Acta* **36**, 35–50.

Lundegardh, H., 1927, Carbon dioxide evolution of soil and crop growth, *Soil Sci.* **23**, 117–153.

Lundgren, D. G., Vestal, J. R., and Tabita, F. R., 1972, The microbiology of mine drainage pollution, in *Water Pollution Microbiology*, R. Mitchell, Ed., Chap. 4, Wiley-Interscience, New York.

Makarov, B. N., 1960, Respiration of soil and composition of soil air on drained peat-bog soils, *Sov. Soil Sci.*, 154–160.

Marval, R. H. A., 1968, Estudio sobre la meteorizacion de la peridotita de Tinaquillo, Edo. Cojedes, Thesis, Universidad Central de Venezuela, Caracas.

Olson, J. S., 1970, Carbon cycle and temperate woodlands, in *Ecological Studies, I. Temperate Forest Ecosystems*, Chap. 15 D. E. Reichle, Ed., Springer-Verlag, New York.

Pačes, T., 1973, Steady state kinetics and equilibrium between ground water and granitic rock, *Geochim. Cosmochim. Acta* **37**, 2641–2663.

Plummer, L. N., Vacher, H. L., Mackenzie, F. T., Bricker, O. P., and Land, L. S., 1976, Hydrogeochemistry of Bermuda: A case history of ground-water diagenesis of biocalcarenites, *Bull. Geol. Soc. Am.* **87**, 1301–1316.

Rabinowitch, E. I., 1951, *Photosynthesis and Related Processes*, Vol. II, Part I, Interscience, New York, 1208 pp.

Robie, R. A. and Waldbaum, D. R., 1968, Thermodynamic properties of minerals and related substances at 298.15°K (25.0°C) and one atmosphere (1.013 bars) pressure and at higher temperatures, U.S. Geological Survey Bulletin 1259, 256 pp.

Russell, E. W., 1961, *Soil Conditions and Plant Growth*, 9th Ed., Wiley, New York.

Schellmann, W., 1964, Zur lateritischen Verwitterung von Serpentinit, *Geol. Jahrb. Hannover* **81**, 645–678.

Silverman, M. P. and Munoz, E. F., 1970, Fungal attack on rock: solubilization and altered infrared spectra, *Science* **169**, 985–987.

Singer, P. C. and Stumm, W., 1970, Acidic mine drainage: the rate determining step, *Science* **167**, 1121–1123.

Spoehr, H. A., 1926, *Photosynthesis*, Chemical Catalog Co., New York.

Stumm, W. and Morgan, J. J., 1970, *Aquatic Chemistry*, Wiley-Interscience, New York, 583 pp.

Tardy, Y., 1971, Characterization of the principal weathering types by the geochemistry of waters from some European and African crystalline massifs, *Chem. Geol.* **7**, 253–271.

Thomas, M. D., 1965, Photosynthesis (Carbon Assimilation): environmental and metabolic relationships, Vol. IVA, in *Plant Physiology, A Treatise*, Chap. 1, F. C. Steward, Ed., Academic Press, New York.

Thomas, M. D. and Hill, G. R., 1949, Photosynthesis under field conditions, in

Photosynthesis in Plants, Chap. 2, J. Franck and W. E. Loomis, Ed., Iowa State College Press. Ames, Iowa.

Thrailkill, J. V., 1964, The excavation of limestone caves and the deposition of speleothems. I. Chemical and hydrological factors in the excavation of limestone caves. II. Water chemistry and carbonate speleothem relations in Carlsbad Caverns, New Mexico, Ph.D. Thesis, Princeton University.

Thrailkill, J., 1970, Solution geochemistry of the water of limestone terrains, University of Kentucky Water Resources Institute, Lexington, Kentucky, Research Report 19, 125 pp.

Thrailkill, J., 1972, Carbonate chemistry of aquifer and stream water in Kentucky, *J. Hydrol.* **16**, 93–104.

Walsh, F. and Mitchell, R., 1972a, a pH-dependent succession of iron bacteria, *Environ. Sci. Technol.* **6**, 809–812.

Walsh, F. and Mitchell, R., 1972b, An acid-tolerant iron-oxidizing Metallogenium, *J. Gen. Microbiol*, **72**, 369–376.

Walsh, F. and Mitchell, R., 1972c, Biological control of acid mine pollution, *J. Water Pollut. Control*, **44**, 763–768.

White, D. E., Hem, J. D. and Waring, G. A., 1963, Chemical composition of subsurface waters, in *Data of Geochemistry*. U. S. Geological Survey Professional Paper 440-F, 6th ed., Chapter F.

Whittemore, D. O. and Langmuir, D., 1975, The solubility of ferric oxyhydroxides in natural waters, *Groundwater* **13**, 360–365.

Wollast, R., 1967, Kinetics of the alteration of K-feldspar in buffered solutions at low temperatures, *Geochim. Cosmochim. Acta* **31**, 635–648.

Woodwell, G. M., Whittaker, R. H., Reiners, W. A., Likens, G. E., Delwiche, C. C., and Botkin, D. B., 1978, The biota amid the world carbon budget, *Science*, **199**, 141–146.

Zajic, J. E., 1969, *Microbial Biogeochemistry*, Academic Press, New York, 345 pp.

Zeissink, H. E., 1969, The mineralogy and geochemistry of a nickeliferous laterite profile (Greenvale, Queensland, Australia), *Miner. Deposita* **4**, 132–152.

Zelitch, I., 1971, *Photosynthesis, Photorespiration, and Plant Productivity*, Academic Press, New York, 347 pp.

CHAPTER III

EVAPORATION, RAINFALL, AND RUNOFF

The combined rate of physical and chemical weathering is geologically rapid. Measurements of stream loads and sediment accumulation in the oceans demonstrate that present-day continental denudation rates are a few centimeters per thousand years (see for instance Judson and Ritter, 1964; Holeman, 1968). The mean life of mountain ranges with respect to erosion to base level is, therefore, measured in tens of millions of years. If this has also been true in the past, the mean erosion rate for any given 100-million-year period of earth history has been virtually identical to the mean rate of uplift during that period. Rates of erosion are coupled to rates of uplift through the current topography of the earth and via the contemporary efficiency of erosional processes. For a particular rate of uplift, the mean elevation during a particular period of earth history is proportional to the contemporary inefficiency of erosional processes.

The chemical composition of eroded matter depends on the mean composition of the material lifted above sea level, on the relative effectiveness of chemical weathering compared to physical weathering, and on the response of the uplifted materials to the chemistry of the atmosphere. In the complete absence of water, weathering at present earth surface temperatures would be inefficient and almost entirely physical. It is a truism that present-day weathering owes much of its efficiency to the large amount of rainfall and runoff from the continents. This chapter is concerned with the relationships among evaporation, rainfall, and runoff, and with the extent to which they may have varied in the past in response to climatic and geographic variables.

THE TOTAL AMOUNT OF EVAPORATION

Rainfall on land depends largely on evaporation at sea. Unfortunately the direct measurement of evaporation rates at sea is rather difficult. Sverdrup (1951) summarized the somewhat unsatisfactory state of current theory and data at that time, but pointed out that the three most promising methods for calculating evaporation rates from the oceans gave surprisingly concordant results. The analysis of the heat balance of the earth has made considerable progress since the early fifties (Budyko, 1956; Budyko et al., 1962; Budyko, 1963; Budyko and Kondratiev, 1964). Sellers (1965) has used these data, together with those of Brooks and Hunt (1930), Meinardus (1934), and Wüst (1954) to derive the figures in Table 3-1 for the annual evaporation of water in successive 10° strips of latitude. Figure 3-1 compares Budyko's (1963) latitudinal distribution of annual evaporation over land and sea with that obtained by Manabe and Holloway (1975) from a mathematical model of the atmosphere; Figure 3-2 compares Manabe and Holloway's (1975) calculated mean annual rate of evaporation for continents plus oceans with Budyko's (1963) observations. The agreement between computation and observation is quite good, and the results of such computations for meteoro-

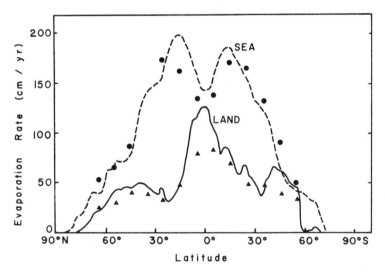

Figure 3-1. The latitudinal distribution of the simulated annual evaporation rate broken down into means over land (solid line) and over sea (dashed line) (Manabe and Holloway, 1975). For comparison, values for evaporation from land and sea areas derived from observed data (Budyko, 1963) are indicated by triangles and dots, respectively. Copyrighted by American Geophysical Union.

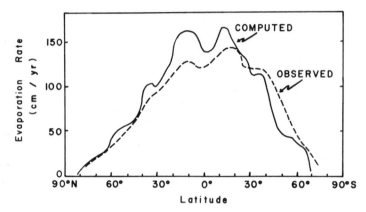

Figure 3-2. Latitudinal distribution of the annual mean rate of evaporation computed by Manabe and Holloway (1975) and derived from observed data (Budyko, 1963). Copyrighted by American Geophysical Union.

logic data in continental configurations other than those of the present day can, therefore, be taken quite seriously.

The distribution of land and sea is well known. Table 3-2 (List, 1958) is based on Kossinna's (1921) figures, which have been confirmed by Menard and Smith (1966). These data and Sellers' (1965) figures for evaporation from the various oceans show that the current annual rate of evaporation of ocean water is approximately 4.4×10^{20} cm^3/yr, and that the total annual evaporation of water from land areas is 0.7×10^{20} cm^3/yr (Table 3-3). Total evaporation is, therefore, near 5.1×10^{20} cm^3/yr, and the mean global evaporation rate is near 100 cm^3/cm^2 yr.

These data can be used to set limits on the probable variation of total evaporation. Mean annual temperatures have normally been higher than at present (see, for instance, Dorf, 1960). Fairbridge (1964) has pointed out (see Figure 3-3) that the greatest temperature differences between past and present climates are to be found in latitudes north of 30°N and south of 30°S, while equatorial temperatures have tended to be rather constant and similar to present-day values. As shown in Figures 3-1 and 3-2 the evaporation rate of ocean water decreases rapidly today with decreasing mean annual temperature in the higher latitudes. A strong maximum can therefore be set for the annual evaporation of water from the oceans during warm periods of earth history, such as the late Cretaceous and the mid-Tertiary periods, by taking the mean rate of evaporation for the entire ocean to have been equal to the current evaporation rate for the latitude belt 30°N to 30°S. Despite such greatly increased evaporation rates, particularly in very

Table 3-1

Some Physical and Climatological Data[a]

Latitude Zone	O	h	T	r	E	w_a	PE	n	α
80–90° N	93.4	137	249.6	12.0	4.2	0.490	6.7	62	61
70–80	71.3	220	257.3	18.5	14.5	0.648	7.8	66	46
60–70	29.4	202	266.0	41.5	33.3	0.852	13.3	65	24
50–60	42.8	296	273.7	78.9	46.9	1.164	18.6	60	14
40–50	47.5	382	280.7	90.7	64.1	1.521	16.3	53	12
30–40	57.2	496	287.2	87.2	100.2	1.895	12.6	46	10
20–30	62.4	366	293.6	79.0	124.6	2.637	8.2	43	10
10–20	73.6	146	298.3	115.1	138.9	3.673	8.6	47	9
0–10	77.2	158	298.7	193.4	123.5	4.107	12.9	52	8
0–90° N.	60.6	284	286.4	100.9	94.4	2.385	12.1	52	14
0–10° S	76.4	154	298.0	144.5	130.4	4.090	9.7	52	7
10–20	78.0	121	296.5	113.2	154.1	3.666	8.5	48	8
20–30	76.9	156	292.0	85.7	141.6	2.986	7.9	48	8
30–40	88.8	106	286.7	93.2	125.6	2.381	10.7	54	8
40–50	97.0	5	281.9	122.6	89.5	1.810	18.6	66	7
50–60	99.2	5	274.4	104.6	52.0	1.261	22.7	72	17
60–70	89.6	388	262.2	41.8	17.4	0.684	16.7	76	21
70–80	24.6	1420	243.7	8.2	4.5	0.287	7.8	65	63
80–90	0.0	2272	225.3	3.0	0.0	0.156	5.3	54	84
0–90° S	80.9	216	284.6	100.0	106.4	2.549	12.1	57	13
Globe	70.8	250	285.5	100.4	100.4	2.467	12.1	54	13

[a] After Sellers, 1965. O = percentage of surface covered by ocean; h = mean elevation in meters; $T(°K)$ = mean annual surface temperature; r = mean annual precipitation in centimeters; E = mean annual evaporation in centimeters; w_a = centimeters of precipitable water vapor; PE = precipitation efficiency; n = percentage of cloud cover; α = short-wave albedo.

high latitudes, the evaporation of water from the oceans calculated on this basis would only have been about 30% greater than at present because only half of the earth's surface lies outside the latitude belt 30°N–30°S. Even this rather modest increase may well be an overestimate. It seems likely that the meridional transport of heat by atmospheric circulation is less rapid during nonglacial periods than during glacial and interglacial periods, and that mean wind velocities are probably smaller during warm periods of earth history. As the rate of evaporation of water depends on wind speed (Sverdrup, 1951), the rate of evaporation during warm periods may have been somewhat smaller at a given temperature than the rate of evaporation at the same temperature during glacial or interglacial times.

Table 3-2

Distribution of Water and Land in Various Latitude Belts[a]

Latitude	Northern Hemisphere				Southern Hemisphere			
	Water	*Land*	*Water*	*Land*	*Water*	*Land*	*Water*	*Land*
90–85°	0.979	—	100.0	—	—	0.978	—	100.0
85–80	2.545	0.384	86.9	13.1	—	2.929	—	100.0
80–75	3.742	1.112	77.1	22.9	0.522	4.332	10.7	89.3
75–70	4.414	2.326	65.5	34.5	2.604	4.136	38.6	61.4
70–65	2.456	6.116	28.7	71.3	6.816	1.756	79.5	20.5
65–60	3.123	7.210	31.2	69.8	10.301	0.032	99.7	0.3
60–55	5.399	6.613	45.0	55.0	12.006	0.006	99.9	0.1
55–50	5.529	8.066	40.7	59.3	13.388	0.207	98.5	1.5
50–45	6.612	8.458	43.8	56.2	14.693	0.377	97.5	2.5
45–40	8.411	8.016	51.2	48.8	15.833	0.594	96.4	3.6
40–35	10.029	7.627	56.8	43.2	16.483	1.173	93.4	6.6
35–30	10.806	7.943	57.7	42.3	15.782	2.967	84.2	15.8
30–25	11.747	7.952	59.6	40.4	15.438	4.261	78.4	21.6
25–20	13.354	7.145	65.2	34.8	15.450	5.049	75.4	24.6
20–15	14.981	6.164	70.8	29.2	16.147	4.998	76.4	23.6
15–10	16.553	5.080	76.5	23.5	17.211	4.422	79.6	20.4
10– 5	16.628	5.332	75.7	24.3	16.898	5.062	76.9	23.1
5– 0	17.387	4.737	78.6	21.4	16.792	5.332	75.9	24.1
90– 0°	154.695	100.281	60.7	39.3	206.364	48.611	80.9	19.1

[a]List, 1958.
All oceans and seas $361.059 \times 10^6 km^2$, 70.8%.
All land $148.892 \times 10^6 km^2$, 29.2%.

Table 3-3

Annual Water Balance of the Oceans[a]

Ocean	E	r	Δf_L	Δf_0	Δf
Atlantic Ocean	104.0	78.0	−20.0	−6.0	−26.0
Indian Ocean	138.0	101.0	−7.0	−30.0	−37.0
Pacific Ocean	114.0	121.0	−6.0	13.0	7.0
Arctic Ocean	12.0	24.0	−23.0	35.0	12.0
All Oceans	125.0	112.0	−13.0	0	−13.0

[a]Sellers, 1965, p. 85. E = total evaporation; r = total precipitation; Δf_L for oceans = inflow from surrounding lands; Δf_0 for oceans = inflow from surrounding oceans, all in centimeters per year. $\Delta f = \Delta f_L + \Delta f_0$.

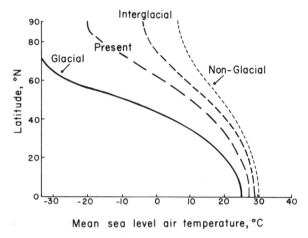

Figure 3-3. Approximate latitudinal variation of mean sea level air temperature for glacial, present, interglacial, and nonglacial, e.g., late Cretaceous or mid-Tertiary time (after Fairbridge, 1964).

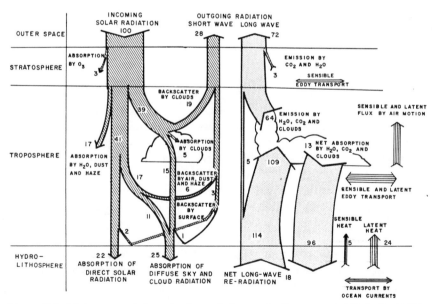

Figure 3-4. Mean annual heat budget of the earth (Rotty and Mitchell, 1974) slightly modified, showing principal items affecting disposition of short-wave solar radiation (fine stippling), long-wave terrestrial radiation (coarse stippling), and heat transport by air motions. Units are the percentage of insolation arriving at the top of the atmosphere, based mostly on estimates for the northern hemisphere by Budyko and Kondratiev (1964). Units of sensible and latent heat flux through the earth's surface (lower right) are estimates for the whole planet.

Strong constraints on total evaporation are also set by the earth's energy balance. The total annual evaporation of water from the oceans is severely limited by the availability of solar energy. Figure 3-4 shows that of 100 units of incoming solar radiation about 28 are backscattered by clouds, air, dust, haze, and the earth's surface. Forty-seven units are absorbed by the hydrosphere and lithosphere. Twenty-four of these are used for evaporation of water. If all of the 47 units now absorbed by the hydrosphere and lithosphere were used for evaporating water, the total evaporation rate would be nearly doubled, but it is difficult to see how the net long-wave reradiation could be changed so drastically. Therefore, it seems unlikely that the total evaporation rate was ever as much as twice its present value, unless the solar constant was considerably greater than it is at present. This seems somewhat unlikely; however, Wetherald and Manabe (1975) estimate that an increase of only 6% in the solar constant would increase the intensity of the hydrologic cycle by 27%, and, in the absence of satisfactory explanations of the lack of detectable solar neutrinos, it is difficult to have confidence in predictions of the solar constant based on the theory of the solar interior (Ulrich, 1975; Bahcall and Davis, 1976).

THE TOTAL AMOUNT OF PRECIPITATION

Precipitation on land is a well-measured quantity, and rainfall at sea is probably most reliably estimated by subtracting the figure for continental rainfall from the total amount of annual evaporation. It is rather interesting to compare the percentage of evaporation from and precipitation on land and sea with the areas covered by continents and oceans. As might have been expected, the data of Table 3-4 show that the percentage of the total annual evaporation from the ocean is proportionately much larger than from the continents. The distribution of precipitation is more nearly in proportion to the relative areal extent of continents and oceans, but average rainfall per unit area of continent is still somewhat smaller than average rainfall per unit area of ocean.

Figure 3-5 shows Gerasimov's (1964) summary of the latitude dependence of precipitation on land. The peak in the tropics is well developed, there is a

Table 3-4

Comparison of Evaporation, Precipitation, and Areal Extent of
Oceanic and Continental Areas

Percentage	Oceans (%)	Continents %
Of total area	71	29
Of annual evaporation	88	12
Of annual precipitation	79	21

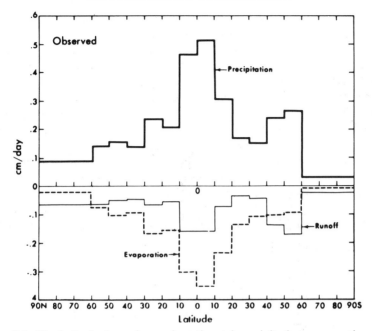

Figure 3-5. The latitude dependence of continental precipitation, evaporation, and runoff (Gerasimov, 1964), adapted from Manabe and Holloway (1975). Copyrighted by American Geophysical Union.

pronounced high in the southern hemisphere at midlatitudes; at high latitudes precipitation is small in both hemispheres.

The total annual rainfall on the continents depends not only on the total global evaporation rate but on a host of meteorologic and physiographic variables. The nature of the global circulation and the distribution, size, and topography of the continents are obviously important. South America receives nearly twice (see Table 3-6) and Australia only a little more than half of the average continental rainfall, but the remaining continents have very nearly the same average rainfall. In fact, the relative uniformity of rainfall over continent-sized areas is rather remarkable, and it is likely that a given increase in the mean evaporation rate of ocean water would be matched by a similar increase in the mean precipitation rate over continental areas.

The Effect of Land Area on Continental Rainfall

The presently exposed area of the continents is abnormally large. In the past the exposed area of the continents has generally been smaller, but probably not less than one-half the present area since mid-Precambrian time. A reduction of the subaerially exposed continental area from its present value of 29 to 15% of the earth's surface, and a corresponding increase of oceanic area from 71 to 85% would affect both the total amount of evaporation and the distribution of precipitation between land and sea. Total evaporation is roughly proportional to the area covered by ocean, and rainfall on land is roughly proportional to land area. A reduction of land area by a factor of 2 would, therefore, probably reduce the total volume of rain falling on the exposed continents by approximately a factor of $(85/71)$ $(15/29)=0.62$.* With decreasing size, continents might be expected to receive a progressively larger relative share of rainfall. At present the ratio of the rainfall fraction to the area fraction of continental areas is 0.72 (see Table 3-4). This ratio probably approaches unity, and may even exceed unity with progressively decreasing size of the continents. If unity is reached when the land area is reduced by a factor of 2, rainfall would be reduced by 17 rather than by 38%. The actual reduction would probably be close to the lower figure, but would depend heavily on physiographic aspects, such as the presence or absence of mountain barriers in the path of prevailing winds.

The Effect of Continental Drift on Continental Rainfall

The formation of a supercontinent such as Gondwanaland would tend to reduce the ratio of the rainfall fraction to the area fraction. The magnitude of this effect is difficult to evaluate. There is no correlation between the size and annual rainfall on continents today, but rather between annual rainfall and location with respect to dry and wet belts. It seems likely that the formation of supercontinents affects the total rainfall on land areas less strongly than do changes in climate and the inundation of the continents by shallow seas.

On the other hand, changes in the latitudinal distribution of the continents could affect rainfall very considerably. If all of the continents were concentrated in the polar regions, the mean rainfall would certainly be much lower than at present. A concentration of continents in a belt encircling the globe in the equatorial regions could also reduce the mean annual rainfall, but for different reasons. Presently available polar wander curves (McElhinny, 1973) suggest that neither of these extremes has prevailed during the Phanerozoic Era, but detailed model calculations are needed to evaluate the effect of continental motions on rainfall in the past.

The formation of mountain ranges probably has only a small positive effect on total rainfall on land areas, but mountain ranges frequently affect the distribution of rainfall on land very considerably. The effect of rainfall distribution on runoff rates is discussed in the next section.

*For a somewhat more detailed calculation, see p. 73.

The rather surprising outcome of the discussion in this section is that the total rainfall received by land areas has probably varied by less than a factor of 3 since mid-Precambrian time. The typically milder climates produced a rainfall 1.15 ± 0.15 times the present rate; the typically greater degrees of inundation, a rainfall 0.80 ± 0.20 times the present rate; and the presence of supercontinents a rainfall something like 0.90 ± 0.10 times the present rate. The sum of all of these variables produces a mean value of rainfall 0.83 times the present value with a somewhat unlikely upper limit of 1.30 and lower limit of 0.48 times the present total rainfall. Only unexpectedly large variations due to the position of the continents seem capable of upsetting these conclusions.

RIVER RUNOFF FROM THE CONTINENTS

Most dissolved materials that reach the oceans arrive as solutes in river water. The flux of river water from the continents is, therefore, of great importance for the chemistry of the oceans. A substantial number of estimates for the total annual river flux have been reported. Most of these fall between 3.0×10^{19} and 5.0×10^{19} cc/yr (see, for instance, Alekhin and Brazhnikova, 1963; Livingstone, 1963; and Holeman, 1968). The most recent figures are from compilations by Lvovitch (1973): 4.4×10^{19} cc/yr, and by Kozoun et al. (1974); 4.7×10^{19} cc/yr (see Tables 3-5; 3-6). The runoff, Δf, from continental areas depends on a large number of variables (see for instance Sellers, 1965; Budyko, 1974; Kozoun et al., 1974). Many of these are interdependent, and it is difficult to predict precisely the runoff from any given watershed. Certain rough approximations are, however, possible, and a reasonable account can be given for the runoff from major drainage basins, which average many of the idiosyncrasies of small watersheds.

The Relationship Between Runoff and Rainfall

The runoff ratio, $\Delta f / r$, that is, the ratio of river runoff to rainfall in a drainage basin, must depend on the rainfall itself. When rainfall is so heavy that evaporation can only remove a small fraction of the annual water accumulation, the runoff ratio must approach unity. Conversely, when rainfall is so sparse that evaporation can easily remove the annual water accumulation, the runoff ratio is apt to be very much less than 1; in extreme cases it is 0. The maximum annual evaporation is, of course, strongly dependent on the intensity of solar radiation. Various functions have been proposed that relate the runoff ratio, $\Delta f / r$, to the rainfall, r, in the manner required by these extremes. The simplest of these functions was introduced

Table 3-5
Data for Major River Basins of the World[a]

	Locality	Area of Drainage Basin (1000 Km²)	Mean Latitude (°)	Average Rainfall r (cm/yr)	Runoff Ratio Δf/r
	Europe				
1.	Danube	817	45	86.3	0.28
2.	Dniepr	504	49	66.0	0.16
3.	Don	422	50	57.5	0.11
4.	Volga	1360	50	65.7	0.27
	Asia				
5.	Ob	2990	58	54.3	0.24
6.	Yenisei	2580	60	56.0	0.42
7.	Lena	2490	62	46.2	0.46
8.	Kolima	647	65	42.0	0.50
9.	Amur	1855	52	60.5	0.32
10.	Yellow	745	38	45.8	0.16
11.	Yangtze	1800	31	110.0	0.50
12.	Pearl	437	23	158.0	0.53
13.	Mekong	810	20	157.0	0.40
14.	Ganges & Brahmaputra	1730	27	146.5	0.48
15.	Indus	960	29	56.8	0.17
16.	Tigris-Euphrates	750	34	38.2	0.16
	Africa				
17.	Nile	2870	15	73.0	0.034
18.	Senegal	441	15	51.0	0.10
19.	Niger	2090	10	79.8	0.16
20.	Congo	3822	3	155.0	0.24
21.	Orange	1020	29	36.7	0.041
22.	Limpopo	440	23	58.0	0.10
23.	Zambezi	1330	15	99.0	0.081
24.	Dschuba	750	3	41.0	0.056
	North America				
25.	Yukon	852	65	50.8	0.48
26.	Columbia	669	45	78.0	0.51
27.	Colorado	635	36	37.7	0.095
28.	Rio Bravo del Norte	570	31	40.6	0.079
29.	Mississippi	3220	40	87.5	0.21
30.	Missouri	1370	43	60.6	0.10
31.	St. Lawrence	1290	44	103.6	0.33
32.	Nelson	1070	56	56.6	0.14
33.	Mackenzie	1800	65	48.0	0.40
	South America				
34.	Amazon	6915	5	215.0	0.46
35.	Panama + Uruguay	2970	25	124.0	0.20
36.	Orinoco	1000	7	199.0	0.46
37.	San Francisco	600	14	105.0	0.15
	Australia				
38.	Murray	1057	35	51.7	0.043
39.	Darling	650	32	44.4	0.030

[a]Kozoun et al. (1974).

Table 3-6

Water Balance of the Continents and of the Land as a Whole[a]

Continent	Area ($10^6 km^2$)	Approximate Mean Latitude (°)	Average Rainfall, r (cm/yr)		Average Runoff, Δf (cm/yr)		Average Runoff ratio, Δf/r		Total Annual Runoff, ΔF ($10^{19} cc/yr$)	
			Lvovitch 1973	Kozoun 1974	Lvovitch 1973	Kozoun 1974	Lvovitch 1973	Kozoun 1974	Lvovitch 1973	Kozoun 1974
Europe[b]	9.8	52	73	79	32	31	0.43	0.39		
Asia	45.0	50	73	74	29	33	0.40	0.45		
Africa	30.3	10	69	74	14	15	0.20	0.20		
North America[c]	20.7	45	67	76	29	34	0.43	0.45		
South America	17.8	15	165	160	58	66	0.43	0.41		
Australia[d]	7.4	25	44		4.8		0.11			
Antarctica	14.4	80		18		16.5		0.92		
Whole Land	149.0	42	83	80	29	31	0.35	0.39	4.4	4.7

[a]Data from Lvovitch, 1973; Kozoun et al. 1974.
[b]Including Iceland.
[c]Excluding the Canadian Archipelago and including Central America.
[d]Australian continent only.

by Schreiber (1904), who found that the runoff ratio for central European rivers was related approximately to the rainfall by the expression

$$\frac{\Delta f}{r} = e^{-\alpha/r} \tag{3-1}$$

When rainfall is sufficiently heavy so that $\alpha/r \ll 1$, equation 3-1 can be expanded to the form

$$\frac{\Delta f}{r} \cong 1 - \frac{\alpha}{r} \tag{3-2}$$

It follows that α is then equal to the difference $(r - \Delta f)$ between rainfall and runoff, that is, to the quantity of water than can be evaporated annually from a completely saturated land surface. If we let this rate of evaporation be E_0, then equation 3-1 becomes

$$\ln\frac{\Delta f}{r} = -\frac{E_0}{r} \tag{3-3}$$

This implies that the runoff ratio depends on the ratio between the maximum amount of annual evaporation and the annual rainfall. Figure 3-6 shows that this relationship holds reasonably well for the major river basins of the world between latitudes 25° and 40°. The rainfall and runoff data were taken from Kozoun et al. (1974); the numbers accompanying the data points identify the rivers in Table 3-5. The observed scatter in the data is

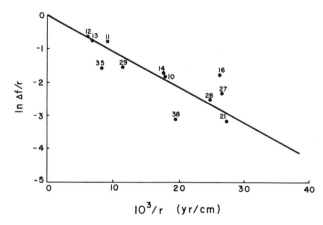

Figure 3-6. Relationship between $\ln\Delta f/r$ and $10^3/r$ for major drainage basins between latitudes 25° and 40° (data from Kozoun et al., 1974).

rather to be expected, since runoff from drainage basins also depends on seasonality of rainfall, topography, vegetation, and on a host of other physical, chemical, and biological variables.

Several other functions relating $\Delta f/r$ to r have been proposed. Examples are the expressions used by Ol'dekop (1911) and Budyko (1974), but these do not seem to improve the fit with river runoff data to any great extent. Sellers (1965) has shown that for many drainage basins in the United States the runoff ratio is linearly related to rainfall, that is,

$$\frac{\Delta f}{r} = br \tag{3-4}$$

However, this expression breaks down when $br > 1$, and is not generally applicable.

The Effect of Latitude on the Runoff Ratio

The Schreiber equation 3-1 predicts that the runoff ratio should be latitude dependent, since the evaporation rate E_0 should decrease rapidly toward lower temperatures at high latitudes. The expected relationship is shown in Figure 3-7. The function $-r\ln(\Delta f/r)$, which is equal to E_0 in the Schreiber equation, has been plotted against the mean latitude of each basin. The decrease of E_0 with increasing latitude is evident. The rather large amount of scatter is due to the individuality of the various basins at each latitude,

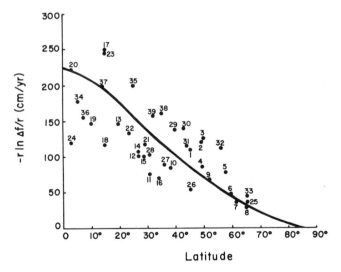

Figure 3-7. The variation of $-r\ln(\Delta f/r)$ of major drainage basins with their mean latitude (data from Kozoun et al., 1974).

and is a measure of the effect of geographic parameters other than latitude and rainfall on the runoff ratio of the major drainage basins.

If the physical basis of the Schreiber equation is approximately correct, there should be a reasonable correlation between the calculated values of E_0, the mean annual temperature, and the mean evaporation rate, E, at various latitudes. These relationships exist, and are shown in Figure 3-8. The error bars are a measure of the scatter in Figure 3-7. The means of the E_0 values rise smoothly with the mean annual temperature and are defined rather closely by the equation

$$E_0 = 1.2 \times 10^9 e^{\frac{-4.62 \times 10^3}{T}} \qquad (3\text{-}4a)$$

The coefficient of the exponent in this expression is surprisingly close to the value of $\Delta H/R$, where ΔH is the heat of vaporization of water in cal/mol and R is the gas constant in cal/mol deg. At latitudes higher than $30°$ the average value of E_0 is nearly the same as the mean annual rate of evaporation from the oceans in the same latitude belt (see Table 3-1). At low latitudes the average value of $E_0 > E$, but the scatter in E_0 for different river basins within each $10°$ belt of latitude includes the values of E at all latitudes.

Table 3-6 summarizes two recent sets of rainfall and runoff ratio data for entire continents and for the continents as a whole. Lvovitch's (1973) data agree reasonably well with those of Kozoun et al. (1974); future refinements will certainly affect the figures somewhat. The function $-r\ln\Delta f/r$ for each continent is plotted against the approximate mean continental latitudes in

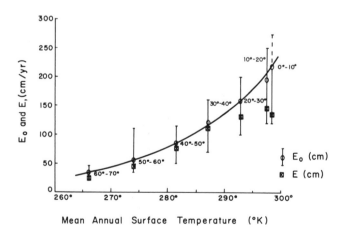

Figure 3-8. The relationship between the average value of E_0, the mean annual evaporation E, and the mean annual surface temperature between the equator and latitude $70°$; data from Figure 3-7 and Sellers (1965, p. 5).

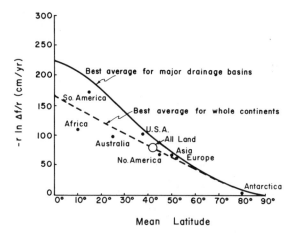

Figure 3-9. The function $-r\ln\Delta f/r$ for whole continents plotted against their approximate mean latitude; data from Lvovitch, 1973; Kozoun et al., 1974, and Sellers, 1965.

Figure 3-9. At high latitudes the continental means agree very well with the data for large river basins. Toward lower latitudes the continental means lie systematically lower than the best fit for large river basins. Much of this divergence is explained below in terms of the effects of heterogeneity in rainfall on a continental scale. It seems that the rather simplified approach to the relationship between runoff and rainfall is adequate for the purposes of the present analysis.

The Effect of Climate on Worldwide Runoff

The climatic and geographic factors that affect rainfall also affect runoff. It was shown above that a return to nonglacial climates may well be accompanied by an increase of roughly 30% in the evaporation rate and in total global rainfall. It is likely that the effect of a milder climate is nearly the same for the total river runoff from the continents. Increased rainfall should increase Δf; however, an increase in mean annual temperature should also increase the value of E_0. Since

$$\Delta f \cong re^{-E_0/r} \tag{3-5}$$

the percentage increase of the runoff, Δf, compared to that of the rainfall, r, depends on the relative increase of E_0 and r during climatic changes.

The data in Figure 3-3 suggest that during nonglacial periods conditions at the poles corresponded to those currently existing at approximately $60°$ latitude. We should, therefore, be able to gain some insight into total river runoff during nonglacial periods by spreading the current variation of the

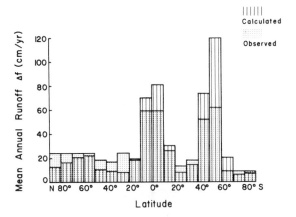

Figure 3-10. Observed and calculated mean annual runoff as a function of latitude.

mean annual runoff between 60°N and 60°S over the entire latitude spectrum from 90°N to 90°S, and by multiplying the new runoff figures for each of the various latitude intervals by the corresponding land areas.

Figure 3-10 shows the variation of the present-day mean annual runoff (Gerasimov, 1964), the values computed from equation 3-5 using the reported values of rainfall in 10° latitude intervals (Kozoun et al., 1974), and the best fit continent-wide values of E_0 along the dashed line in Figure 3-9. The agreement between the observed and calculated runoff values is reasonable, although not spectacular. The most serious discrepancy occurs between 50 and 60°S latitude, where the calculated runoff rate is twice the value reported by Gerasimov (1964). The annual runoff in 10°-latitude strips has been computed using the land areas in Table 3-2. The results are shown in Figure 3-11. The difference between the total observed runoff (3.8×10^{19} cm^3/yr) and the calculated runoff (3.4×10^{19} cm^3/yr) is rather smaller than might have been expected from the discrepancies in Figure 3-10; this happy circumstance is due to the small amount of land area between 50 and 60°S.

If the data in Figure 3-10 are adjusted for a nonglacial climate by extending the present day variation in river runoff between 60°N and 60°S to cover the entire range from 90°N to 90°S, a new set of runoff figures can be calculated for various latitude intervals. The sum of these figures should give at least a rough indication of total river runoff during periods of nonglacial climate. The sum computed using the observed present-day runoff values is 4.7×10^{19} cm^3/yr; the sum computed with the calculated values of the runoff is 4.9×10^{19} cm^3/yr. It seems likely, then, that during nonglacial times the global runoff is approximately 25% greater than the present value. The similarity between the expected increase in the annual

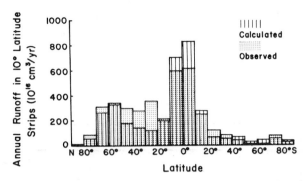

Figure 3-11. The observed and calculated annual runoff at various latitudes.

runoff and the increase in the annual rainfall indicates that the ratio E_0/r remains virtually constant during climatic changes of this kind. During glacial periods runoff was probably smaller than it is today (Manabe and Hahn, 1976).

The Effect of Land Area on Worldwide Runoff

Inundation of continental areas by shallow seas probably increases mean annual rainfall slightly. If as much as half of the continents were covered by shallow seas, the area of ocean water would be increased by ca. 20%; the effect of this increase on the total evaporation rate depends on the latitudinal distribution of the inundated continental areas. If the distribution were about average, then the increase in rainfall would depend largely on the difference between the average evaporation rate from the oceans and land areas. The mean evaporation rate from the oceans today is ca. 130 cm/yr; the mean evaporation rate from land is ca. 54 cm/yr. Thus

$$\Delta E \approx \frac{+130(\Delta A) - 54 \cdot \Delta A}{A_\oplus} \, \text{cm/yr} \qquad (3\text{-}6)$$

where ΔA is the increase in the area of the oceans and $A_\oplus$ is the area of the whole earth. The maximum value of $\Delta A / A_\oplus$ seems to be about 0.15. Thus

$$\Delta E_{\text{max}} \approx +11 \, \text{cm/yr}$$

which is approximately 10% of the present global evaporation rate.

Rainfall per unit area on the exposed land would therefore be greater by approximately 10%. If all other things remained constant, the mean runoff would increase by somewhat more than 10% since the ratio of the new value, $\Delta f'$, to the old value, Δf, of the mean annual runoff would be approximately

$$\frac{\Delta f'}{\Delta f} \approx \frac{r' \exp(-E_0/r')}{r \exp(-E_0/r)} \tag{3-7}$$

$$\approx 1.1 \frac{\exp(-80/1.1 \times 83.4)}{\exp(-80/83.4)} = 1.2 \tag{3-8}$$

The ratio of the new value of the total runoff, $\Delta F'$, to the old value, ΔF, would then be approximately

$$\frac{\Delta F'}{\Delta F} \cong \frac{\Delta f' \cdot A_c'}{\Delta f \cdot A_c} = 1.2 \times 0.5 = 0.6$$

where A_c'/A_c is the ratio of the continental area after inundation to the present area of the continents. The reduction of the total runoff due to inundation of half the area of the continents is therefore about 40%

The continents have rarely been more exposed than at present. An increase of 20% above the present level in the world-wide annual runoff from the continents due to additional exposure, therefore, seems a safe upper limit to excursions in ΔF unless the ratio of the mass of continental crust to the mass of sea water differed markedly from its present value during the early history of the earth.

The Effect of Continental Drift on Worldwide Runoff

Continental drift could have had quite major effects on world-wide runoff. The large effects on rainfall of positioning the continents either in the polar regions or in a circumequatorial belt have been discussed earlier. Polar wander data suggest that neither configuration existed during Phanerozoic time. Since the early Cretaceous polar wander seems to have been no more than a few degrees, and the latitude distribution of the continents during the past 100 million yr has not been very different from that of the present day (Jurdy and Van der Voo, 1975). However, less than drastic changes in the distribution of the continents might have sizable effects on the mean value of rainfall and runoff. If the continents were repositioned so that the mean runoff ratio for all continents were as small as that of Africa, the total runoff would be reduced by a factor of 2 (see Table 3-6); if the mean world runoff ratio could be set equal to that of Australia, the world runoff would be reduced by a factor of 4.

Internal drainage tends to increase with continental size, and internal drainage in supercontinents such as Gondwana, may have reduced the runoff ratio significantly. However, the runoff ratio of Asia, the largest continent at present, is 15% greater than the present world average. It is likely that this is a reasonable maximum for the upper limit of excursions of the world-wide runoff ratio due to continental drift. A very strong upper limit is set by the unlikely event that all the water evaporated on the globe is deposited on land. If the mean evaporation rate is taken to be 100 cm/yr, then the maximum rainfall on land is

$$\frac{100}{29} \times 100 \text{ cm/yr} = 345 \text{ cm/yr}$$

Thus

$$\Delta f_{\max} \approx r e^{-E_0/r} = 345 e^{-80/345}$$

and

$$\frac{\Delta f_{\max}}{\Delta f_{\text{present}}} \approx \frac{275}{83} = 3.3$$

The geometry required to produce such intense rainfall on the continents is quite unrealistic; a more reasonable value for the ratio $\Delta f_{\max}/\Delta f_{\text{present}}$ is probably about 2.

The Effect of Heterogeneities in Rainfall on Worldwide Runoff

Physiographic features such as mountain ranges are apt to affect runoff even when they do not affect total rainfall, because heterogeneity in rainfall distribution per se increases the effective runoff ratio. This can be shown as follows: If an area A_T receives an evenly distributed rainfall rate, r_A, the total runoff ΔF is given approximately by the expression

$$\Delta F = \Delta f \cdot A_T \cong r_A e^{-E_0/r_A} \cdot A_T \tag{3-9}$$

If such a region is divided into two equal areas, and if one of these receives a rainfall $(r_A + \delta)$ while the other receives a rainfall $(r_A - \delta)$, the total amount of rain falling within the region remains unchanged. However, the total runoff, ΔF_1, is somewhat different,

$$\Delta F_1 \cong \tfrac{1}{2} A_T \left[(r_A + \delta) e^{-E_0/(r_A + \delta)} + (r_A - \delta) e^{-E_0/(r_A - \delta)} \right] \tag{3-10}$$

The difference between ΔF and ΔF_1 diminishes as r_A becomes progressively larger than E_0. This can be shown by expanding the exponential terms in equation 3-10; when $r_A \gg E_0$ only the first term of the expansion needs to be considered; thus

$$\Delta F_1 \cong \frac{A_T}{2}\left\{(r_A+\delta)\left[1-\frac{E_0}{(r_A+\delta)}\right]+(r_A-\delta)\left[1-\frac{E_0}{(r_A-\delta)}\right]\right\} \quad (3\text{-}11)$$

$$\cong \frac{A_T}{2}\left[(r_A+\delta-E_0)+(r_A-\delta-E_0)\right] \quad (3\text{-}12)$$

$$\cong A_T(r_A-E_0) \quad (3\text{-}13)$$

The magnitude of the effect of heterogeneities in actual rainfall distribution is neither negligible nor overwhelming. Figure 3-12 shows a cumulative rainfall–area plot for the United States. If r_i and E_0^i are the rainfall and mean maximum evaporation rate for each fractional area A_i/A_T, then the total runoff should be approximately

$$\Delta F_1 \cong \sum_i r_i \cdot e^{-E_0^i/r_i} \cdot A_i \quad (3\text{-}14)$$

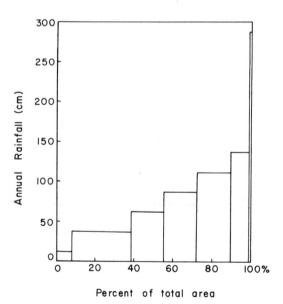

Figure 3-12. Cumulative rainfall–area plot for the United States (Moritz and Larson, 1946).

and the mean annual runoff should be

$$\Delta f_1 \cong \frac{1}{A_T} \sum_i r_i \cdot e^{-E_0^i/r_i} \cdot A_i \qquad (3\text{-}15)$$

If a mean value of 120 cm/yr for E_0 is taken for the entire United States, the calculated value of Δf_1 using the data in Figure 3-12 is 0.27; this compares favorably with the reported value of 0.26 (Sellers, 1965). On the other hand, if the mean annual runoff had been computed for the United States on the basis of an average rainfall of 75 cm/yr, without considering regions of differing rainfall separately, the calculated mean annual runoff ratio, again using 120 cm/yr as the value of E_0, would have been 0.20. Conversely, if we accept a value of 0.27 for the mean annual runoff ratio, a mean maximum evaporation rate E_0 of 120 cm/yr is required to produce agreement with the runoff ratio calculated by means of equation 3-15, whereas a mean maximum evaporation rate of 100 cm/yr gives a fit between the observed runoff ratio and the value calculated without subdividing the United States into several regions of different rainfall. The best value of E_0 for the United States based on data for individual river basins in Figure 3-9, is 110 cm/yr. The best value for E_0, based on continental averages in Figure 3-10, is 90 cm/yr.

These calculations show that heterogeneities in rainfall distribution can affect runoff by several tens of percent. Some of the scatter in the runoff–rainfall plots for major river basins is probably due to different degrees of heterogeneity in the rainfall within the various basins. More important, the effects of heterogeneities in rainfall probably account for much of the difference in Figure 3-9 between the E_0-latitude plot for individual river basins and for continents as a whole.

Little is known about past variations in the heterogeneity of rainfall. Based on the foregoing analysis, it seems unlikely that changes in the global rainfall heterogeneity could have increased or decreased runoff by more than 50%, but more data are needed to check this estimate.

Summation of the Estimated Effects on Worldwide Runoff

Table 3-7 summarizes the potential magnitude of the variation in world-wide river runoff that could be produced by changes in the mean annual temperature, the area of the exposed continents, the position of continental land areas, and the heterogeneity of rainfall distribution. Although the probable effect of the individual parameters is modest, their combined effect as measured by the product $\Pi_i(\Delta F_i/\Delta F^0)$ could be very large. The estimated minimum runoff is only one-twentieth of the present-day value; the

Table 3-7

Summary of the Estimated Effects of Climatic and Geographical Factors
on World-Wide River Runoff, Δf

	$\Delta F / \Delta F^{\circ}$		
Change Factor	*Minimum*	*Present*	*Maximum*
Changes in mean annual temperature	0.7	1.0	1.3
Changes in land area	0.6	1.0	1.2
Changes in position of continents	0.25	1.0	2
Changes in the heterogeneity of rainfall distribution	0.5	1.0	1.5
Product of all four effects $\Pi_i (\Delta F_i / \Delta F^0)$	0.05	1.0	4.7

estimated maximum is nearly five times the present value. It is unlikely that all the changes in climate and geography would combine to yield either of these extremes. However, the present assessment of the individual effects is by no means certain, and the effects of other variables, such as changes in the land flora, have not been included. It seems very likely that the annual flow of river water to the oceans has varied considerably during geologic history. The next chapter deals with the effects of these changes on the annual flux of solutes to the oceans.

REFERENCES

Alekhin, O. A. and Brazhnikova, L. V., 1963, Removal of solutes from continents by rivers and the relationship of this process to the mechanical erosion of the Earth's surface, in *Chemistry of the Earth's Crust*, A. P. Vinogradov, Ed., Israel Program for Scientific Translation, 1966, Jerusalem.

Bahcall, J. N. and Davis, R., Jr., 1976, Solar neutrinos: a scientific puzzle, *Science* **191**, 264–267.

Brooks, C. E. P. and Hunt, T. M., 1930, The zonal distribution of rainfall over the Earth, *Mem. R. Meteorol. Soc.* **3**, 139–158.

Budyko, M. I., 1956, Teplovoi Balans Zemnoi Poverkhnosti, Gidrometeorologicheskoe Izdatel'stvo, Leningrad, English Translation: Stepanova, 1958, *The Heat Balance of the Earth's Surface*, Office of Technical Services, U.S. Dept. of Commerce, Washington, D.C., 259 pp.

Budyko, M. I., 1963, *Atlas of the Heat Balance of the Earth* (in Russian), U.S.S.R. Glavnaia Geofizicheskaia Observatoriia, Moscow.

Budyko, M. I., 1974, *Climate and Life*, English Edition, D. H. Miller, Ed., Academic Press, N.Y., 508 pp.

Budyko, M. I., Yefimova, N. A., Aubenok, L. I., and Strokina, L. A., 1962, The heat balance of the surface of the earth, *Sov. Geogr.* **3**, 3–16.

Budyko, M. I. and Kondratiev, K. Y., 1964, The heat balance of the earth, in *Researches in Geophysics*, Vol. 2, H. Odishaw, Ed., Massachussets Institute of Technology Press, Cambridge, Mass., pp. 529–554.

Dorf, E., 1960, Climatic changes of the past and present, *Am. Sci.* **48**, 341–364.

Fairbridge, R. W., 1964, The importance of limestone and its Ca/Mg content to palaeoclimatology, in *Problems in Palaeoclimatology*, A. E. M. Nairn, Ed., Interscience, N.Y., pp. 431–477.

Gerasimov, I. P., 1964, *Physical-Geographical Atlas of the World* (in Russian), Academy of Sciences, U.S.S.R. and Department of Geodesy and Cartography, State Geodetic Commission, Moscow.

Holeman, J. N., 1968, The sediment yield of major rivers of the world, *Water Resour. Res.* **4**, 737–747.

Judson, S. and Ritter, D. F., 1964, Rates of regional denudation in the United States, *J. Geophys. Res.* **69**, 3395–3401.

Jurdy, D. M. and Van der Voo, R., 1975, True polar wander since the early Cretaceous, *Science* **187**, 1193–1196.

Kossinna, E., 1921, Die Tiefen des Weltmeeres, Inst. Meereskunde Veröff. Geographisch-naturwissenschaftliche Reihe, Heft 9, 70 pp.

Kozoun, V. I., Sokolov, A. A., Budyko, M. I., Voskresensky, K. P., Kalinin, G. P., Konoplyantsev, A. A., Korotkevich, E. S., Kuzin, P. S., and Lvovitch, M. I., 1974, *World Water Balance and Water Resources of the Earth*, prepared by the U.S.S.R. National Committee for the International Hydrological Decade, V. I. Kozoun, Editor-in-Chief, Leningrad.

List, R. J., 1958, *Smithsonian Meteorological Tables*, 6th rev. ed., first reprint, Smithsonian Institution, Washington, D.C., p. 484.

Livingstone, D. A., 1963, Chemical composition of rivers and lakes, in *Data of Geochemistry*, 6th ed., M. Fleischer, Ed., U.S. Geological Survey Professional Paper 440–G, 64 pp.

Lvovitch, M. I., 1973, The global water balance, *EOS Trans. Am. Geophys. Union* **54**, 28–42.

Manabe, S. and Holloway, J. L., Jr., 1975, The seasonal variation of the hydrologic cycle as simulated by a global model of the atmosphere, *J. Geophys. Res.* **80**, 1617–1649.

Manabe, S. and Hahn, D. G., 1976, Simulation of the tropical climate of an ice age, unpublished manuscript.

McElhinny, M. W., 1973, *Paleomagnetism and Plate Tectonics*, Cambridge Univ. Press, Cambridge, 357 pp.

Meinardus, 1934, Niederschlagsverteilung auf der Erde, *Meteorol. Z.* **51**, 345–350.

Menard, H. W. and Smith, S. M., 1966, Hypsometry of ocean basin provinces, *J. Geophys. Res.* **71**, 4305–4325.

Moritz, E. A. and Larson, E. O., 1946, *The Colorado River*, Dept. of the Interior, Washington, D.C.

Ol'dekop, E. M., 1911, On evaporation from the surface of river basins, Tr. IUr'ev Obs., Leningrad.

Rotty, R. M., and Mitchell, J. M., Jr., 1974, Man's energy and the world's climate, 67th Annual Meeting of the American Institute of Chemical Engineers, Washington, D.C.

Schreiber, P., 1904, Über die Beziehungen zwischen dem Niederschlag und der Wasserführung der Flüsse in Mitteleuropa, *Meterol. Z.* 21, 441–452.

Sellers, W. D., 1965, *Physical Climatology*, Univ. of Chicago Press, Chicago, Ill. 272 pp.

Sverdrup, H. U., 1951, Evaporation from the oceans, in *Compendium of Meteorology*, T. F. Malone, Ed., American Meteorological Society, pp. 1071–1081.

Ulrich, R. K., 1975, Solar neutrinos and variations in the solar luminosity, *Science* 190, 619–624.

Wetherald, R. T. and Manabe, S., 1975, The effects of changing the solar constant on the climate of a general circulation model, *J. Atmos. Sci.* 32, 2044–2059.

Wüst, G., 1954, Gesetzmässige Wechselbeziehungen zwischen Ozean und Atmosphäre in der zonalen Verteilung von Oberflächensalzgehalt, Verdunstung, und Niederschlage, *Arch. Meteorol. Geophys. Bioklimatol. Ser. A.* 7, 305–328.

CHAPTER IV

PHYSICAL AND CHEMICAL TRANSPORT IN RIVER SYSTEMS

The load of particulate matter and dissolved salts carried by rivers is quantitatively the most important input to the oceans. It is the product of physical and chemical weathering on the continents, and its quantity and composition is a measure of the transfer of CO_2, O_2, H_2O, and other weathering agents from the atmosphere to terrestrial rocks. It therefore plays a major role in the operation of the atmosphere–ocean–crust system. The flux of particulate and dissolved matter in river systems is a complex function of all of the parameters influencing the physics and chemistry of weathering. A precise formulation of the response of river loads to geographic, climatic, petrologic, and biologic factors is not possible. Nevertheless, rough relationships can be derived between these factors and river loads, and the response of the system to variations in the critical parameters can be predicted at least approximately.

THE PARTICULATE LOAD OF RIVERS

A very large body of literature deals with the variables that control the particulate load of rivers, especially the relationship between rates of erosion and their dependence on a variety of topographic and climatic factors. On a small scale, efforts to parameterize erosion rates have been reasonably successful (see, for instance, the summaries by Leopold, Wolman, and Miller, 1964; Gregory and Walling, 1973). Langbein and Schumm (1958) have shown that the sediment yield in many drainage basins increases with effective annual precipitation as shown in Figure 4-1, passes through a

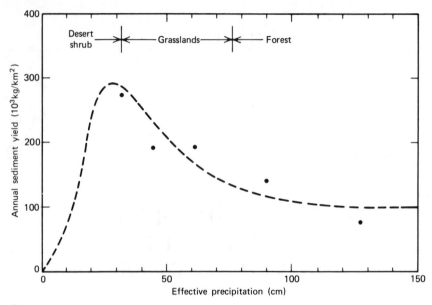

Figure 4-1a. Climatic variation of sediment yield as determined from records at sediment stations; adapted from Langbein and Schumm (1958). Copyrighted by American Geophysical Union.

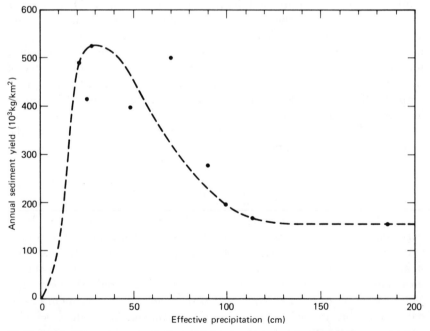

Figure 4-1b. Climatic variation of sediment yield as determined from reservoir surveys; adapted from Langbein and Schumm (1958). Copyrighted by American Geophysical Union.

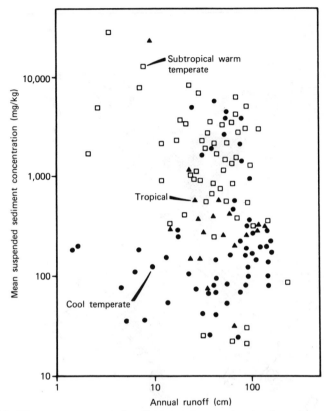

Figure 4-2. The mean suspended sediment concentration in rivers from cool temperate, warm temperate and subtropical, and tropical regions as a function of annual runoff, based on data by Fournier (1969); from Gregory and Walling (1973).

maximum near an annual precipitation of 30 cm, then decreases, and finally becomes nearly insensitive to the annual precipitation rate. In extremely arid regions the rate of erosion is certainly very small, but the plot in Figure 4-2 of the mean suspended sediment concentration in rivers from many parts of the world as a function of annual river runoff shows that there is no simple relationship between these two parameters on a world-wide basis, and that Langbein and Schumm's (1958) data are not generally applicable. A similar data plot for the conterminous United States (Curtis, Culbertson, and Chase, 1973) yields much the same result. Attempts by Strakhov (1967), Fournier (1960), Douglas (1967), and Fleming (1969) to explain the world-wide pattern of suspended yield in Figure 4-3 have produced no consistent or generally accepted set of conclusions. The rate of erosion clearly depends at least on the interplay of total rainfall and its seasonality, on mean annual temperature, on elevation differences, and on the soil and rock types present within a given drainage basin. Holeman (1968) has most recently compiled

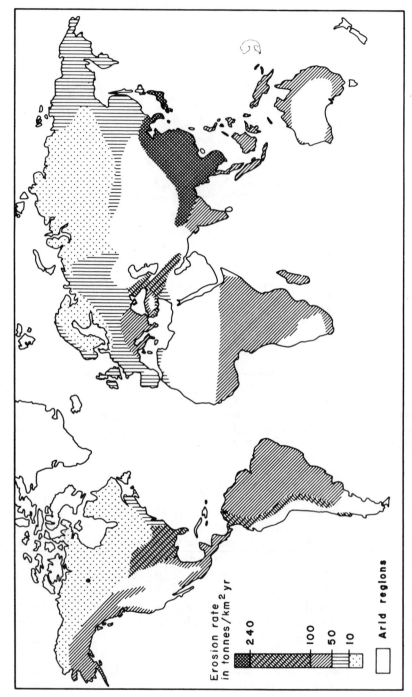

Erosion rate
in tonnes/km² yr

240
100
50
10

Arid regions

Figure 4-3. Worldwide erosion rates based on Strakhov's (1967) data; figure from Gregory and Walling (1973).

the suspended load data for the major rivers of the world. Many of his numbers are reproduced in Table 4-1. The data are clearly incomplete, and many are based on observations covering rather short time periods. They do, however, confirm the importance of the variables mentioned above. Thus, the much higher sediment yield of the Missouri River basin than that of the Ohio River basin is probably due largely to differences in rainfall and topography. The very large suspended load of the Colorado River prior to the construction of reservoir systems, compared to that of the Missouri carries this comparison even further. On the other hand the very high sediment load of the Yellow River in China is largely a consequence of the rapid erosion of extensive loess deposits in its drainage basin, whereas the very high sediment yield of the rivers draining the Himalayas undoubtedly is due to the combination of extreme topographic relief and a monsoon climate in this part of Asia. By contrast, the relatively low sediment yield of the Amazon basin is due to the very heavy rainfall within the topographically low part of the basin (Gibbs, 1967). Holeman's (1968) estimated suspended sediment yield of the continents is summarized in Table 4-2. As pointed out by Garrels and Mackenzie (1971), the correlation seems more than fortuitous. Asia, the highest continent, is being eroded much more rapidly than the other continents. Europe, the lowest continent, is being eroded at the second-lowest rate. Figure 4-4 shows that the correlation between sediment yield and mean elevation of the continents is rather better than might be expected. Only Africa falls well off the simplest curve of best fit. This may reflect in part the effects of a very large desert area, a low mean annual runoff, and the topography of the continent. The rather good fit of Australia may be somewhat fortuitous. Sediment yield data for this continent are very limited, and it would not be surprising if Holeman's (1968) estimated sediment yield for Australia turns out to be considerably too high.

The rough correlation between sediment yield and continental elevation suggests that on a continent-wide scale many of the parameters that are responsible for the large variations in the sediment yield of individual river basins average out fairly well, and that much of the variance in continental sediment yield data is explained in terms of differences in mean continental elevation. The curve in Figure 4-4 is of the form

$$y = \alpha \left(e^{\beta h} - 1 \right). \qquad (4\text{-}1)$$

If the sediment yield figures are converted into a rate of denudation, ϕ, in units of centimeters per year, then the data are best fitted by the equation

$$\phi = 2.0 \times 10^{-4} \left(e^{5.0 \times 10^{-5} h} - 1 \right)$$

where h is the mean height of the continents in centimeters. There is

Table 4-1

The 20 Largest Rivers Arranged in Order of their Water Discharge[a]

	Drainage Area (10³ km²)	Discharge (10³ m³/sec)	Runoff Δf (cm/yr)	Suspended Sediment Yield (kg/km² yr × 10³)	Suspended Sediment Concentration (mg/kg)	Dissolved Solids Yield (kg/km² yr × 10³)	Dissolved Solids Concentration (mg/kg)
1. Amazon[3][4]	5,930	175	92	55.3	66		53
2. Congo[4]	4,000	40	31	14.5	51		
3. Orinoco[4]	950	23	72	82.5	77		
4. Yangtze[4]	1,030[b,c]	22	67	444	700		
5. Bramaputra[4]	560	20	110	1179	1070		
6. Mississippi[1][2]	3,268	18.4	18	82.5	510	43.7	245
7. Yenisei[4]	2,480	17.5	22	3.8	190		
8. Mekong[4]	390[b,c]	15.0	120	395	365		
9. Parana[4]	2,300	14.9	20	31.7	175		
10. St. Lawrence[4]	1,300	14.2	34	2.5	8		
11. Ganges[4]	1,060[b]	14.2	42	1270	3400		
12. Irrawaddy[4]	370[b]	13.6	115	744	710		
13. Ob[4]	2,440	12.5	16	5.4	37		
14. Volga[4]	1,350	8.0	19	12.7	73		
15. Columbia[1][2]	669	8.0	38	19.0	56	29.0	85
16. Pearl-West[4]	310	7.9	80	79.8	110		
17. Mackenzie[5]	1,700	7.4	14	2.7	21		
18. Indus[4]	1,050	6.8	20	408	2200		
19. Danube[4]	810[c]	6.2	24	21.8	100		
20. Niger[4]	1,100[c]	6.1	17	3.8	25		

[a]Sources of data: (1) Leifeste, 1974; (2) Curtis, Culbertson, and Chase, 1973; (3) Gibbs, 1972; (4) Holeman, 1968; (5) Livingstone, 1963.
[b]Holeman gives conflicting data for these figures.
[c]Serious discrepancy between this figure and that in Kozoun et al. (1974).

Table 4-2

Summary of Measured Annual Sediment Yields of Selected Rivers to Oceans[a]

Continent	Measured Drainage Area (km²)	Annual Suspended Sediment Discharge	
		(10^6 kg)	(10^3 kg/km²)
North America	6,380,976	547,908	86
South America	9,890,938	552,872	56
Africa	8,146,755	196,195	24
Australia	1,073,425	42,956	40
Europe	3,514,197	110,622	31
Asia	10,907,016	5,819,303	534
Total	39,913,307	7,269,856	182

[a]Holeman (1968) after conversion to metric units.

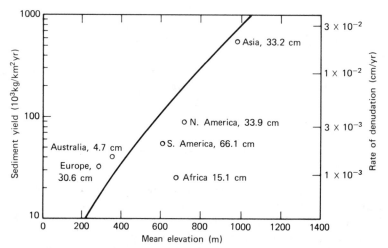

Figure 4-4. The relationship between total sediment yield and mean elevation of the continents; the mean annual runoff for each continent is included; data from Holeman (1968) and Chapter III.

considerable evidence that the rate of sediment transport by rivers today is not very different from the rate of sediment transport during the Tertiary era. Curray and Moore (1971), for instance, have shown that the growth of the Bengal deep-sea fan is commensurate with the current rate of sediment discharge from the Ganges and Brahmaputra.

Discrepancies do exist between current sediment transport rates and the estimated Cenozoic erosion rates in eastern North America (Mathews, 1975), but these erosion rates are based on possibly faulty assumptions about the position of the watershed between the Atlantic drainage and the Mississippi drainage basin during the Cenozoic era. If it is true that the continents are currently in a state of near balance between uplift and erosion, then

$$\frac{dh}{dt} \approx \phi - \alpha(e^{\beta t} - 1) - \eta \approx 0 \qquad (4\text{-}2)$$

where η is the rate of chemical denudation. If the contribution of η is neglected for the moment, then in general the rate of change of mean continental elevation with time depends on the time variation of the rate of elevation, ϕ, and on the response of the erosion rate to h.

A particularly simple case is instructive. Let a continent have reached a steady state elevation h_1 in equilibrium with a rate of uplift ϕ_1. Now let the rate of uplift suddenly change from ϕ_1 to a new, constant value, ϕ_2, then

$$\frac{dh}{dt} \cong \phi_2 - \alpha(e^{\beta h} - 1) \qquad (4\text{-}3)$$

This expression can be integrated readily to yield the expression

$$h = \frac{1}{\beta} \ln \frac{(\phi_2 + \alpha)}{A e^{-\beta(\phi_2 + \alpha)t} + \alpha} \qquad (4\text{-}4)$$

where

$$A = \frac{(\phi_2 + \alpha) - \alpha e^{\beta h_1}}{e^{\beta h_1}} \qquad (4\text{-}5)$$

If erosion really follows these equations, the mean elevation of a continent of an initial height of 1000 m decreases very rapidly due to mechanical erosion during the first 10 million yr after cessation of uplift. As shown in Figure 4-36 the mean elevation decreases very slowly indeed after h has been reduced to 200 m. It is shown later that the inclusion of a term for chemical erosion in equation 4-2 influences the form of the response curve of the mean elevation h to changes in the uplift rate ϕ, and is largely responsible for the continued erosion of low-standing continents.

The sum of the annual sediment discharge of the major rivers of the world can be extrapolated to yield a figure for the total quantity of sediment transported annually to the oceans by rivers. Holeman's (1968) estimate of 20.2×10^9 short tons seems to refer to suspended sediment only. Since bedload accounts for approximately 10% of the total particulate load of many rivers, the total annual sediment transport by rivers is probably close to 2.0×10^{16} g. Estimates by six other authors since 1950 have been as low as 1.4×10^{16} g and as high as 6.4×10^{16}g (Fournier, 1960). However, Fournier's value is almost twice that of the next highest estimate. It is obviously difficult to estimate the uncertainty in Holeman's figure; a value of $(2.0 \pm 1.0) \times 10^{16}$ g/yr is adopted here.

Erosion rates have increased enormously in many areas since settlement by man. There is, therefore, considerable room for doubt that present-day erosion rates are in any sense characteristic of the "undisturbed" system. Nevertheless the similarity between the present-day sediment input rates into the Gulf of Mexico and the Bay of Bengal and the mean rate calculated by dividing the total amount of sediment by the accumulation time suggests that the effect of man has been modest, perhaps as large as a factor of 2.

A solution to this paradox is suggested by Trimble's (1975) study of 10 river basins in the southeastern United States. Apparently, only about 5% of the material eroded from upland slopes since European settlement has been exported. The eroded material not transported from the basins has been deposited as colluvium and alluvium, and the streams are not in steady state. Recently, the agricultural use of land in these drainage basins has declined, and soil conservation practices have improved. The streams have, therefore, regained their transport ability, and some of the modern alluvial deposits are being dissected. Export of these sediments will, however, be hampered by water reservoirs, which act as efficient sediment traps. Predictions of sediment transport rates during the next 1000 years surely will have to include estimates of the effects of changes in the rates of erosion and entrapment, and of the response of the capacity of streams to transport sediment to the oceans. Holeman's (1968) sediment transport figure is certainly not unaffected by man, but it seems likely that the magnitude of the human effect is smaller than the uncertainty of the estimate itself.

THE TOTAL DISSOLVED LOAD OF RIVERS

The concentration and composition of dissolved solids in river waters are highly variable. Figure 4-5 summarizes the average concentration of total dissolved solids* in rivers of the United States as reported in a series of U.S Geological Survey Water–Supply Papers (Anderson and George, 1966; Billingsley et al., 1957; Irelan and Mendieta, 1964; Love, 1958, 1960a–b, 1961a–c, 1964a–e, 1965a–d, 1966a–e, 1967a–d, 1970; McCarthy and Keighton, 1964) and in reports by Gladwell and Mueller (1967) and Sylvester and Rambow (1967). The concentration of dissolved solids ranges from 37 to 8000 ppm. There is a large variability of total dissolved solids among rivers that fall within a particular range of runoff (Δf) values, but the increase in concentration with decreasing runoff is unmistakable. In the

*Normally these figures refer to solid residues on heating to 180°C.

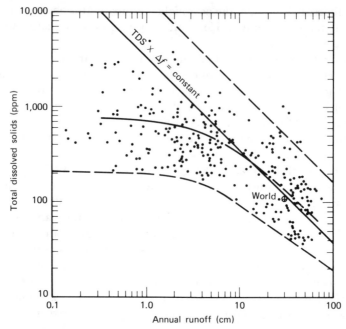

Figure 4-5. Total dissolved solids (TDS) in river waters of the United States as a function of the runoff Δf.

upper range of runoff values the concentration of total dissolved solids is nearly inversely proportional to Δf. Toward the lower values of the runoff the concentration of dissolved solids increases less rapidly. The value 117 ppm for the world mean of total dissolved solids was obtained by reducing Livingstone's (1963) value of 120 ppm as recommended by Gibbs (1972). These data suggest that the total annual dissolved load per unit area $(\text{TDS} \times \Delta f)$ in the United States is virtually independent of runoff for runoff values in excess of 10 cm/yr, and that the total annual flux of dissolved solids per unit area decreases rapidly at Δf values below 10 cm/yr. Durum, Heidel, and Tison (1961) and Van Denburgh and Feth (1965) reached similar conclusions on the basis of a somewhat smaller sample.

Many of the points in Figure 4-5 represent the composition of river water from relatively small areas. Large rivers tend to integrate the contributions from their tributaries. The composition of major rivers should, therefore, fall well within the limits set by the data points for their tributaries. If all rivers behaved like those within the United States, the spread in their position within a TDS-Δf diagram should be smaller than the spread in Figure 4-5. This is not the case. Figure 4-6 shows, however, that a number of major

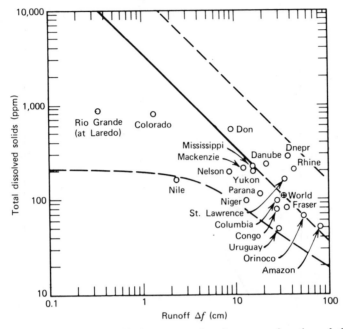

Figure 4-6. Total dissolved solids in some major rivers as a function of the mean annual runoff.

rivers fall quite close to the limits set by the data points in Figure 4-5; the Nile actually falls outside these boundaries. Rivers within the United States clearly do not span the entire world range of TDS-Δf relationships.

It has been proposed that the computation of the total input of dissolved solids into the oceans requires only a knowledge of the chemistry of the 20 largest rivers of the world. Unfortunately this does not turn out to be true. Table 4-1 ranks the 20 most important rivers according to mean annual runoff. The combined runoff is 450×10^3 m^3/sec or 1.4×10^{19} cc/yr. This is only 31% of the estimated mean annual world-wide runoff of 4.6×10^{19} cc/yr. As might be expected, the rivers with largest runoff tend to drain the wettest areas of the globe. This is shown clearly in Figure 4-7. The cumulative area is proportionately smaller than the cumulative runoff and approaches the world mean only as progressively smaller rivers and river-free desert areas are included.

The output of total dissolved salts has been used traditionally in calculations of the relative intensity of chemical and physical weathering (see for instance Leopold, Wolman, and Miller, 1964). This is a questionable procedure, as a good deal of the chloride, sulfate, and bicarbonate in river water is

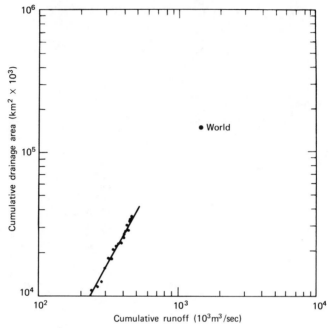

Figure 4-7. Cumulative drainage area and cumulative runoff for major rivers of the world (for data see Table 4-1).

ultimately atmospheric in origin and should be subtracted from the total dissolved solids to define a meaningful removal rate of material in solution from the continents. There is no a priori reason to assume that the concentration of the various dissolved components of river water responds similarly to climatic changes. In fact, they turn out to behave quite individually (Hem, 1959).

CHLORIDE IN RIVERS

A good deal of chloride in river water is derived from the oceans, has been cycled through the atmosphere, and has been brought back to earth in rain water. Mairs (1967), for instance, has shown that the chloride content of lakes in Maine decreases from values in excess of 10 ppm near the coast to < 1 ppm some 150 miles inland. Such a pattern almost certainly owes its origin to the atmospheric transport of chloride from the Atlantic Ocean. Junge and Gustafson (1957) found that the chloride content of precipitation is in excess of 0.5 ppm only in coastal areas, and that over most of the United States the average chloride content of rain water is less than 0.3 ppm. Cadle (1973), however, found considerably higher Cl concentrations in

Table 4-3

Mean Composition of River Waters of the World[a]

Continent	HCO_3^-	SO_4^{2+}	Cl^-	NO_3^-	Ca^{2+}	Mg^{2+}	Na^+	K^+	Fe	SiO_2	Sum
North America	68	20	8	1	21	5	9	1.4	0.16	9	142
South America	31	4.8	4.9	0.7	7.2	1.5	4	2	1.4	11.9	69
Europe	95	24	6.9	3.7	31.1	5.6	5.4	1.7	0.8	7.5	182
Asia	79	8.4	8.7	0.7	18.4	5.6	9.3		0.01	11.7	142
Africa	43	13.5	12.1	0.8	12.5	3.8	11		1.3	23.2	121
Australia	31.6	2.6	10	0.05	3.9	2.7	2.9	1.4	0.3	3.9	59
World	58.4	11.2	7.8	1	15	4.1	6.3	2.3	0.67	13.1	120
Anions[b]	0.958	0.233	0.220	0.017							1.428
Cations[b]					0.750	0.342	0.274	0.059			1.425

[a]Livingstone (1963); concentrations in ppm.
[b]Millequivalents of strongly ionized components.

rain in the arid West and in the vicinity of industrial centers. The mean runoff ratio, $\Delta f/r$, for North America is 0.40 (see Table 3-3). If all the chloride in North American river water were atmospherically derived, the mean chloride concentration of river water would be about 1–4 ppm. Livingstone's (1963) mean for the chloride content of North American rivers is 8 ppm (see Table 4-3). Other sources of chloride, therefore, must also be of importance. This conclusion has been strengthened by an analysis of the chloride balance in 11 river basins of the western United States by Van Denburgh and Feth (1965), who showed that chloride brought into these basins by rain and snow accounted for only 1.6–17% of the total chloride removed in runoff.

The compilation of data in Figure 4-8 for the chloride content of United States rivers is based on the same sources used to construct Figure 4-5. The enormous spread in chloride concentration both in streams of high and low runoff is impressive. Except in areas of heavy industrial pollution, atmospheric chloride concentration by evaporation cannot account for chloride contents in excess of 20 ppm in rivers from areas with a runoff in excess of 20 cm/yr. A good deal of the nonatmospheric chloride in these rivers is apparently derived from sedimentary rocks, possibly by the addition of subsurface connate brines (see, for instance, Hitchon, Levinson, and Reeder, 1969). At least some of the salts in these brines come from the solution of evaporites, and may be responsible for the rapid differential cycling of this rock type (Garrels and Mackenzie, 1971, p. 272). Sodium and chloride are usually the most abundant ions in subsurface brines (White, Hem, and Waring, 1963; White, 1965), although calcium and sulfate are often important and sometimes dominant. A plot of the concentration of sodium

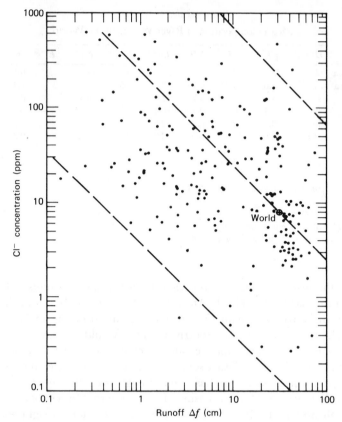

Figure 4-8. The chloride concentration in rivers of the United States. Data largely from U.S. Geological Survey Water Supply Papers (listed in references); world mean after Livingstone (1963).

versus chloride in the analyses of river water by Livingstone (1963) showed that in 98 of 120 analyses m_{Na^+} was equal to or greater than m_{Cl^-}, and that m_{Na^+} was less than 0.75 m_{Cl^-} in only four analyses. This result does not prove, but is consistent with the fact that chloride is normally added to river water largely balanced by sodium.

Sugawara (1964) has suggested that the consumption of industrial products accounts for approximately a quarter of the chloride content of river waters on the islands of Honshu, Hokkaido, Kyushu, and Shikoku. This must surely be a strong maximum for the world as a whole. Although the relative contribution of chloride from recycled atmospheric, industrial, and evaporite sources is not well known, it is clear that they far exceed the

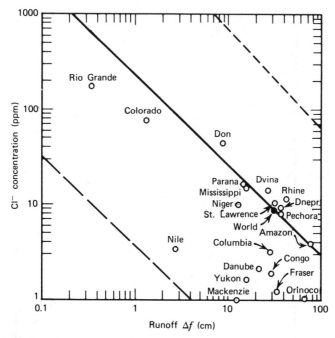

Figure 4-9. The chloride concentration in some major rivers; data largely from Livingstone (1963).

contributions from the weathering of igneous rocks and from volcanic emanations. Virtually all chloride in river waters is, therefore, recycled in the sense that it has been in the oceans at least once before (Hem, 1959, p. 110).

The data of Figures 4-8 and 4-9 show that the mean chloride concentration of river water increases crudely with decreasing mean annual runoff. A change in the mean annual world runoff would, therefore, tend to produce only small changes in the total input of chloride into the oceans since the product $\bar{m}_{Cl^-} \times \Delta f$ is roughly constant.

SULFATE IN RIVERS

Figures 4-10 and 4-11 show that the distribution of sulfate in rivers is similar to that of chloride. At each annual runoff value there is a sizable scatter in the SO_4^{2-} concentration; however, it is clear that toward lower Δf values the mean value of the sulfate concentration in river water rises, and that the product $\bar{m}_{SO_4^{2-}} \times \Delta f$ is approximately constant. Sulfate in rivers is derived

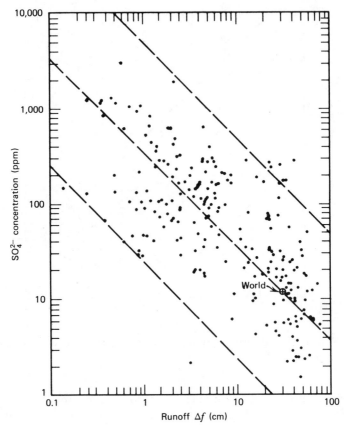

Figure 4-10. The sulfate concentration in rivers of the United States. Data largely from U.S. Geological Survey Water–Supply Papers (see references); world mean after Livingstone (1963).

from a variety of sources. Weathering of sulfide and sulfate minerals and sulfate derived from the sea are the major sources of nonanthropogenic sulfate. Most of the anthropogenic sulfur is produced during fossil fuel burning (see Table 4-4). If we accept Livingstone's estimate for the SO_4^{2-} content of average river water in Table 4-3 and the best estimate for the total annual runoff from Chapter III, then the total SO_4^{2-} transport in rivers is currently

$$11.2 \times 10^{-6} \frac{\text{g } SO_4^{2-}}{\text{g water}} \times 4.6 \times 10^{19} \frac{\text{g water}}{\text{yr}} \cong 520 \times 10^{12} \text{ g/yr}$$

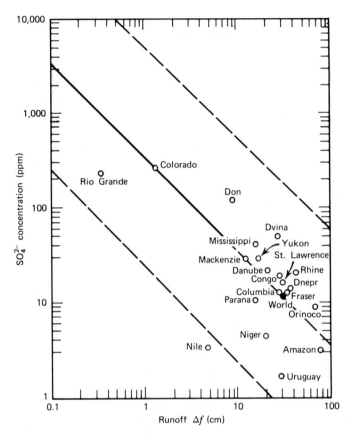

Figure 4-11. The sulfate concentration in some major rivers; data largely from Livingstone (1963).

Table 4-4

Anthropogenic Sources of Atmospheric Sulfur[a]

Coal burning	45.4
Oil burning	5.5
Industrial smelting and petroleum refining	13.6
Transportation	0.4
Total	64.9

[a] In units 10^{12} g/yr; Friend (1973).

97

Anthroprogenic sulfur in the atmosphere is virtually all oxidized and rains out as sulfate, largely on land. During oxidation the 65×10^{12}g of sulfur are converted to 195×10^{12}g of SO_4^{2-}. Thus, roughly 40% of the river-borne SO_4^{2-} is probably anthropogenic in origin. Berner (1971) has arrived at a figure of 28%. Garrels, Mackenzie, and Hunt's (1975, p. 82) data indicate an anthropogenic contribution of 45%. The differences between these figures are probably not significant.

The data in Figure 4-11 for the major rivers of the world corroborate the suggestion that fossil fuel burning still does not overshadow the SO_4^{2-} content of river waters on a world-wide basis. The relationship between $m_{SO_4^{2-}}$ and Δf in rivers from industrialized areas is not dramatically different from the $m_{SO_4^{2-}}$–Δf relationship for rivers in nonindustrialized areas. Most of the SO_4^{2-} in rivers like the Amazon must be derived from weathering and from the rain-out of marine sulfate. It is quite difficult to obtain precise estimates for the average sulfide and sulfate content of crustal rocks (see Chap. VI). The best estimate of the S^{2-} content of average rock undergoing weathering is probably $0.3 \pm 0.1\%$ (Holland, 1973).

Ericksson (1960, 1963) has estimated that sea spray introduces approximately 44×10^{12} g of sulfur/yr into the atmosphere, and that roughly 10% of this quantity is transported to continental regions and rained out. This flux is small compared to the sulfur flux due to weathering and fossil fuel burning, and is probably balanced, at least roughly, by the atmospheric transport of sulfur from fossil fuel burning from the continents to the oceans.

The best estimate of the sulfate–sulfur content of average rock undergoing weathering is probably $0.3 \pm 0.1\%$ (see Chap. VI). During weathering virtually all of this sulfur is released and enters stream water as SO_4^{2-}. If we accept Holeman's (1968) figure for the total amount of sediment transported annually to the oceans, then the accompanying complement of river sulfate is approximately

$$(0.6 \pm 0.2) \times 10^{-2} \frac{\text{g S}}{\text{g rock}} \times \frac{3.0 \text{ g } SO_4^{2-}}{\text{g S}} \times 2 \times 10^{16} \text{ g rock/year}$$

$$\cong (360 \pm 120) \times 10^{12} \text{ g } SO_4^{2-} / \text{year}$$

The sum of the estimated rate of sulfur release during weathering and the anthropogenic flux agree quite well with the best estimate of the river flux. This agreement may, however, be somewhat fortuitous, since neither the release of sulfur during weathering nor the total river flux of sulfate is particularly well known. A backward extrapolation suggests that the river flux of SO_4^{2-} in the preindustrial era was roughly 60% of the present value, and that this flux was not particularly sensitive to variations in the mean annual runoff. A forward extrapolation suggests that fossil fuel burning is apt to become the dominant source of river sulfate during the next 50 yr, but that the importance of anthropogenic SO_4^{2-} is likely to decrease in 100–150 years as the stock of fossil fuels approaches exhaustion.

BICARBONATE IN RIVERS

A glance at Figures 4-12 and 4-13 shows that the distribution of bicarbonate on a log concentration–log Δf diagram is quite different from that of chloride and sulfate. The rather wide concentration range at runoff values in excess of 15 cm/yr narrows significantly toward lower runoff values, and the bicarbonate concentration increases slowly from a mean near 150 ppm at a runoff of 10 cm/yr to a mean near 300 ppm at a runoff of 0.1 cm/yr. Between 0.1 and 10 cm/yr the bicarbonate concentration therefore increases much less rapidly than the concentration of either sulfate or chloride. River waters in areas of high rainfall normally contain more bicarbonate

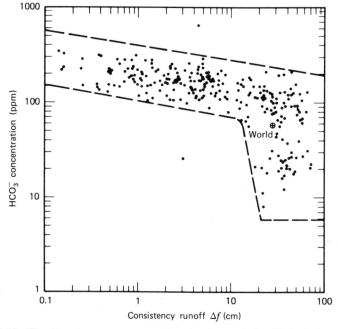

Figure 4-12. The bicarbonate concentration in rivers of the United States. Data largely from U.S. Geological Survey Water–Supply Papers (see references); world mean after Livingstone (1963).

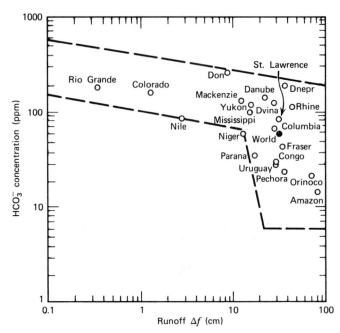

Figure 4-13. The bicarbonate concentration in some major rivers; data largely from Livingstone (1963).

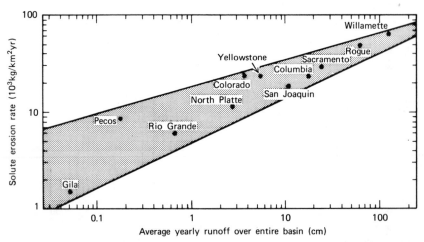

Figure 4-14. The relationship between solute erosion rate and average yearly runoff for 11 river basins of the western United States; adapted from Van Denburgh and Feth (1965). Copyrighted by American Geophysical Union.

than sulfate or chloride. In arid regions the concentration of sulfate and chloride often exceed that of bicarbonate. The sum of the concentration of the three anions largely determines the concentration of dissolved salts in rivers. Were it not for the small slope of the log $m_{HCO_3^-}$–log Δf curve in the low-Δf region, the product of the total dissolved solids concentration and the runoff would be nearly constant. As it is, the rate of removal of dissolved salts per unit area increases rather regularly with increasing runoff; this is illustrated very well by the data in Figure 4-14 for the 11 river basins studied by Van Denburgh and Feth (1965).

Three aspects of the data in Figures 4-12 and 4-13 require explanation: The large range of the bicarbonate concentration in river systems with runoff in excess of 15 cm/yr, the relatively small range of bicarbonate concentrations at Δf values below 15 cm, and the small slope of the log $m_{HCO_3^-}$–log Δf curve for annual runoff values between 10 and 0.1 cm.

Bicarbonate Concentrations in the High-Δf Region

The Amazon is by far the largest river in the world, and is an excellent example of a river with a bicarbonate content well below the world average in a region of high runoff. Gibbs (1967, 1972) has pointed out that the low concentration of dissolved solids in the Amazon River is due in large part to the distribution of rainfall in relation to the topography and lithology in the Amazon River basin. The lower parts of the basin receive much of the rainfall and are underlain by highly leached lake sediments that contribute silica but very little in the way of cations, and hence indirectly bicarbonate, to the dissolved solids of the river. Although the Amazon is apparently a rather extreme example, the other major South American rivers in Figure 4-13 are also distinguished by bicarbonate concentrations well below the world mean of about 58 ppm (Livingstone, 1963).

Other distributions of rock and sediment types in river basins can produce entirely different patterns of river chemistry. The total dissolved solids and the bicarbonate content of rivers draining crystalline rocks is usually considerably smaller than those of rivers draining sediments and sedimentary rocks under equivalent conditions. Table 4-5 illustrates this point for several tributaries of the St. Lawrence River. Tributaries from the north drain largely crystalline terrain, tributaries from the south drain largely carbonate terrain. The average HCO_3^- content of tributaries from the north is 12 ppm, that of tributaries from the south 114 ppm. The differences in the SiO_2 content of water from the two sets of tributaries is minor, but the Ca^{2+} concentration is closely correlated with the HCO_3^- concentration of water in several St. Lawrence tributaries. Clearly, the solution of carbonates, mainly

Table 4-5

The Composition of Some Rivers and Lakes in the St. Lawrence River Drainage.[a]

	A	B	C	D	E	F	G	H	I	J	K	L	M	N	O	P	Q
HCO_3^-	26.2	126.1	134.2	210.5	229.3	102.1	15.3	27.5	120	30.5	43.3	16.9	6.7	6.1	100.7	52.1	100.0
SO_4^{2-}	8.5	16.8	13.6	58.8	124.2	13.1	8.9	8.0	82	7.2	12.0	5.3	5.0	3.1	5.3	6.3	13.1
Cl^-	1.0	3.4	3.0	36.2	14.8	1.8	1.5	1.6	32	1.1	1.6	1.0	0.8	0.5	1.5	2.1	4.2
F^-									0.2								
NO_3^-	1.33	0.50	0.08	2.13	1.58	1.5	1.10	0.62	4.3	0.26	0.52	0.72	5.8	0.60	0.40	0.61	1.32
Ca^{2+}	9.0	36.0	39.5	66.6	87.3	31.8	7.5	10.5	54	16.0	15.6	6.2	4.8	3.6	28.2	14.6	27.6
Mg^{2+}	3.6	4.8	6.3	20.1	24.1	4.8	2.8	3.0	12	3.2	4.1	1.6	1.3	1.5	4.8	4.2	7.5
Na^+	3.8	5.3	1.8	21.2	15.7	2.7	1.1	3.6	21 ⎫	5.2	3.6	2.4	3.3		6.8	2.6	2.4
K^+									3.4 ⎭								
Fe	0.15	None	0.05	0.25	0.06	0.7	0.08	0.22	0.06	0.02	0.05	0.24	0.15	0.20	0.5	0.06	0.04
SiO_2	3.0	4.8	6.0	9.5	9.9	6.5	1.4	5.8	6.7	6.6	6.6	13.5	4.2	3.0	9.4	5.4	5.6
Total dissolved solids	56.6	198	205	425	507	165	39.7	60.8	336	70.1	87.4	47.9	32.1	>18.6	158	88.0	162
Rock type in drainage	C	S	S	S	S	S	C	C+S	S	S	S	C	C+S	C	S	S	S

[a] Data from Livingstone (1963). C: Crystallines only; C+S: Crystallines + sedimentary rocks; S: Sedimentary rocks only.
A. Lake Nipissing at North Bay, Ontario; depth sample 2 miles from shore. May 26, 1939. B. Lake Couchiching at Orillia, Ontario; depth sample 3 miles from shore. July 17, 1934. C. Lake Simcoe, Ontario. Depth sample at mouth of Kampenfeldt Bay. Aug. 12, 1935. D. Thames River at Chatham, Ontario. Mean of six analyses from 1934–40. E. Grand River at Brantford, Ontario. Mean of nine analyses from 1934–42. F. Trent River at Trenton, Ontario. Mean of four analyses, 1934–37. G. Lake Temiskaming at Haileybury, Ontario. Depth sample 1 mile from shore, Aug. 27, 1937. H. Ottawa, River at Hawkesbury, Ontario. Mean of eight analyses, 1934–38. I. Cuyahoga River at Botzum, Ohio. U.S. Geological Survey (1952). J. Magog River at Sherbrooke, Quebec. June 26, 1942. K. Richelieu River at St. Johns, Quebec. Mean of four samples, 1935–42. L. St. Charles River at Chateau d'Eau, Quebec. Mean of four samples, 1934–39. M. St. Maurice River at Three Rivers, Quebec. Mean of five samples, 1934–41. N. Saguenay River at Riverbend, Quebec. July 12, 1935. O. Nipigon River at Nipigon, Ontario. Aug. 2, 1937. P. St. Mary's River at Sault Ste Marie, Ontario. Mean of five analyses, 1936–38. Q. St. Clair River at Point Edward, Ontario. Mean of four analyses, 1934–37; concentrations in ppm.

calcite and dolomite, is responsible for the much larger HCO_3^- concentration of the tributaries draining sedimentary terrains. The relatively high HCO_3^- concentration of the St. Lawrence and its position in Figure 4-13 as compared to the Amazon reflects the influence of the tributaries from carbonate terrains. If these were absent, the HCO_3^- content of the St. Lawrence would be nearly identical to that of the Amazon.

Although the St. Lawrence is a fairly typical river, the concentration of bicarbonate and of total dissolved solids in water from crystalline terrains is not always small compared to world mean concentrations. White, Hem, and Waring (1963) have shown that bicarbonate concentrations of several hundred parts per million are not uncommon in ground water in some igneous terrains; Garrels (1967) has examined the chemical systematics of such waters. Drever (1971) has found bicarbonate concentrations of up to 375 ppm in the basin of the Rio Ameca on the Pacific slope of Mexico in rivers draining an area largely underlain by easily altered extrusive and intrusive igneous rocks and by some metamorphic rocks. Nevertheless, it is likely that in Figures 4-12 and 4-13 much of the large spread in the HCO_3^- concentration of rivers in areas where the runoff exceeds 20 cm/yr is due to differences in the proportion of carbonates within individual drainage systems.

Mean annual temperature seems to play a surprisingly subordinate role in determining the bicarbonate content of river water. This is best illustrated by the similarity of the bicarbonate concentration of the Mississippi, the Yukon, and the Mackenzie. Studies of the chemistry of ground water in basaltic terrains in Hawaii (Visher and Mink, 1964) and of surface water in vegetated areas of Iceland (Cawley, Burruss, and Holland, 1969) have shown that the bicarbonate content of water from heavily vegetated parts of both areas is normally between 50 and 120 ppm, and is apparently not influenced to any great extent by the large differences between the mean annual temperature in these two climatically disparate areas.

Even in areas of northern Iceland where the Skjálfandafljót flows across a basaltic lava desert devoid of vegetation, the bicarbonate concentration in water reaches values as high as 25 ppm (0.4 meq/kg) (Cawley, Burruss, and Holland, 1969). Unless CO_2 is supplied by an unsuspected, buried peat layer, the HCO_3^- in this river system reaches values about one-third as high as those in the more vegetated lower reaches of the river without the benefit of CO_2 supplied by plant decay in a soil horizon (see Figure 4-15). Reynolds and Johnson (1972) have found a similar situation in the northern Cascade Mountains. At least a partial explanation for the surprisingly small effect of the absence of organic matter decay in a soil zone is offered by a comparison of the data for the partial pressure of CO_2 in Icelandic river water and those from the fairly heavily vegetated Ameca River basin (Drever, 1971). The CO_2 pressure in the water of a river is defined by its pH and by its

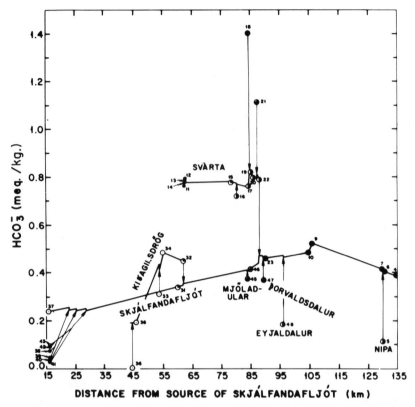

Figure 4-15. The bicarbonate concentration in water from the Skjálfandafljót River and its tributaries in northern Iceland. Open circles indicate stations free of vegetation. Solid circles indicate stations, such as farmland, that are heavily vegetated. Partially filled circles indicate intermediate degrees of plant cover; data from Cawley, Burruss, and Holland (1969).

bicarbonate concentration

$$a_{H^+} \cdot a_{HCO_3^-} = K_1 B P_{CO_2} \qquad (4\text{-}6)$$

Figure 4-16 shows that the CO_2 pressure in river water from the Ameca River basin almost invariably exceeds that of the atmosphere. This is rather to be expected in areas where high-CO_2 waters draining vegetated areas pass into a river system. The persistence of CO_2 pressures in excess of $10^{-3.5}$ atm is perhaps somewhat surprising. Apparently, the loss of CO_2 from river water to the atmosphere is not sufficiently rapid to erase the effects of the influx of high-P_{CO_2} ground water and the decay of organic matter within the river itself (see also Garrels and Mackenzie, 1971).

In the Skjálfandafljót drainage basin the situation is quite different. Figure 4-17 shows that in the areas free of vegetation the effective CO_2

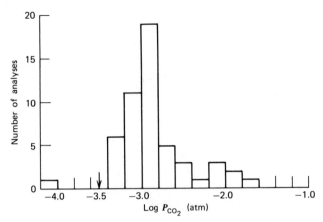

Figure 4-16. Effective CO$_2$ pressure in river waters from the Rio Ameca drainage basin; data from Drever (1971).

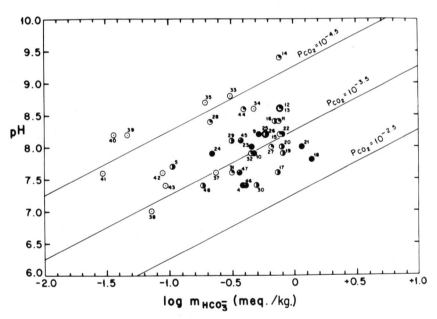

Figure 4-17. The effective CO$_2$ pressure in river waters from the drainage basin of the Skjálfandafljót, northern Iceland; Cawley, Burruss, and Holland (1969).

pressure in river water is normally considerably below that in the atmosphere. With increasing degree of plant cover the P_{CO_2} value of the river water increases and finally surpasses atmospheric P_{CO_2} in the heavily vegetated areas. In the absence of plants, dissolved CO_2 is removed from river water by reaction with the basaltic rocks of this area, so that P_{CO_2} drops to values below $10^{-3.5}$ atm. However, CO_2 is apparently replenished sufficiently rapidly from the air, so that HCO_3^- concentrations as high as 25 ppm are attained even in the absence of plants. This single study does not establish with certainty the ratio of chemical weathering rates in the absence of higher land plants to rates in the presence of such plants; the results do, however, suggest that the effect of CO_2 pickup by river water in times before the advent of higher land plants tended to compensate for the ready availability of CO_2 in soil atmospheres today.

Bicarbonate Concentrations in the Low-Δf Region

In arid areas, where runoff is less than 10 cm/yr, the range of the bicarbonate content of river waters at a given value of Δf decreases markedly, and the median value of $m_{HCO_3^-}$ increases slowly (see Figure 4-12). One possible reason for this behavior is that the bicarbonate concentration in low-Δf rivers tends to be limited by the solubility of one or more mineral phases. Calcite is an obvious candidate. Figures 4-18 and 4-19 show that in rivers within the United States and in most of the major rivers of the world the bicarbonate concentration in moles per liter is slightly more than twice the concentration of calcium up to calcium concentrations of ca. 50 ppm. At higher calcium concentrations the bicarbonate concentration is nearly constant.

It was shown in Chapter II that in solutions saturated with respect to calcite

$$a_{Ca^{2+}} \cdot a_{HCO_3^-}^2 = \frac{K_1 K_C B}{K_2} P_{CO_2} \qquad (4\text{-}7)$$

and that

$$m_{Ca^{2+}} \cdot m_{HCO_3^-}^2 = \left(\frac{K_1 K_C B}{K_2 \gamma_{Ca(HCO_3)_2}^3} \right) P_{CO_2} \qquad (4\text{-}8)$$

In very dilute solutions the activity coefficient $\gamma_{Ca(HCO_3)_2}^3$ is nearly equal to unity. Solutions saturated with respect to calcite at a given temperature and CO_2 pressure therefore have compositions that lie on lines of slope $-\frac{1}{2}$ in

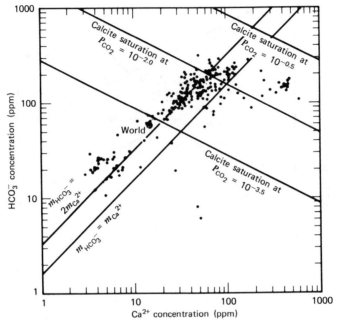

Figure 4-18. The relationship between the bicarbonate and calcium concentrations in rivers of the United States. Data largely from U.S. Geological Survey Water–Supply Papers (see references); world mean after Livingstone (1963).

$\log m_{Ca^{2+}} - \log m_{HCO_3^-}$ diagrams. Lines of this kind have been drawn in Figures 4-18 and 4-19 for pressures of $10^{-3.5}$, $10^{-2.0}$, and $10^{-0.5}$ atm at 25°C. The data in these figures suggest that many rivers contain a concentration of CO_2 in excess of the concentration in equilibrium with the partial pressure of CO_2 in the atmosphere. This agrees with Garrels and Mackenzie's observation (1971, p. 126) that the average CO_2 pressure in many rivers is about 10 times that of the atmosphere.

The concentration of dissolved solids in some rivers in arid terrains approaches, and may even exceed, 10,000 ppm. In these solutions $\gamma_{Ca(HCO_3)_2}^3$ may drop to values as small as 0.15. The effective CO_2 pressures in such rivers may, therefore, be as much as a factor of 7 times smaller than the value suggested by the lines of constant P_{CO_2} in Figures 4-18 and 4-19; this effect probably explains the unreasonably high values of P_{CO_2} suggested for a few rivers by the data of Figure 4-19.

The $HCO_3^- - Ca^{2+}$ plots in Figures 4-18 and 4-19 indicate that most dilute rivers in well-watered areas are undersaturated with respect to calcite, but that in arid areas saturation is reached at the CO_2 pressures commonly encountered in river water. It remains to show how this observation can

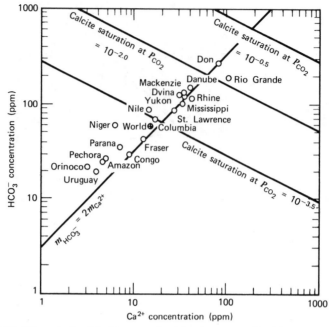

Figure 4-19. The relationship between the bicarbonate and calcium concentrations in some major rivers; data from Livingstone (1963).

explain the $HCO_3^- - \Delta f$ relationship at low runoff values in Figures 4-12 and 4-13. If a river water that is undersaturated with calcite is allowed to evaporate at constant P_{CO_2}, the concentrations of Ca^{2+} and HCO_3^- increase by essentially the same factor until saturation with calcite is reached. At this point the conditions in equation 4-8 are satisfied. Further evaporation requires the precipitation of $CaCO_3$ via the reaction

$$Ca^{2+} + 2HCO_3^- \rightarrow CaCO_3 + CO_2 + H_2O$$

If the concentrations of Ca^{2+} and HCO_3^- at initial saturation are $m_{Ca^{2+}}^0$ and $m_{HCO_3^-}^0$, respectively, their concentration at any later time is

$$m_{Ca^{2+}} = \sigma \left(m_{Ca^{2+}}^0 - \Delta \right)$$

$$m_{HCO_3^-} = \sigma \left(m_{HCO_3^-}^0 - 2\Delta \right)$$

where Δ equals the number of moles of $CaCO_3$ precipitated from a volume of 1 liter of river water at the time of first saturation with respect to $CaCO_3$, and where σ equals V^0/V, the evaporative concentration of the river water since first saturation with $CaCO_3$. Thus, during evaporative concentration

$$\sigma^3\left(m^0_{Ca^{2+}} - \Delta\right)\left(m^0_{HCO_3^-} - 2\Delta\right)^2 = \left(\frac{K_1 K_C B}{K_2 \gamma^3_{Ca(HCO_3)_2}}\right) P_{CO_2} \qquad (4\text{-}9)$$

The reaction path of the solutions depends on the Ca^{2+}/HCO_3^- ratio. Three cases can be distinguished

1. $2m^0_{Ca^{2+}} > m^0_{HCO_3^-}$
2. $2m^0_{Ca^{2+}} = m^0_{HCO_3^-}$
3. $2m^0_{Ca^{2+}} < m^0_{HCO_3^-}$

Calcium-rich river waters of type 1 tend to move downward to the right along the lines of $CaCO_3$ saturation at constant P_{CO_2} in Figure 4-18; type 2 river waters stay more or less at the point of saturation; type 3 river waters tend to move upward to the left along the lines of $CaCO_3$ saturation at constant P_{CO_2}.

As shown in Figures 4-18 and 4-19, many river waters are essentially of the second kind. On reaching saturation with calcite, the Ca^{2+}/HCO_3^- ratio of such rivers remains nearly constant, and the actual concentration of the ions increases only slightly, in response to changes in $\gamma_{Ca(HCO_3)_2}$ during evaporative concentration. This probably explains the cluster of points along the line $m_{HCO_3^-} = 2m_{Ca^{2+}}$ between the calcite saturation line at CO_2 pressures of $10^{-3.5}$ and $10^{-2.0}$ atm. Most of the rivers not close to this line contain excess Ca^{2+}. During evaporative precipitation these waters should and do scatter toward the high-Ca^{2+} side of the diagram as shown in Figure 4-18.

The scheme outlined above is clearly too simple, because the evolution of river water does not follow a path of simple evaporative concentration (see pp. 141–143). However, the agreement between the observations and the predictions is very satisfactory, and it seems likely that a part of the variance of the concentration of $m_{HCO_3^-}$ in $\log m_{HCO_3^-}$–$\log \Delta f$ diagrams is due to $CaCO_3$ precipitation during evaporative concentration.

There are three major sources for the HCO_3^- in river water:

1. atmospheric CO_2 that has interacted either directly with soils and rocks or has been "pumped" by photosynthesis followed by plant decay and root respiration.
2. CO_2 that has been produced by the oxidation of organic matter and elemental carbon already present in rocks prior to their elevation into the soil zone.
3. CO_3^{2-} in carbonate minerals, largely in calcite, aragonite, and dolomite, but also in siderite, ankerite, rhodochrosite, and other carbonate minerals.

In the weathering of igneous and high-grade metamorphic rocks atmospheric CO_2, as defined here, is by far the major source. In the weathering of carbonaceous shales the second source is apt also to be important. In the weathering of limestones and dolomites CO_2 from the atmosphere and CO_3^{2-} from $CaCO_3$ and $CaMg(CO_3)_2$ contribute equally to the HCO_3^- content of ground and river water. The relative importance of the three sources for rock weathering as a whole is considered in the next sections and in Chapter VI.

CALCIUM IN RIVERS

The variation of the calcium concentration in river water with runoff in Figures 4-20 and 4-21 is reminiscent of the variation of bicarbonate with runoff in Figures 4-12 and 4-13. This strong similarity was to be expected, as Figures 4-18 and 4-19 had demonstrated a considerable degree of correlation between the concentration of calcium and bicarbonate in rivers. The

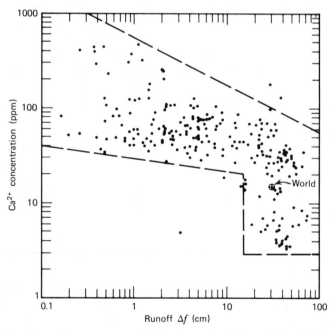

Figure 4-20. The calcium concentration in rivers of the United States. Data largely from U.S. Geological Survey Water–Supply Papers (see references); world mean after Livingstone (1963).

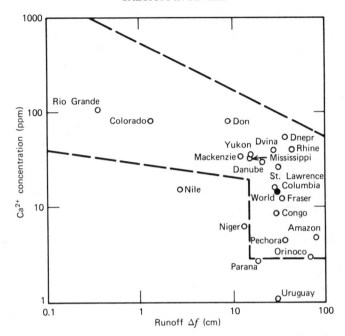

Figure 4-21. The calcium concentration in some major rivers; data largely from Livingstone (1963).

major difference between the log concentration–log Δf relationship of the two ions is found in arid regions, where the concentration range of calcium is larger than that of bicarbonate; this seems to be due largely to the solution of gypsum and/or anhydrite.

Calcium in river water is derived largely from the weathering of carbonates, sulfates, and silicates. Calcite, aragonite, and dolomite are by far the most important carbonates, and gypsum and anhydrite by far the most important sulfates. The relative proportion of these sources of calcium must be known before the acid consumption of the continents can be estimated. This, in turn, is a vital component of the carbon cycle, and is basic for a quantitative formulation of the relationships that determine the CO_2 content of the atmosphere (see Chap. VI).

Most of the calcium in the world average river water of Table 4-2 must be derived from the weathering of carbonates and silicates. It was shown above that ca. 40% of the mean sulfate concentration of 0.23 meq/kg in world average river water is due to fossil fuel burning. The remainder is probably divided fairly evenly between the contribution from sulfate and from sulfide weathering. Thus ca. 0.07 meq of SO_4^{2-} should be balanced by Ca^{2+} due to gypsum and anhydrite solution. The remaining 0.16 meq of sulfur reacts with surface rocks and sediments as H_2SO_4, and, therefore, tends to become balanced by a mixture of the major cations Ca^{2+}, Mg^{2+}, Na^+, and K^+. The proportion of these ions must vary from place to place. Pyrite that is oxidized in a carbonate environment yields H_2SO_4 which is then neutralized largely by Ca^{2+} and Mg^{2+} carbonates. On the other hand, pyrite that is oxidized in a schist or gneiss environment yields H_2SO_4 which is neutralized by salts of all four of the major cations in proportions that depend on their abundance in the area and on the relative ease with which their host minerals are attacked by dilute sulfuric acid. As a first approximation, it seems likely that the relative proportions of Ca^{2+}, Mg^{2+}, Na^+, and K^+ released by sulfuric acid attack are approximately the same as their proportions in world average river water. Since roughly half of the milliequivalents of cations in the average world river waters of Table 4-3 are due to Ca^{2+}, half of 0.16, or about 0.08 meq/kg of Ca^{2+} are probably released during attack by H_2SO_4 from the atmosphere and from the weathering of pyrite. The total amount of Ca^{2+} balanced by SO_4^{2-} in world average river water is therefore roughly 0.15 meq/kg, that is, about 20% of the calcium content of average world river water.

A small fraction of the remainder comes from the solution of chlorides and from brines in which Ca^{2+} is balanced by Cl^-. The bulk of the remaining river calcium must, however, be derived from the weathering of carbonate and silicate minerals. Carbonates can be considered minerals in which metal oxides have already been neutralized by CO_2. To this extent their weathering and redeposition does not involve additional acid consumption. Solution and redeposition of calcite via the reactions

$$CaCO_3 + CO_2 + H_2O \underset{\text{redeposition}}{\overset{\text{solution}}{\rightleftharpoons}} Ca^{2+} + 2HCO_3^- \qquad (4\text{-}10)$$

involves no overall gain or loss of CO_2. From a chemical point of view the process is identical to the mechanical transport of $CaCO_3$ from a continent to the oceans. On the other hand, carbonate minerals can be considered to be reservoirs of CO_2. The solution of limestone and the redeposition of the dissolved calcium as a constituent of a calcium silicate releases CO_2 to the ocean–atmosphere system. The solution of dolomite and the precipitation of the released Mg^{2+} as a constituent of Mg-silicates has the same effect. Carbonate minerals can therefore play the role of CO_2 banks (see Chap. VI). For these reasons it is of considerable interest to estimate the fraction of Ca^{2+} and Mg^{2+} in river water that has been released by the weathering of silicates and the fraction that has been released by the weathering of carbonate minerals.

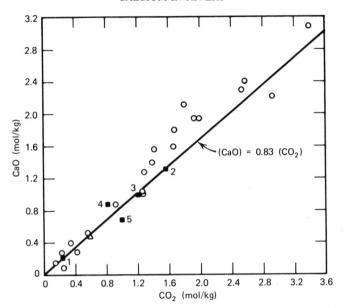

Figure 4-22. The relationship between the corrected CaO and the combined CO_2 content of shales and slates. The numbered analyses refer to (1) Late Proterzoic sediments of the Russian Platform; (2) Paleozoic sediments of the Russian Platform; (3) Paleozoic sediments of the North American Platform; (4) Mesozoic and Cenozoic sediments of the Russian Platform, and (5) Mesozoic and Cenozoic sediments of the North American Platform. ○ Rocks of the Russian Platform, Vinogradov and Ronov (1956); ■ Averages for the Russian and North American Platforms, Ronov and Migdisov (1971); △ Average shale, Clarke (1924).

Most of the calcium in sedimentary rocks is present as a constituent of carbonate minerals. This is obvious for carbonate rocks, but it is also true for shales as demonstrated, for instance, by the data in Figure 4-22. Total CO_2 and total CaO corrected for CaO bound as $CaSO_4$ are well correlated, and the line of best fit goes through the origin. The quantity of calcium present in shales as a constituent of silicate minerals must, therefore, be quite small compared to the quantity present as a constituent of carbonates.

In most shales the number of moles of CO_2 per kilogram of rock exceeds the number of moles of CaO. The excess CO_2 is bound largely as a constituent of dolomite and siderite (see, for instance, Van Moort, 1973). Figures 4-23 and 4-24 show plots of MgO versus bound CO_2 for the same rock units as those in Figure 4-22. These data show that much of the Mg in most shales is not present as a constituent of carbonates. This comes as no surprise, since chlorite, montmorillonite, and other layer silicates containing Mg are prominent constituents of shales.

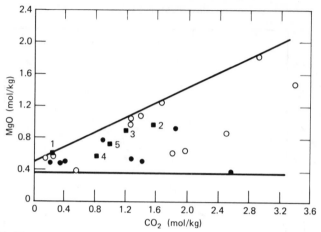

Figure 4-23. The relationship between the MgO and the CO_2 content of shales; see Fig. 4-22 for sources of data.

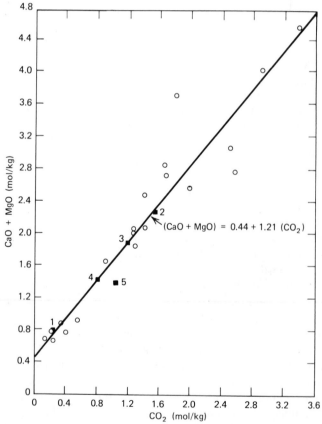

Figure 4-24. The relationship between the sum of the MgO and the CaO content (corrected for SO_3) and the CO_2 content of shales; see Fig. 4-22 for sources of data.

114

If we let

$$w = \text{mol } CaCO_3 \text{ per kilogram of rock}$$

$$x = \text{mol Ca in silicates per kilogram of rock}$$

$$y = \text{mol } CaMg(CO_3)_2 \text{ per kilogram of rock}$$

$$z = \text{mol Mg in silicates per kilogram of rock}$$

then, if we neglect the contribution of iron carbonates,

$$CaO \cong w + x + y \tag{4-11}$$

$$MgO \cong y + z, \quad \text{and} \tag{4-12}$$

$$CO_2 \cong w + 2y \tag{4-13}$$

The data in Figure 4-22 indicate that $x \approx 0$ and that

$$w + y \approx 0.83 \; CO_2 \tag{4-14}$$

The data in Figure 4-24 indicate that

$$CaO + MgO = (w + x + y) + (y + z) \approx 0.44 + 1.21 \; CO_2 \tag{4-15}$$

From equations 4-13 and 4-14 it follows that

$$\frac{w}{y} = \frac{\text{mol } CaCO_3/\text{mol rock}}{\text{mol } CaMg(CO_3)_2/\text{mol rock}} \approx 3.9 \tag{4-16}$$

From equations 4-13 and 4-15 it follows that

$$z \approx 0.44 + 0.21 \text{ mol } CO_2 \tag{4-17}$$

The calculated calcite–dolomite ratio, w/y, is close to that for carbonate rocks as a whole. The average Mg-silicate content of shales apparently consists of 0.44 mol/kg whose presence is independent of carbonates, and, rather curiously, of a portion that increases slowly with the total carbonate content of the shale. The proportion of Mg that is present as a constituent of a carbonate phase in shales, ψ_{Mg}, is equal to

$$\psi_{Mg} = \frac{y}{y + z} \cong \frac{0.17 \; CO_2}{0.44 + 0.38 \; CO_2} \tag{4-18}$$

The world average shale contains roughly 5% CO_2 by weight. In such a shale

$$\bar{\psi}_{Mg} = \frac{(0.17)(1.1)}{0.44 + [(0.38)(1.1)]} = 0.22$$

Thus, whereas calcium in shales is largely present in carbonates, most magnesium is present in silicate minerals. The mean value of ψ, defined as

$$\bar{\psi} = \frac{CO_2}{CaO + MgO}$$

in shales is approximately 0.5 as shown by the means of the values in Tables 4-6 and 4-7.

Table 4-6

Average Analyses of Shales from the Russian and American Platforms[a]

	1 *Late Proterozoic Russian Platform*	*2* *Paleozoic Russian Platform*	*3* *Paleozoic North American Platform*	*4* *Mesozoic & Cenozoic Russian Platform*	*5* *Mesozoic & Cenozoic North American Platform*
SiO_2	57.65	47.94	53.68	55.61	60.10
TiO_2	0.86	0.78	0.88	0.72	0.62
Al_2O_3	17.04	14.26	15.98	14.45	14.71
Fe_2O_3	4.26	4.10	4.02	3.65	3.55
FeO	3.17	1.85	1.40	1.95	0.45
MnO	0.113	0.064	—	0.057	—
MgO	2.38	3.85	3.61	2.20	3.00
CaO	1.20	7.53	5.48	4.90	3.74
K_2O	4.18	3.70	4.33	2.39	2.16
Na_2O	0.93	0.65	0.51	1.10	0.87
P_2O_5	0.090	0.112	0.101	0.107	0.093
$C_{org.}$	0.35	0.70	—	0.94	—
CO_2	1.10	6.80	5.27	3.62	4.35
$S_{sulfate}$	0.100	0.351	—	0.137	—
S_{pyr}	0.300	0.473	—	0.377	—
Cl	0.118	0.148	—	0.101	—
H_2O	6.29	6.07	4.61	7.67	6.52
	100.78	100.73	100.44	100.87	100.71
$\bar{\psi}$	0.32	0.68	0.64	0.57	0.75

[a]Ronov and Migdisov (1971).

Table 4-7

Analyses of Common Sedimentary Rocks[a]

	Sandstones		Shales		Limestones	
	1	2	3	4	5	6
SiO_2	78.66	84.86	60.15	55.43	5.19	14.09
TiO_2	0.25	0.41	0.76	0.46	0.06	0.08
Al_2O_3	4.78	5.96	16.45	13.84	0.81	1.75
Fe_2O_3	1.08	1.39	4.04	4.00	}0.54	}0.77
FeO	0.30	0.84	2.90	1.74		
MnO	trace	trace	trace	trace	0.05	0.03
CaO	5.52	1.05	1.41	5.96	42.61	40.60
SrO	trace	none	—	—	—	—
BaO	0.05	0.01	0.04	0.06	—	—
MgO	1.17	0.52	2.32	2.67	7.90	4.49
Na_2O	0.45	0.76	1.01	1.80	0.05	0.62
K_2O	1.32	1.16	3.60	2.67	0.33	0.58
H_2O-	0.31	0.27	0.89	2.11	0.21	0.30
H_2O^+	1.33[b]	1.47[b]	3.82	3.45	0.56[b]	0.88[b]
P_2O_5	0.08	0.06	0.15	0.20	0.04	0.42
CO_2	5.04	1.01	1.46	4.62	41.58	35.58
SO_3	0.07	0.09	0.58	0.78	0.05	0.07
$C_{org.}$	—	—	0.88	0.69	—	—
Cl	trace	trace	—	—	0.02	0.01
S	—	—	—	—	0.09	0.07
	100.41	99.86	100.46	100.48	99.98	100.34
$\bar{\psi}$	0.90	0.75	0.41	0.65	0.99	0.97

[a]Clarke (1924).

[b]Includes organic matter.

1. Composite analysis of 253 sandstones.
2. Composite analysis of 371 sandstones used for building purposes.
3. Composite analysis of 51 Paleozoic shales.
4. Composite analysis of 27 Mesozoic and Cenozoic shales.
5. Composite analysis of 345 limestones.
6. Composite analysis of 498 limestones for building purposes.

Tables 4-6 to 4-11 contain what are probably the best available data for the average composition of sedimentary rocks. The composite analyses of sandstones, shales, and limestones reported by Clarke (1924) and listed in Table 4-7 have been used in numerous texts since their publication. The most extensive set of new analyses since 1924 has been supplied by Vinogradov, Ronov, and their co-workers, who have carried out large-scale studies of the chemical composition of sedimentary rocks within the Soviet

Table 4-8

Sediments of the Russian Platform[a] and of Precambrian Lutites[b]

	Sands + Aleurites[a]	Clays + Shales[a]	Carbonates[a]	Average Sediment[a]	Precambrian Lutites[b]
SiO_2	70.03	52.15	8.27	39.90	58.32
TiO_2	0.49	0.77	0.13	0.45	0.78
Al_2O_3	8.02	14.69	1.65	7.76	18.44
Fe_2O_3	2.80	3.95	0.81	2.39	3.75
FeO	1.49	2.06	0.42	1.27	4.13
MnO	0.07	0.07	0.04	0.06	0.08
MgO	1.81	3.04	7.44	3.98	2.21
CaO	4.29	5.71	39.74	16.99	0.78
K_2O	2.16	3.25	0.57	1.90	4.42
Na_2O	0.71	0.85	0.29	2.04	0.96
P_2O_5	0.12	0.11	0.05	0.07	0.13
C_{org}	0.24	0.74	0.30	0.41	0.69
CO_2	3.40	4.84	35.67	14.15	0.74
Cl	0.08	0.13	0.07	1.80	—
H_2O	3.76	6.50	2.32	4.10	3.86
$S_{sulfate}$	0.23	0.24	1.07	1.05	0.14 (SO_3)
S_{pyr}	0.20	0.41	0.20	0.26	0.54 (FeS_2)
	100.54	100.22	100.97	100.45	99.97
$\bar{\psi}$	0.64	0.62	0.91	0.80	0.24

[a]Ronov and Migdisov (1971).
[b]Nanz (1953)

Union. Their revised data for the average composition of sedimentary rocks of the Russian Platform (Ronov and Migdisov, 1971) are summarized in Table 4-8, and Ronov and Yaroshevsky's (1967) proposed compositions for platform sediments, geosynclinal sediments, and continental sediments as a whole are reproduced in Table 4-9. It is interesting to compare these averages with those of Poldervaart (1955) in Table 4-10. Although the two sets of figures agree reasonably well, there are some rather unexpectedly large differences, particularly in the SiO_2 content of platform sediments and in the CO_2 content of geosynclinal sediments. These, and less pronounced differences, are due to the choice of chemical compositions for the component rock types and to the proportions chosen for each particular mix. Poldervaart (1955) used the data reported by Clarke (1924) for the composition of average sandstones, limestones, and shales, and was heavily influenced by North American data in arriving at the proportion of the various rock types in platform and geosynclinal sediments.

Table 4-9
Chemical Composition of Continental Sedimentary Rocks[a]

	Platform Sediments						Geosynclinal Sediments						Average Continental Sediment[d]
	1 Sandstones	2 Shales	3 Carbonates	4 Evaporites	5 Effusives	6 Average[b] Composition	7 Sandstones	8 Shales	9 Carbonates	10 Evaporites	11 Effusives	12 Average[c] Composition	
SiO_2	75.75	55.09	9.80	3.02	49.22	49.21	62.93	55.76	13.30	3.02	55.62	50.00	49.82
TiO_2	0.49	0.86	0.18	0.05	1.53	0.65	0.52	0.71	0.14	0.05	1.00	0.65	0.65
Al_2O_3	6.90	16.30	2.54	0.76	15.74	10.88	12.12	17.56	2.70	0.76	16.12	13.69	12.97
Fe_2O_3	2.55	4.17	0.85	0.27	3.33	2.97	2.30	3.61	0.43	0.27	4.17	2.98	2.97
FeO	1.31	1.87	0.54	0.13	8.02	1.65	3.20	3.35	0.94	0.13	4.52	3.21	2.81
MnO	0.06	0.05	0.06	0.01	0.18	0.06	0.11	0.08	0.09	0.13	0.24	0.13	0.11
CaO	3.40	4.75	38.93	28.25	10.00	12.28	5.69	4.08	42.40	28.25	6.91	11.42	11.64
SrO	—	—	—	—	—	—	—	—	—	—	—	—	—
BaO	—	—	—	—	—	—	—	—	—	—	—	—	—
MgO	1.43	2.46	6.83	6.16	6.11	3.37	2.28	2.52	2.92	6.16	4.22	2.99	3.09
Na_2O	0.58	0.75	0.24	13.06	2.51	0.81	1.92	1.27	0.56	13.06	3.33	1.83	1.57
K_2O	1.83	3.01	0.76	0.28	0.73	2.11	1.69	2.76	0.49	0.28	2.02	2.00	2.03
H_2O^-	—	—	—	—	—	—	—	—	—	—	—	—	—
H_2O^+	2.11	5.16	1.40	0.64	1.81	3.42	2.47	4.37	1.05	0.64	1.46	2.73	2.91
P_2O_5	0.16	0.11	0.07	0.01	0.18	0.12	0.11	0.15	0.10	0.01	~0.34	0.18	0.17
CO_2	2.75	3.92	35.05	17.55	~0.01	10.31	4.22	2.80	34.44	17.55	~0.01	7.56	8.26
SO_3	0.29	0.43	2.36	18.15	0.03	1.13	0.12	0.11	0.03	18.15	0.03	0.12	0.38
C_{org}	0.30	0.99	0.33	0.09	~0.01	0.63	0.25	0.78	0.35	0.09	~0.01	0.42	0.47
Cl	0.09	0.12	0.08	14.78	0.005	0.40	0.09	0.12	0.08	14.78	0.007	0.12	0.19
S	—	—	—	—	—	—	—	—	—	—	—	—	—
	100.00	100.04	100.02	103.21	99.42	100.00	100.02	100.03	100.02	103.21	100.01	100.03	100.04
$\bar{\psi}$	0.68	0.64	0.96	0.93	<0.01	0.81	0.61	0.48	0.94	0.93	<0.01	0.62	0.67

[a] Ronov and Yaroshevsky (1967); for slightly revised figures see Ronov and Yaroshevsky (1976).
[b] Computed on basis of 23.6% sandstones, 49.5% shales, 21.0% carbonates, 2.0% evaporites, 3.9% effusives.
[c] Computed on basis of 18.7% sandstones, 39.4% shales and slates, 16.3% carbonates, 0.3% evaporites, 25.3% effusives.
[d] Computed on basis of 27% platform sediments, 73% geosynclinal sediments.

Table 4-10

Comparison of Ronov and Yaroshevsky's (1967) and Poldervaart's (1955)
Composition of Average Platform and Average Geosynclinal Sediments

	Platform		Geosynclinal	
	1 *Ronov and Yaroshevsky* *(1967)*	*2* *Poldervaart* *(1955)*	*3* *Ronov and Yaroshevsky* *(1967)*	*4* *Poldervaart* *(1955)*
SiO_2	49.21	60.6	50.00	51.9
TiO_2	0.65	0.4	0.65	0.5
Al_2O_3	10.88	8.9	13.69	11.4
Fe_2O_3	2.97	2.4	2.98	2.6
FeO	1.65	1.2	3.21	2.0
MnO	0.06	trace	0.13	trace
CaO	12.28	10.6	11.42	12.6
SrO	—	—	—	—
BaO	—	—	—	—
MgO	3.37	2.9	2.99	3.8
Na_2O	0.81	0.8	1.83	1.3
K_2O	2.11	2.1	2.00	2.4
H_2O^-	—	—	—	—
H_2O^+	3.42	—	2.73	—
P_2O_5	0.12	0.1	0.18	0.1
CO_2	10.31	10.0	7.56	11.4
SO_3	1.13	—	0.12	—
C_{org}	0.63	—	0.42	—
Cl	0.40	—	0.12	—
S	—	—	—	—
	100.00	100.00	100.03	100.00
$\bar{\psi}$	0.81	0.87	0.62	0.81

Ronov and Yaroshevsky's (1967) data appear to be similarly biased by the
Russian data. An average of the two sets of data in Table 4-10 is, therefore,
meaningful, and is probably a small step toward true world-wide averages,
rather than simply another exercise in mathematical inbreeding.

The number of moles of CaO, SO_3, MgO, and CO_2 per kilogram of rock
in Ronov and Yaroshevsky's (1967) and Poldervaart's (1955) average platform and geosynclinal sediments are listed in Table 4-12. CaO_{corr}, the
number of moles of CaO not associated with sulfate, and the sum of CaO_{corr}
and MgO are also shown. Together these data can set meaningful limits to

Table 4-11

Components of Average Platform and Geosynclinal Sediments in Table 4-7

Rock Types	Platform Sediments		Geosynclinal Sediments	
	1 *Ronov and Yaroshevsky* (*1967*)	*2* *Poldervaart* (*1955*)	*3* *Ronov and Yaroshevsky* (*1967*)	*4* *Poldervaart* (*1955*)
Sandstone	23.6	43	18.7	13
Graywacke	0.	0	0.	5
Shale	49.5	41	39.4	52
Carbonates	21.0	16	16.3	22
Evaporites	2.0	0	0.3	0
Effusives	3.9	0	25.3	8
	100.0	100.0	100.0	100.0

Table 4-12

Data for and Values of ψ, ψ_{Ca}, and ψ_{Mg} for Platform and Geosynclinal Sediments

Compound	Platform		Geosynclinal	
	1 *Ronov and Yaroshevsky* (*1967*)	*2* *Poldervaart* (*1955*)	*3* *Ronov and Yaroshevsky* (*1967*)	*4* *Poldervaart* (*1955*)
CaO (mol/kg)	2.19	1.89	2.04	2.25
SO_3 (mol/kg)	0.14	0	0.1	0
$CaO-SO_3 \equiv CaO_{corr}$ (mol/kg)	2.05	1.89	2.03	2.25
MgO (mol/kg)	0.84	0.72	0.74	0.94
CO_2 (mol/kg)	2.34	2.27	1.72	2.59
$CaO_{corr} + MgO$ (mol/kg)	2.89	2.61	2.77	3.19
$\bar{\psi}$	0.81	0.87	0.62	0.81
ψ_{Ca}^{max}	1.00	1.00	0.85	1.00
ψ_{Ca}^{min}	0.81	0.87	0.62	0.81
ψ_{Mg}^{max}	0.81	0.87	0.62	0.81
ψ_{Mg}^{min}	0.35	0.53	0.00	0.36

the distribution of calcium and magnesium between carbonate and silicate phases in sedimentary rocks. In all except analysis 3, the number of moles of CO_2 exceeds the number of moles of CaO per kilogram of rock. The fraction of CaO_{corr} present in carbonate phases, ψ_{Ca}, could therefore be 1.00 in analyses 1, 2, and 4. In analysis 3 the value of ψ_{Ca}^{max} is defined by the ratio

$$\psi_{Ca}^{max} = \frac{CO_2 \, (mol/kg)}{CaO_{corr} \, (mol/kg)} = \frac{1.72}{2.03} = 0.85$$

It is virtually certain that the fraction of calcium present in carbonate phases in sedimentary rocks is greater than the fraction of magnesium present in carbonate phases. The minimum value of ψ_{Ca} is, therefore, set quite reasonably by the condition that $\psi_{Ca} = \psi_{Mg}$ since, by definition,

$$\psi_{Ca} = \frac{(CaO)_{carb}}{(CaO)_{total}}$$

and

$$\psi_{Mg} = \frac{(MgO)_{carb}}{(MgO)_{total}}$$

it follows that when $\psi_{Ca} = \psi_{Mg}$

$$(CaO)_{carb} + (MgO)_{carb} = \psi_{Ca}(CaO)_{total} + \psi_{Mg}(MgO)_{total}$$

$$= \psi_{Ca}\left[(CaO)_{total} + (MgO)_{total}\right]$$

and

$$\psi_{Ca} = \frac{(CaO)_{carb} + (MgO)_{carb}}{(CaO)_{total} + (MgO)_{total}} = \psi_{Mg} \equiv \bar{\psi} \qquad (4\text{-}19)$$

Thus the value of $\bar{\psi}$ is both the minimum value of ψ_{Ca} and the maximum value of ψ_{Mg}. The minimum value of ψ_{Mg} is defined by the condition that all of the CaO_{corr} is present in carbonate phases, or that all of the CO_2 is present in $CaCO_3$. Thus in analyses 1, 2, and 4

$$\psi_{Mg}^{min} = \frac{(CO_2 - CaO_{corr}) \, (mol/kg)}{MgO(mol/kg)} \qquad (4\text{-}20)$$

and in analysis 3

$$\psi_{Mg}^{min} = 0$$

For comparison with the data of Table 4-12 values of $\bar{\psi}$ have been calculated for the analyses in Tables 4-7 to 4-10 and are shown as the last figure in each column. The carbonates of cations other than calcium and magnesium have been neglected in this calculation, as they almost certainly play a subordinate role in the mineralogy of sedimentary rocks. Pettijohn (1957) has pointed out that siderite and ankerite are both very minor constituents of sedimentary rocks, although siderite may be more common as a cement in sandstones and as a component of shales than is generally believed.

In virtually all types of sedimentary rocks the value of ψ_{Ca} lies between 0.8 and 1.0. Perhaps the best mean value to take for ψ_{Ca} is the average of Ronov and Yaroshevsky's (1967) and Poldervaart's (1955) values for platform and geosynclinal sediments weighted according to the proportion of the mass of these sediment types

$$\bar{\psi}_{Ca} \cong \left[(0.92 \pm 0.08)(0.27) \right] + \left[(0.82 \pm 0.10)(0.73) \right]$$

$$= 0.85 \pm 0.08$$

A similar calculation for $\bar{\psi}_{Mg}$ yields

$$\bar{\psi}_{Mg} = \left[(0.74 \pm 0.20)(0.27) \right] + \left[(0.45 \pm 0.20)(0.73) \right]$$

$$= 0.50 \pm 0.20$$

The uncertainty in $\bar{\psi}_{Mg}$ is obviously much greater than the uncertainty in $\bar{\psi}_{Ca}$. Both functions are known less precisely than the mean value of $\bar{\psi}$.

$$\bar{\psi} = \left[(0.84 \pm 0.03)(0.27) \right] + \left[(0.72 \pm 0.09)(0.73) \right]$$

$$= 0.75 \pm 0.07.$$

The results of these calculations show that roughly 85% of all calcium and roughly 50% of all magnesium in sedimentary rocks is present in carbonate minerals. It seems reasonable to expect that in rivers draining sedimentary terrains the proportion of calcium and magnesium derived from the weathering of carbonate minerals exceeds the proportion of carbonate minerals in the parent rocks. The assessment of the relative importance of this effect requires a look at the magnesium content of river waters.

MAGNESIUM IN RIVERS

Figures 4-25 and 4-26 show that the relationship between runoff and the magnesium content of river waters is quite similar to that for calcium. This impression is reinforced by the good correlation between the concentration

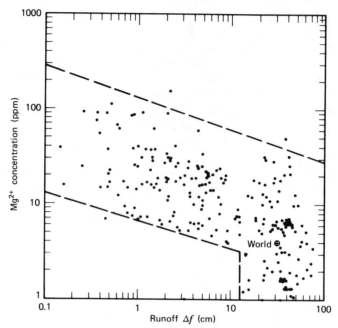

Figure 4-25. The magnesium concentration in rivers of the United States. Data largely from U.S. Geological Survey Water–Supply Papers (see references); world mean after Livingstone (1963).

of calcium and that of magnesium in Livingstone's (1963) compilation of river water analyses, at least up to calcium concentrations of 100 ppm. Lines of best fit in such plots invariably pass through or near the point defined by the concentration of calcium and magnesium in Livingstone's (1963) mean river water. Clarke's (1924) average river water appears to have a calcium–magnesium ratio somewhat higher than the best value; Conway's (1942) calcium–magnesium ratio appears to be high by about a factor of 2.

The molar ratio of calcium to magnesium in mean river water is probably close to Livingstone's (1963) value of

$$\frac{m_{Ca^{2+}}}{m_{Mg^{2+}}} \cong \frac{15/40.08}{4.1/24.33} = 2.2$$

This value is determined by the average composition of river water draining sedimentary terrains, by the average composition of river water draining igneous and metamorphic terrains, and by the proportion of river water draining these two types of terrains. Clarke's (1924) oft-quoted figures for

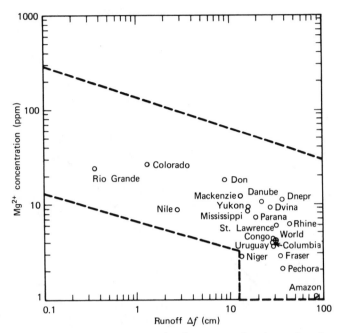

Figure 4-26. The magnesium concentration in some major rivers, data largely from Livingstone (1963).

the proportion of land surface covered by sedimentary rocks (75.7%) and by igneous and metamorphic rocks (24.3%) are apparently quite precise (Blatt and Jones, 1975). The concentration of total dissolved solids in rivers draining igneous and metamorphic terrains was shown to be roughly one-half the total dissolved solids concentration in rivers draining sedimentary terrains (see, also, Conway, 1942). The dissolved solids from sedimentary terrains, therefore, make up approximately 86% of the total dissolved load of rivers; those from igneous and metamorphic terrains constitute about 14% of the total dissolved load of rivers.

The ratio of calcium to magnesium in underground water from igneous and metamorphic terrains is quite variable. Figure 4-27 shows that water in basalts and gabbros has a Ca/Mg ratio of about one-third the value for water in granitic rocks and in many schists and gneisses. The mean Ca/Mg mole ratio in rivers draining igneous and metamorphic terrains may, therefore, be slightly smaller than that of Livingstone's (1963) mean river water, but it seems unlikely that the ratio $m_{Ca^{2+}}/m_{Mg^{2+}}$ is smaller than 1.5.

We are now in a position to return to the question of the preferential solution of carbonates from sedimentary rocks. The Ca/Mg ratio in average

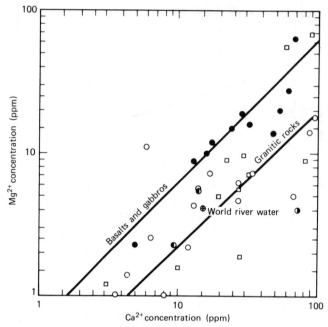

Figure 4-27. The calcium and magnesium concentration in underground waters in igneous terrains; data from White, Hem, and Waring (1963). ☉ Granitic rocks; ◑ intermediate rocks; ● basalts and grabbos; ⊡ schists and gneisses.

sedimentary rock is

$$\frac{\text{CaO (mol/kg)}}{\text{MgO (mol/kg)}} \cong \frac{(2.04)(0.27) + (2.14)(0.73)}{(0.78)(0.27) + (0.84)(0.73)} \cong 2.6$$

whereas the ratio of CaO in the carbonates of sedimentary rocks to MgO in these carbonates is

$$\frac{\text{CaO (carb)}}{\text{MgO (carb)}} \cong 2.6 \times \frac{(0.85 \pm 0.08)}{(0.50 \pm 0.20)} \cong 4.4^{+2.7}_{-1.8}.$$

These figures show that if preferential leaching of carbonates in sedimentary rocks were important, the ratio $m_{Ca^{2+}}/m_{Mg^{2+}}$ in mean river water would probably be greater than the present value, 2.2. Apparently, the removal of magnesium from silicates during the weathering of sedimentary rocks also contributes a good deal to the magnesium content of river water. This somewhat surprising result is, however, in agreement with the data of Figure 4-28 that suggest that the Ca/Mg ratio in underground water from non-

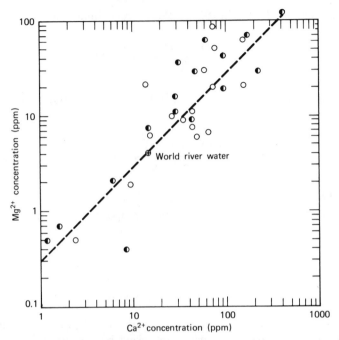

Figure 4-28. The calcium and magnesium content of underground waters in non-carbonate sedimentary rocks; data from White, Hem, and Waring (1963). ⊙ Sandstone, arkose, greywacke, ◐ siltstone, clay, and shale.

carbonate sedimentary rocks is similar to that in mean river water. When all these data are combined, it can be shown that approximately $74 \pm 10\%$ of the calcium in river water and $40 \pm 20\%$ of the magnesium in river water are derived from the solution of carbonate minerals; the remainder is derived largely from the solution of silicate minerals. Ronov and Yaroshevsky's (1967) data for the SO_3 content of sedimentary rocks can account for only a small fraction of river water sulfate assigned above to the solution of gypsum and anhydrite. It is possible that sulfate minerals are dissolved preferentially from parts of the sedimentary section not undergoing mechanical weathering, and that such preferential solution accounts for the virtual absence of evaporites in pre-Cambrian sediments (Garrels and Mackenzie, 1969).

The analyses of average sedimentary rock in Table 4-10 as well as Gehman's (1962) data show that carbon in carbonates is much more abundant than elemental carbon. In Ronov and Yaroshevsky's (1967) platform sediments

$$\frac{CO_2(mol/kg)}{C_{org}(mol/kg)} \cong \frac{2.35}{0.52} = 4.5$$

while in Ronov and Yaroshevsky's (1967) geosynclinal sediments

$$\frac{CO_2(\text{mol}/\text{kg})}{C_{\text{org}}(\text{mol}/\text{kg})} \cong \frac{1.72}{0.35} = 4.9$$

If we weight the platform and geosynclinal rocks as above, then for continental sediments as a whole

$$\frac{CO_2(\text{mol}/\text{kg})}{C_{\text{org}}(\text{mol}/\text{kg})} \cong 4.7$$

It is shown in Chapter VI that the value for this ratio is consistent with the relationships between the isotopic composition of organic carbon, carbonate carbon, and the probable composition of carbon in the earth's mantle.

SODIUM IN RIVERS

The close relationship between sodium and chloride in river waters was pointed out above, and is further emphasized by the similarity of the log concentration–$\log \Delta f$ plots of the two ions. The concentration of sodium in rivers of the United States (Figure 4-29) has a dispersion not quite as large as that of the chloride concentration (see Figure 4-8), but the logarithm of the concentration of both ions increases approximately linearly with decreasing $\log \Delta f$.

It seems likely that most of the chloride in river water is derived from the atmospheric cycling of sea salt and from the solution of evaporites. Sodium is the dominant but not the only cation in both these sources. If we subtract the 0.22 meq/kg of sodium required to balance all the chloride in Livingstone's (1963) average river water from the total sodium concentration, there remain 0.054 meq/kg that must be derived largely from the weathering of silicate minerals in igneous, metamorphic, and sedimentary rocks. It is interesting to see whether the supply of such a quantity of sodium is consistent with the data for the average chemical composition of these rock types and with the proposed apportionment of calcium to the weathering of sediments and the weathering of crystalline rocks. Various estimates for the proposed composition of continental shield crystalline surface rocks are reproduced in Table 4-13. The mole ratio of calcium to sodium in these rocks is approximately 0.60. During weathering of silicate rocks calcium is removed more completely than sodium (see Chapter II). This is consistent with the data of Figure 4-30, which show that the Ca^{2+}/Na^+ ratio in

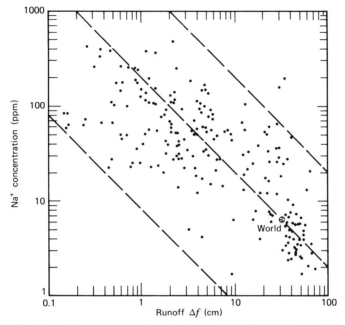

Figure 4-29. The sodium concentration in rivers of the United States. Data largely from U.S. Geological Survey Water–Supply Papers (see references); world mean after Livingstone (1963).

underground waters from igneous and metamorphic terrains is quite variable, and that the mean value of the ratio is approximately

$$\frac{m_{Ca^{2+}}}{m_{Na^+}} \approx \frac{1}{0.7} \times \frac{23}{40} \approx 0.8$$

that is, somewhat greater than the Ca^{2+}/Na^+ ratio in the parent rocks. It seems likely that the average Ca^{2+}/Na^+ ratio in river water draining igneous and metamorphic terrains has a similar value.

In the average sedimentary rock of Table 4-9 the mole ratio of Ca/Na is close to 4.0. During weathering of these rocks Ca is almost certainly removed preferentially, at least to the degree that Ca is removed preferentially during the weathering of igneous and metamorphic rocks. The ratio of the concentration of Ca^{2+} to that of sodium derived from the weathering of silicates in rivers draining sedimentary terrains is, therefore, probably about 6.

The Ca/Na ratio derived from silicate weathering in world average river water depends both on the $(Ca/Na)_{sil}$ ratio in water draining sedimentary and nonsedimentary terrains and on the fraction of total dissolved salts

Table 4-13

Some Estimates of the Composition of Crystalline Shield Rocks[a]

	1	2	3	4	5
SiO_2	66.4	67.45	63.08	59.14	65.03
TiO_2	0.6	0.41	0.81	1.05	0.50
Al_2O_3	15.5	14.63	16.75	15.34	15.15
Fe_2O_3	1.8	1.27	2.38	3.08	1.82
FeO	2.8	3.13	2.91	3.80	3.46
MnO	0.1	0.04	0.02	0.12	0.07
CaO	3.8	3.39	4.07	5.08	3.32
SrO	—	—	—	0.02	—
BaO	—	—	0.02	0.05	—
MgO	2.0	1.69	1.78	3.49	2.27
Na_2O	3.5	3.06	3.64	3.84	3.15
K_2O	3.3	3.55	3.07	3.13	3.03
H_2O^-	—	} 0.79	—	} 1.15	} 1.76
H_2O^+	—		0.79		
P_2O_5	0.2	0.11	0.22	0.30	0.13
CO_2	—	0.12	0.39	0.10	—
SO_3	—	—	—	—	0.14
C_{org}	—	—	—	—	—
Cl	—	—	—	0.05	—
S	—	—	0.12	0.05	—
	100.0	99.64	100.05	99.79	99.83

[a](1) Average composition of continental shield crystalline surface rocks (Poldervaart, 1955); (2) Average composition of the rocks of Finland (Sederholm, 1925); (3) Preliminary estimate of the composition of the earth's crust in the Canadian Shield (Grout, 1938); (4) Composition of average igneous rocks (Clarke, 1924, p. 29); (5) Average composition of the shields (Ronov and Migdisov, 1971).

contributed by rivers draining these two types of terrains. In the previous section it was concluded that approximately 86% of the dissolved salts were derived from sedimentary, and 14% from igneous and metamorphic terrains. On this basis the ratio of $(m_{Ca^{2+}}/m_{Na^+})_{sil}$ in world average river water should be approximately

$$\left(\frac{m_{Ca^{2+}}}{m_{Na^+}}\right)_{sil} \cong (6)(0.86) + (0.8)(0.14) \approx 5$$

This ratio should be nearly the same as the ratio $m_{Ca^{2+}}/(m_{Na^+} - m_{Cl^-})$ if all the chloride in world average river water is balanced by sodium. From

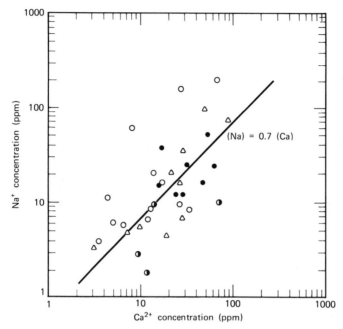

Figure 4-30. The sodium and calcium concentration in underground waters from metamorphic and igneous terrains. Data from White, Hem, and Waring (1963). ⊙ Granitic rocks; ○ intermediate igneous rocks; ● basic igneous rocks; △ schists and gneisses.

the data in Table 4-3 it follows that

$$\overline{\left(\frac{m_{Ca^{2+}}}{m_{Na^+} - m_{Cl^-}}\right)} = \frac{0.375}{0.274 - 0.220} = 6.9$$

The uncertainty in this ratio and in the calculated value for $(m_{Ca^{2+}}/m_{Na^+})_{sil}$ is so large, that the difference is probably not significant. If the difference does turn out to be significant, it is in the right direction in that it suggests that not all of the river chloride is balanced by river sodium.

POTASSIUM IN RIVERS

The concentration of potassium is frequently not reported in analyses of river waters. Despite the relatively small number of data points in Figures 4-31 and 4-32 it is clear, however, that the $\log m_{K^+} - \log \Delta f$ line has a slope

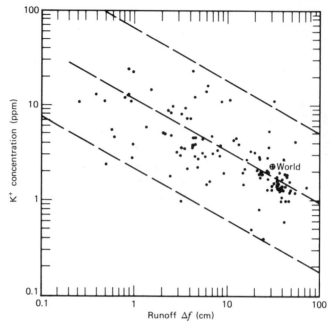

Figure 4-31. The potassium concentration in rivers of the United States. Data largely from U.S. Geological Survey Water–Supply Papers (see references); world mean after Livingstone (1963).

of about $-\frac{1}{2}$. This is significantly smaller than the slope of -1 found for the log concentration $-\log \Delta f$ relationships of Na^+, Cl^-, and SO_4^{2-}. Thus most potassium in rivers must have a different origin than these ions. The solution of evaporite minerals containing potassium and atmospheric cycling contribute a rather small but poorly known fraction of the total potassium in rivers. Most potassium in rivers is derived from the decomposition of silicate minerals during weathering. Potassium is usually not released completely during the weathering of silicates; a good deal of potassium is frequently retained within illites and mixed-layer minerals and on cation exchange sites. If 50% of the potassium and none of the calcium were retained during the weathering of crystalline and sedimentary rocks, the $m_{Ca^{2+}}/m_{K^+}$ ratio in average river water would be near 6.1. This value is very close to Livingstone's (1963) $m_{Ca^{2+}}/m_{K^+}$ ratio of 6.3 in world average river water. The agreement is encouraging, but the analytical data for mean river water and figures for the percentage retention of potassium during weathering are sufficiently uncertain that the similarity between the two ratios may be somewhat fortuitous.

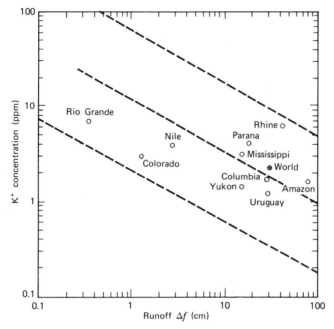

Figure 4-32. The potassium concentration in some major rivers; data largely from Livingstone (1963).

SILICA IN RIVERS

Silica in river water is present largely as H_4SiO_4 derived from the weathering of silicate minerals other than quartz. The variation of SiO_2 as a function of Δf in Figures 4-33 and 4-34 is quite similar to that of bicarbonate. The similarity is particularly striking in the Δf range 0.1–10 cm, where the slope of the $\log m_{SiO_2} - \log \Delta f$ line is almost identical to that of the $\log m_{HCO_3^-} - \log \Delta f$ line. In the high-Δf region of the diagram the range of the bicarbonate concentration tends to be somewhat larger than that of the silica concentration. There is certainly no necessary connection between the concentration of the two components in a given river. This is demonstrated most clearly by the chemistry of the large South American rivers. The Amazon, for instance, carries some 11 ppm of SiO_2, only slightly less than Livingstone's world average of 13.1 ppm. On the other hand the bicarbonate concentration of 22 ppm (Gibbs, 1972) is only one-third of the bicarbonate content of average river water. This is a consequence of the distribution of rainfall and rock types in the Amazon River basin. In the lower reaches of the river the decomposition of aluminosilicates, such as kaolinite, continues to contribute silica to the river but virtually no cations balanced by bicarbonate.

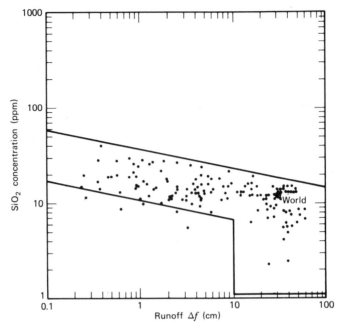

Figure 4-33. The silica concentration in rivers of the United States. Data largely from U.S. Geological Survey Water–Supply Papers (see references); world mean after Livingstone (1963).

Normally, however, the weathering of silicate minerals does release both silica and cations balanced by bicarbonate. In fact, a crude test of the parentage of the ions in average river water can be built on this observation. If all of the calcium, magnesium, sodium, and potassium are released from 100 g of Clarke's (1924) average igneous rock (Table 4-13, column 4), about 0.37 mol of cations are freed. Of the 0.98 mol of SiO_2 present in the original 100 g of average igneous rock about 25% remain as quartz. A part of the other 0.74 mol combine with Al_2O_3, Fe_2O_3, and MgO to form a variety of clay minerals. The mole ratio SiO_2/Al_2O_3 in the common clays is between 2.0 and 4.0; a figure of 3.0 ± 0.7 for the mean value of the SiO_2/Al_2O_3 ratio is quite reasonable. As 100 g of average igneous rock contain 0.15 mol of Al_2O_3, about 0.45 ± 0.10 mol of SiO_2 will be combined in aluminosilicates. This leaves roughly 0.29 ± 0.10 mol of SiO_2 to be transported away in solution. In river water produced in this fashion the ratio of SiO_2 to total cations in solution is, therefore, approximately $0.29 \pm 0.10/0.37 = 0.78 \pm 0.30$. The release of cations during chemical weathering is rarely complete. The actual ratio of released SiO_2 to released cations is therefore rarely equal to

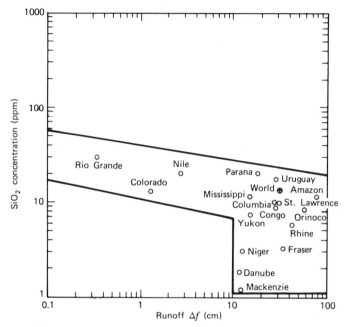

Figure 4-34. The silica concentration in some major rivers; data largely from Livingstone (1963).

the calculated ratio for complete release (see Chap. II), but should not be very different. Silicate weathering contributes about 0.35 mmol/kg of cations (see Table 4-14) to the composition of average river water. The dissolved silica content of average river water is 0.22 mmol/kg (13.1 ppm) (see also Davis, 1964). The ratio of SiO_2 to cations contributed by silicate weathering is, therefore, approximately $0.22/0.35 = 0.63$. Agreement between the predicted and observed values of this ratio is probably better than might have been expected, and shows that the proposed model for the evolution of mean river water is quite reasonable.

The excellent correlation between the behavior of bicarbonate and dissolved silica in rivers draining arid terrains is rather intriguing. At runoff values below 10 cm the ratio of the bicarbonate to the silica concentration is about 10. Garrels (1967) has observed a similar value of the HCO_3^-/SiO_2 ratio in high-carbonate waters from a variety of igneous rocks. It is possible that the value of this ratio in arid regions is related to the preponderance of montmorillonite in the weathering products of arid zones (Barshad, 1966), and to the effects of repeated wetting and drying (J. I. Drever, personal communication, 1977).

Table 4-14

The Concentration of Some Minor Elements Dissolved
in River Water

Element	Median (μg/kg)	Range (μg/kg)	Source of Data[a]
Ag	0.09	0–0.94	(1)
	0.3	0.1–0.55	(2)
	0.3		(12)
Al	238	12–2550	(1)
	360		(3)
	400		(12)
As	1.7		(3)
	2		(12)
Au	0.002		(7)
	0.002		(12)
B	10	1.4–58	(1)
	13		(4)
	10		(12)
Ba	45	9–152	(1)
	16		(5)
	20		(12)
Br	19		(4)
	20		(12)
Ce	0.06		(12)
Co	0.19	0.06–0.54	(2)
	—	0–5.8	(1)
	0.1		(12)
Cr	5.8	0.72–84	(1)
	1.4	0.1–4.1	(2)
	1		(12)
Cs	0.020	0.011–0.044	(2)
	0.02		(12)
Cu	5.3	0.83–105	(1)
	1.4		(3)
	10		(4)
	7		(12)
Dy	0.05		(12)
Er	0.05		(12)
Eu	0.007		(12)
F	150		(3)
	88		(4)
	100		(12)
Fe	300	31–1670	(1)
	480		(3)
	670		(12)
Ga	0.089		(6)
	0.09		(12)

Table 4-14 (continued)

Element	Median (µg/kg)	Range (µg/kg)	Source of Data[a]
Gd	0.07		(12)
Hg	0.074		(8)
	0.04		(12)
Ho	0.01		(12)
I	2.2		(3)
	7.1		(4)
La	0.2		(7)
	0.19		(9)
	0.2		(12)
Li	1.1	0.075–37	(1)
	3		(12)
Lu	0.008		(12)
Mn	12		(4)
	20	0–185	(1)
	7		(12)
Mo	0.35	0–6.9	(1)
	1.8	0.49–6.83	(2)
	0.6		(3)
	0.6		(12)
Nd	0.2		(12)
Ni	10	0–71	(1)
	0.3		(12)
P	19		(9)
	20		(12)
Pb	4.0	0–55	(1)
	3.9		(8)
	3		(12)
Pr	0.03		(12)
Rb	1.5	0–8.0	(1)
	1.1	0.55–1.90	(2)
Sb	1.1	0.27–5.09	(2)
	2		(12)
Sc	0.004		(12)
Se	0.20	0.116–0.409	(2)
	0.2		(12)
Sm	0.03		(12)
Sr	60	6.3–802	(1)
	57		(3)
	70		(12)
Tb	0.008		(12)
Th	0.096		(10)
	0.1		(12)
Ti	8.6	0–107	(1)
	3		(12)

137

Table 4-14 (continued)

Element	Median ($\mu g/kg$)	Range ($\mu g/kg$)	Source of Data[a]
Tm	0.009		(12)
U	0.06		(7)
	0.043		(10)
	0.026		(11)
	0.3		(12)
V	0	0–6.7	(1)
	1		(3)
	0.9		(12)
W	0.03		(7)
	0.03		(12)
Y	0.07		(12)
Yb	0.05		(12)
Zn	—	0–215	(1)
	5.0		(3)
	45		(4)
	16		(9)
	20		(12)

[a]Sources of data: (1) Durum and Haffty (1963) (U.S.A.), and Durum (1971); (2) Kharkar, Turekian, and Bertine (1968); (3) Sugawara (1964) (Japan); (4) Konovalov (1959) (U.S.S.R.); (5) Turekian and Johnson (1966) (Eastern U.S.A.); (6) Heide and Ködderitzsch (1964) (Germany); (7) Landström and Wenner (1965); (8) Heide, Lerz, and Böhm, (1957) (Germany); (9) Silker (1964) (Columbia R.); (10) Moore (1967) (Amazon R.); (11) Rona and Urry (1952) (North America); (12) Garrels, Mackenzie, and Hunt (1975) from Turekian (1969).

BRIEF COMMENTS ON SOME OTHER COMPONENTS OF RIVERS

Table 4-14 summarizes a good deal of the presently available data bearing on the concentration of the minor constituents of river water. For most elements these data are still very fragmentary, but they turn out to be of considerable importance in calculations of residence times of elements in the oceans. Among the elements listed in the table only nitrate, iron, and aluminum have concentration ranges which extend well into the parts per million range. Feth (1966) has reported nitrate analyses in excess of 100 ppm in some ground water, although the mean concentration of dissolved nitrate is about 1 ppm (see Table 4-2). The relative contribution of atmospheric nitrate and nitrate produced by plants and plant decay to the nitrate content of rivers is not well defined.

A good deal of nitrogen in river water may be present as a constituent of dissolved organic matter. Normally the concentration of such material is a small fraction of the total dissolved loads of rivers. However, in tropical streams, especially in those draining swampy areas, the dissolved organic matter may account for more than half the total dissolved solids (Clarke, 1924, p. 110; Beck, Reuter, and Perdue, 1974), and is an important carrier of carbon from the land to the oceans. The large concentration of dissolved organic matter in the Amazon has been reconfirmed by Williams (1968). In such rivers the pH is often abnormally low, and the concentration of iron in true solution is abnormally high. Hem (1960) has pointed out the probable importance of organic complexing agents in the transport of iron in such rivers. Coonley, Baker, and Holland (1971) have shown that in the Mullica River, New Jersey, iron must be complexed in some such fashion, and that iron is precipitated quantitatively in the river estuary. Boyle et al. (1974) have shown that this is apparently true for most rivers.

The concentration of aluminum, like that of iron, is normally well below 1 ppm, but may rise above 1 ppm in some relatively acid waters (Hem, 1959). Iron, aluminum, and titanium together form a group of elements that are present in major amounts in the earth's crust, but whose transport in rivers takes place almost entirely as constituents of particulate matter.

SUMMATION

A good deal of this chapter has been devoted to the definition of the composition of average river water and its dissection into its several component parts. Table 4-15 summarizes the most important results. It seems unnecessary to recapitulate all of the arguments contributing to the present form of this table, but a discussion of the salient features may help to bring into focus aspects of the chapter that are of particular importance for the chemical evolution of the atmosphere and of ocean water. The figures of Table 4-15 are, of course, quite rough. Present estimates of the composition of average river water will undoubtedly be revised. Even if present-day estimates were to remain virtually unchanged by additional data, particularly for some of the large, poorly sampled South American, Asian, and African rivers, it would take considerable luck or an unexpected degree of prescience to prevent a redistribution of the solutes in river water among their various sources. Nevertheless, the numbers in Table 4-15 are almost certain to be approximately right.

Nearly two-thirds of the carbon in river bicarbonate is derived from the atmosphere either directly as CO_2 or via photosynthesis followed by plant decay. The remaining bicarbonate is derived largely from the weathering of

Table 4-15

The Provenance of Solutes in Average River Water

Source	Anions (meq/kg)			Cations (meq/kg)				Neutral Species (mmol/kg)
	HCO_3^-	SO_4^{2-}	Cl^-	Ca^{2+}	Mg^{2+}	Na^+	K^+	SiO_2
Atmosphere[c]	0.58[a]	0.09[b]	0.06	0.01	≤ 0.01	0.05	≤ 0.01	≤ 0.01
Weathering or solution of								
Silicates	0	0	0	0.14	0.20	0.10	0.05	0.21
Carbonates	0.31	0	0	0.50	0.13	0	0	0
Sulfates	0	0.07	0	0.07	0	0	0	0
Sulfides	0	0.07	0	0	0	0	0	0
Chlorides	0	0	0.16	0.03	≤ 0.01	0.11	0.01	0
Organic Carbon	0.07	0	0	0	0	0	0	0
Sum	0.96	0.23	0.22	0.75	0.35	0.26	0.07	0.22

[a]Largely as atmospheric CO_2.
[b]Much of this is apparently balanced by H^+.
[c]These figures do not include soil-derived material.

calcite, aragonite, and dolomite. Only about 7% come from the oxidation of organic carbon in sediments. Although oxidation of organic carbon plays a minor role in the chemistry of rivers, its effect on the geochemical cycle of oxygen is quite marked (see Chap. VI).

The source of sulfate in average river water is still somewhat uncertain. The data available at present suggest that roughly 40% are cycled through the atmosphere, and that the other 60% are derived from the weathering of sulfides and the solution of sulfate minerals. Gypsum and anhydrite are by far the most important sulfate minerals. The contribution of their solution to the sulfate content of average river water is apparently quite out of proportion to their abundance in average sedimentary rock; this may imply that the recycling of evaporites is more rapid than that of other types of sedimentary rocks. The oxidation of sulfides is restricted to near-surface rocks, and probably contributes sulfate to river water roughly in proportion to the concentration of sulfides in rocks undergoing weathering.

The recycling of chloride through the atmosphere apparently does not dominate the chloride content of river water to the extent advocated by some authors. If atmospheric chloride really contributes less than half of the chloride in average river water, the solution of evaporitic chlorides must contribute the bulk of the river chloride. This suggestion is certainly in harmony with the proposed rapid cycling of evaporitic sulfates, but both proposals depend on the validity of the figures accepted for the atmospheric contribution of these ions to average river water.

Only a small fraction of calcium in river water is derived from recycled sea salt. The solution of gypsum and anhydrite contributes about 9% of the total. A small contribution from the solution of chlorides has been assumed, but the bulk of the calcium is derived from the solution of carbonate minerals and the weathering of silicate minerals. Arguments detailed above indicate that the solution of carbonates is currently contributing about four times as much calcium as does the weathering of silicates. Magnesium owes its presence in river water to similar processes. However the weathering of silicates contributes about 50% more magnesium than does the solution of carbonates.

The geochemistry of sodium in river water is closely linked to that of chloride. The available data suggest that only about 35% of the sodium in rivers owes its origin to the weathering of silicate minerals; the remaining 65% are due to atmospheric recycling and to the solution of halite in evaporite sediments. At present it appears that halite solution contributes about twice as much sodium as does atmospheric recycling, but this estimate depends heavily on the rather incomplete data for the concentration of NaCl in average rain water.

Potassium and dissolved silica are largely contributed by silicate weathering, although atmospheric recycling is surely not a negligible source of potassium. If the solution of evaporites is as important in the geochemistry of sodium as suggested above, then some potassium must be similarly recycled.

The relationship between runoff and the concentration of the major dissolved species in rivers varies from element to element. Elements in river water that are largely introduced by atmospheric cycling and by the solution of highly soluble compounds have a slope close to -1 in $\log m - \log \Delta f$ diagrams. Na^+, Cl^-, and SO_4^{2-} are river water constituents of this type. Their removal from the continents is largely independent of continental runoff, because the product $m \cdot \Delta f$ is nearly independent of Δf. The slope of the line of best fit for the $\log m - \log \Delta f$ relationship of the other major ions and of silica in river waters is less than 1. A reduction in the worldwide value of Δf would therefore reduce the rate of removal of these ions and of silica in solution from the continents. However, their removal rate would decrease much less than the total river runoff from the continents.

The small negative slope of the Ca^{2+} and HCO_3^- concentrations with decreasing Δf can be explained in terms of the saturation of soil, ground, and river water with respect to calcite in arid terrains. The similarly small slope of the $\log m_{Mg^{2+}} - \log \Delta f$ and $\log m_{H_4SiO_4} - \log \Delta f$ lines, and the intermediate slope of the $\log m_{K^+} - \log \Delta f$ line are more difficult to explain. Kennedy (1971) has shown that in the Mattole River of northern California and in several other drainage basins the variation of the concentration of dissolved silica with river discharge is related to the proportion of water that has penetrated deeply into the weathering zone. During storm runoff there is apparently not sufficient time for chemical equilibrium between soils and dissolved electrolytes to be established in soil water. The various slopes of $\log m$–$\log \Delta f$ relationships are, therefore, almost certainly a complex function of both the path taken by rain water from the point of impact to its entrance into a river system and the biological processes and chemical interactions between river water and the particulate load of rivers. Models of the evolution of river water chemistry tend to be highly oversimplified; they are nevertheless of some use. It was shown in Chapter III that runoff is related approximately to rainfall by the equation

$$\Delta f = re^{-E_0/r} \tag{4-21}$$

If rain water, shortly after falling, acquires a concentration m_i^0 of an ion i, then the concentration of this ion after evaporative concentration is

$$m_i = m_i^0 \cdot \frac{r}{\Delta f} \cong m_i^0 \cdot e^{+E_0/r} \tag{4-22}$$

It follows that the product $m_i \cdot \Delta f$ is constant only if $m_i^0 \cdot r$ is constant, that is, if the concentration, m_i^0, prior to evaporation, is inversely proportional to the rainfall.

It seems likely, however, that in many soils the value of m_i^0 for several ions is essentially independent of r. The relationship between $\log m_i$ and $\log \Delta f$ is then a curve of the type shown in Figure 4-35; if a straight line is forced through such a curve, its slope is approximately -0.5.

A closer approximation to reality demands a separation of runoff into its surface, and subsurface components. The relative proportion of these two components, the concentration of dissolved species in each, and their progressive evaporative concentration largely determine the composition of the resulting river water. In an extreme case such a process can yield a slope of 0 in $\log m_i$–$\log \Delta f$ plots. If all of the surface water evaporates completely and, therefore, contributes nothing to the river runoff, and if evaporation of subsurface waters is minimal, the concentration of dissolved salts in the resulting river water depends only on their concentration in the subsurface waters, while the runoff depends on the fraction of rainfall that penetrates into the subsurface. If the concentration in subsurface waters is constant while Δf is variable, the slope of the $\log m_i$–$\log \Delta f$ curves is, of course, 0 or nearly 0.

It is premature, however, to conclude that these processes alone determine the small slope of the $\log m$–$\log \Delta f$ relationship of dissolved constituents such as H_4SiO_4. The mineralogy and texture, as well as the biota supported by soils, are also functions of rainfall and its seasonality, and all these parameters must also contribute to the shaping of $\log m$–$\log \Delta f$ relationships.

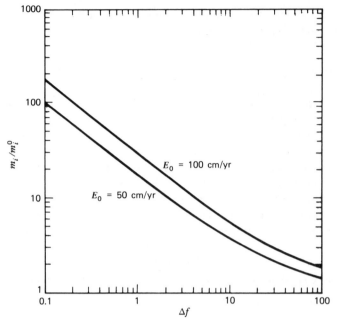

Figure 4-35. The relationship between $\log m_i / m_i^{\circ}$ and $\log \Delta f$, when $m_i / m_i^{\circ} = e^{+E_{\circ}/r}$.

THE COMBINED RATE OF PHYSICAL AND CHEMICAL EROSION

The rate of chemical erosion of the continents can be calculated from the data in Table 4-15 by summing the contribution from the weathering and solution of the various mineral groups. This has been done, and the results are summarized in Table 4-16. All cations derived from silicate weathering have been calculated as the oxides, and silica release has been computed directly from the concentration of dissolved silica in average river water. The contribution of recycled sea salts to the dissolved salts in river water has not been counted, but the sulfur corresponding to 0.08 meq/kg of river SO_4^{2-} has been counted, since this is derived indirectly from man-induced continental "weathering." The continental contribution to the composition of average river water turns out to be ca. 72.6 meq/kg. This accounts for 60% of the total dissolved load of average river water. The remaining 40% consists largely of cycled sea salts and of atmospheric CO_2, O_2, and H_2O that has participated in chemical weathering and has become part of the dissolved salts in river water.

Since the average world runoff is close to 30 cm/yr, the total rate of chemical erosion is approximately

$$72.6 \times 10^{-6} \frac{\text{g solids}}{\text{g river water}} \times \frac{30 \text{ g river water}}{\text{cm}^2/\text{yr}} \approx 2.2 \times 10^{-3} \frac{\text{g solids}}{\text{cm}^2/\text{yr}}$$

At a mean density of 2.7 g/cc this corresponds to a mean denudation rate of ca. 8.1×10^{-4} cm/yr. Garrels' and Mackenzie's values (1971, p. 122 ff) are somewhat higher, apparently because they did not subtract the atmospheric contribution to the total dissolved load of rivers.

The calculated mean chemical erosion rate is nearly independent of land elevation. This is best demonstrated by the lack of correlation between the rate of chemical erosion and the mean elevation of the continents. The data in Table 4-3 indicate that the concentration of total dissolved solids of rivers in North America is similar to that in Asian rivers. The runoff is nearly the

Table 4-16

The Contribution of Continental Material to the Solute Content of Average River Water

	Concentration (meq/kg)	Equivalent solid weight (mg/kg)
Calcium		
CaO	0.14	3.9
$CaCO_3$	0.50	25.0
$CaSO_4$	0.07	4.7
$CaCl_2$	0.03	1.7
Magnesium		
MgO	0.20	4.0
$MgCO_3$	0.13	5.5
Sodium		
$NaO_{0.5}$	0.10	3.1
$NaCl$	0.11	6.4
Potassium		
$KO_{0.5}$	0.05	2.3
KCl	0.01	0.7
Silica	0.22 (mmol/kg)	13.2
Elemental carbon	0.07	0.8
Sulfur from fossil fuel burning	0.08	1.3
Total		72.6

same for the two continents. Thus the rate of chemical erosion of the two continents is nearly the same in spite of the very large difference in their mean elevation above sea level. Chemical erosion is clearly much more a function of atmospheric composition, climate, and exposed rock type than of topography. The term η in equation 4-2 must, of course, also approach 0 with decreasing mean continental elevation, but the approach must be rapid at small values of h, and it is probably best to set η equal to a constant for $h > 0$ and equal to 0 for $h = 0$. The integration of equation 4-3 for a constant rate of uplift, ϕ_2, then yields the expression

$$\eta = \frac{1}{\beta} \ln \frac{(\phi_2 + \alpha - \eta)}{Ae^{-\beta(\phi_2 + \alpha - \eta)t} + \alpha} \qquad (4\text{-}23)$$

where

$$A = (\phi_2 + \alpha - \eta)e^{-\beta h_1} - \alpha. \qquad (4\text{-}24)$$

Negative values of h are excluded by the condition that $\eta = 0$ when $h = 0$.

If the values for α and β derived in the early part of this chapter and the value of η derived above are used in equation 4-3, then

$$\frac{dh}{dt} \approx \phi - 2.0 \times 10^{-4}(e^{5.0 \times 10^{-5}h} - 1) - 8.1 \times 10^{-4} \text{cm/yr} \qquad (4\text{-}25)$$

The rate of chemical erosion is equal to the rate of mechanical erosion when

$$2.0 \times 10^{-4}(e^{5.0 \times 10^{-5}h} - 1) = 8.1 \times 10^{-4}$$

This is true when $h \approx 0.32 \times 10^5$ cm, that is, 320 m. Toward greater elevations the rate of mechanical erosion becomes progressively more dominant. For the world as a whole the rate of mechanical erosion currently exceeds the rate of chemical erosion by a factor of about 6.

These effects are illustrated in Figure 4-36. The upper curve shows the reduction of a continent with an initial mean elevation of 1000 m by mechanical erosion alone. The parameters α and β were given their present-day values, and the uplift rate, ϕ, was set equal to 0. Initially, the mean elevation decreases rapidly and drops to 500 m in 8.5 million years. The reduction to a mean elevation of 250 m requires 24 million years. The decrease in elevation with time slows greatly beyond that point.

The lower curve in Figure 4-36 shows the decrease in mean elevation due to the combined effect of mechanical and chemical erosion. The values of α and β are the same as in the upper curve; the uplift, ϕ, is again 0, and the

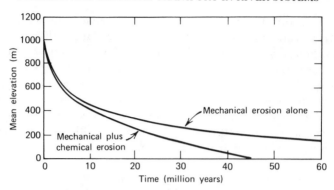

Figure 4-36. The reduction of a continent by mechanical erosion and by combined mechanical and chemical erosion; for a description of the parameters, see text.

rate of chemical denudation η is set at

$$\eta = 8.1 \times 10^{-4} \ \text{cm/yr}$$

Initially, the total rate of erosion is determined essentially by the rate of mechanical erosion. However, when the mean elevation has dropped to ca. 320 m, the rate of chemical denudation equals the rate of mechanical erosion, and below 200 m chemical erosion is quite dominant. The elevation goes to 0 after 46 million years since the rate of chemical erosion has been taken to be a step function of the elevation as described above.

The model developed above is obviously crude. It fails to take into account the real topography of continents, the effects of climate, of disturbances such as scouring and transport by ice sheets, and the changes in η when chemical weathering is extreme. Nevertheless the model does demonstrate quite clearly that the mean rate of uplift largely determines mean elevation, and that the degree of chemical weathering of eroding continental rocks decreases rapidly with increasing mean continental elevation.

The relative importance of chemical and mechanical weathering in the past has depended on the coupling between the rate of injection of acids, largely CO_2, HCl, and H_2SO_4, into the atmosphere and world-wide rates of uplift due to tectonic processes. The consequences of this theorem for the chemistry of the oceans and atmosphere are explored in the next two chapters.

REFERENCES

Anderson, P. W. and George, J. R., 1966, Water-Quality Characteristics of New Jersey Streams, U.S. Geological Survey Water–Supply Paper 1819-G.

Barshad, I., 1966, The effect of a variation in precipitation on the nature of clay mineral formation in soils from acid and basic igneous rocks, in *Proceedings of the International Clay Conference*, Vol. 1, 1966, Jerusalem, Israel, pp. 167–173.

Beck, K. C., Reuter, J. H., and Perdue, E. M., 1974, Organic and inorganic geochemistry of some coastal plain rivers of the southeastern United States, *Geochim. Cosmochim. Acta* **38**, 341–364.

Berner, R. A., 1971, Worldwide sulfur pollution in rivers, *J. Geophys. Res.* **76**, 6597–6600.

Billingsley, G. A., Fish, R. E., and Schipf, R. G., 1957, Water Resources of the Neuse River Basin, U.S. Geological Survey Water–Supply Paper 1414.

Blatt, H. and Jones, R. L., 1975, Proportions of exposed igneous, metamorphic, and sedimentary rocks, *Geol. Soc. Am. Bull.* **86**, 1085–1088.

Boyle, E., Collier, R., Dengler, A. T., Edmond, J. M., Ng, A. C., and Stallard, R. F., 1974, On the chemical mass-balance in estuaries, *Geochim. Cosmochim. Acta* **38**, 1719–1728.

Cadle, R. D., 1973, Particulate matter in the lower atmosphere, in *Chemistry of the Lower Atmosphere*, Chap. 2, S. I. Rasool, Ed., Plenum Press, New York.

Cawley, J. L., Burruss, R. C., and Holland, H. D., 1969, Chemical weathering in Central Iceland: An analog of pre-Silurian weathering, *Science* **165**, 391–392.

Clarke, F. W., 1924, *The Data of Geochemistry*, 5th ed., U.S. Geological Survey Bulletin 770.

Conway, E. J., 1942, Mean geochemical data in relation to oceanic evolution. *R. Irish Acad. Proc.* **48**, 119–159.

Coonley, L. S., Jr., Baker, E. B., and Holland, H. D., 1971, Iron in the Mullica River and Great Bay, New Jersey, *Chem. Geol.* **7**, 51–63.

Curray, J. R. and Moore, D. G., 1971, Growth of the Bengal deep-sea fan and denudation in the Himalayas, *Geol. Soc. Am. Bull.* **82**, 563–572.

Curtis, W. F., Culbertson, J. K., and Chase, E. B., 1973, Fluvial sediment discharge to the oceans from the conterminous United States, U.S. Geological Survey Circular 670.

Davis, S. N., 1964, Silica in streams and ground water, *Am. J. Sci.* **262**, 870–891.

Douglas, I., 1967, Man, vegetation, and the sediment yield of rivers, *Nature* **215**, 925–928.

Drever, J. I., 1971, Chemical weathering in a subtropical igneous terrain, Rio Ameca, Mexico, *J. Sediment Pet.* **41**, 951–961.

Drever, J. I., 1977, personal communication.

Durum, W. H., 1971, Chemical, physical, and biological characteristics of water resources, in *Water and Water Pollution Handbook*, Vol. I, Chap. 1, L. L. Ciaccio, Ed., Marcel Dekker, New York.

Durum, W. H., Heidel, S. G., and Tison, L. J., 1961, Worldwide runoff of dissolved solids, U.S. Geological Survey Professional Paper 424-C, C326–C329.

Durum, W. H. and Haffty, J., 1963, Implications of the minor element content of some major streams of the world, *Geochim. Cosmochim. Acta* **27**, 1–11.

Ericksson, E., 1960, The yearly circulation of chloride and sulfur in nature: meteorological, geochemical and pedological implications, 2, *Tellus* **12**, 63–109.

Ericksson, E., 1963, The yearly circulation of sulfur in nature, *J. Geophys. Res.* **68**, 4001–4008.

Feth, J. H., 1966, Nitrogen compounds in natural water—a review, *Water Resour. Res.* **2**, 41–58.

Fleming, G., 1969, Design curves for suspended load estimation, *Proc. Inst. Civ. Eng.* **43**, 1–9.

Fournier, F., 1960, *Climat et Érosion: La Relation Entre L'érosion du Sol par L'eau et les Précipitations Atmosphèriques*, Presses Univ., Paris, 201 pp.

Fournier, F., 1969, Transports solides effectués par les cours d'eau, *Int. Assoc. Sci. Hydrology Bull.* **14**, 7–47.

Friend, J. P., 1973, The global sulfur cycle, in *Chemistry of the Lower Atmosphere*, Chap. 4, S. I. Rasool, Ed., Plenum Press, New York.

Garrels, R. M., 1967, Genesis of some ground waters from igneous rocks, in *Researches in Geochemistry*, Vol. 2, P. H. Abelson, Ed., Wiley, New York, pp. 405–420.

Garrles, R. M. and Mackenzie, F. T., 1969, Sedimentary rock types: Relative proportions as a function of geologic time, *Science* **163**, 570–571.

Garrels, R. M. and Mackenzie, F. T., 1971, *Evolution of Sedimentary Rocks*, Norton, New York, 397 pp.

Garrels, R. M., Mackenzie, F. T., and Hunt, C., 1975, *Chemical Cycles and the Global Environment*, W. Kaufmann, Los Altos, Calif., 206 pp.

Gehman, H. M., Jr., 1962, Organic matter in limestones, *Geochim. Cosmochim. Acta* **26**, 885–897.

Gibbs, R. J., 1967, The geochemistry of the Amazon River System, Part I, The factors that control the salinity and the composition and concentration of the suspended solids, *Bull. Geol. Soc. Am.* **78**, 1203–1232.

Gibbs, R. J., 1972, Water chemistry of the Amazon River, *Geochim. Cosmochim. Acta* **36**, 1061–1066.

Gladwell, J. S. and Mueller, A. C., 1967, Report No. 2, An Initial Study of the Water Resources of the State of Washington, Vol. 2, Part A, Water Resources Atlas of the State of Washington, State of Washington Water Research Center, Pullman, Wash.

Gregory, K. J. and Walling, D. E., 1973, *Drainage Basin Form and Process, A Geomorphological Approach*, Wiley, New York, 456 pp.

Grout, F. F., 1938, Petrographic and chemical data on the Canadian shield, *J. Geol.* **46**, 486–504.

Heide, F., Lerz, H., and Böhm, G., 1957, Gehalt des Saalewassers an Blei und Quecksilber, *Naturwissenschaften*, **44**, 441–442.

Heide, F. and Ködderitzsch, H., 1964, Der Galliumgehalt des Saale und Elbewassers, *Naturwissenschaften* **51**, 104.

Hem, J. D., 1959, Study and interpretation of the chemical characteristics of natural water, U.S. Geological Survey Water-Supply Paper 1473.

Hem, J. D., 1960, Complexes of ferrous iron with tannic acid, U.S. Geological Survey Water-Supply Paper 1459-D.

Hitchon, B., Levinson, A. A., and Reeder, S. W., 1969, Regional variations of river water composition resulting from halite solution, Mackenzie River drainage basin, Canada, *Water Resour. Res.* **5**, 1395–1403.

Holeman, J. N., 1968, The sediment yield of major rivers of the world, *Water Resour. Res.* **4**, 737–747.

Holland, H. D., 1973, Ocean water, nutrients and atmospheric oxygen, in *Hydrogeochemistry*, of *Symposium on Hydrogeochemistry and Biogeochemistry*, Vol. 1, Clarke Co., Washington, D.C., pp. 68–81.

Irelan, B. and Mendieta, H. B., 1964, Chemical quality of surface waters in the Brazos River basin in Texas, U.S. Geological Survey Water-Supply Paper 1779-K.

Junge, C. E. and Gustafson, P. E., 1957, On the distribution of sea salt over the United States and its removal by precipitation, *Tellus* **9**, 164–173.

Kennedy, V. C., 1971, Silica variation in stream water with time and discharge, in *Nonequilibrium Systems in Natural Water Chemistry*, Chap. 4, J. D. Hem, Ed., Advances in Chemistry Series 106, American Chemical Society, Washington, D.C.

Kharkar, D. P., Turekian, K. K., and Bertine, K. K., 1968, Stream supply of dissolved silver, molybdenum, antimony, selenium, chromium, cobalt, rubidium, and cesium to the oceans, *Geochim. Cosmochim. Acta* **32**, 285–298.

Konovalov, G. S., 1959, The transport of microelements by the most important rivers of the U.S.S.R., *Dokl. Akad. Nauk USSR* **129**, 912–915.

Kozoun, V. I., Sokolov, A. A., Budyko, M. I., Voskresensky, K. P., Kalinin, G. P., Konoplyantsev, A. A., Korotkevich, E. S., Kuzin, P. S., and Lvovitch, M. I., 1974, *World Water Balance and Water Resources of the Earth*, U.S.S.R. National Committee for the International Hydrological Decade, Leningrad.

Landström, O. and Wenner, C. G., 1965, Neutron-activation analysis of natural water applied to hydrology, Aktiebolaget Atomenerg, Sweden, AE-204.

Langbein, W. B. and Schumm, S. A., 1958, Yield of sediment in relation to mean annual precipitation, *Trans. Am. Geophys. Union* **39**, 1076–1084.

Leifeste, D. K., 1974, Dissolved solids discharge to the oceans from the conterminous United States, U.S. Geological Survey Circular 685.

Leopold, L. B., Wolman, M. G., and Miller, J. P., 1964, *Fluvial Processes in Geomorphology*, W. H. Freeman, San Francisco, 522 pp.

Livingstone, D. A., 1963, Chemical composition of rivers and lakes, U.S. Geological Survey Professional Paper 440G, 64 pp.

Love, S. K., 1958, Quality of surface waters for irrigation Western United States, 1954, U.S. Geological Survey Water-Supply Paper 1430.

Love, S. K., 1960a, Quality of surface waters for irrigation Western United States, 1956, U.S. Geological Survey Water-Supply Paper 1485.

Love, S. K., 1960b, Quality of surface waters of the United States, 1957; Parts 1–4, U.S. Geological Survey Water-Supply Paper 1520.

Love, S. K., 1961a, Quality of surface waters of the United States, 1957; Parts 5 and 6, U.S. Geological Survey Water-Supply Paper 1521.

Love, S. K., 1961b, Quality of surface waters of the United States, 1957; Parts 7 and 8, U.S. Geological Survey Water-Supply Paper 1522.

Love, S. K., 1961c, Quality of surface waters of the United States, 1957; Parts 9-14, Colorado River Basin to Pacific Slope Basins in Oregon and Lower Columbia River Basin, U.S. Geological Survey Water-Supply Paper 1523.

Love, S. K., 1964a, Quality of surface waters of the United States, 1962; Parts 1 and 2, U.S. Geological Survey Water-Supply Paper 1941.

Love, S. K., 1964b, Quality of surface waters of the United States, 1962; Parts 3 and 4, U.S. Geological Survey Water-Supply Paper 1942.

Love, S. K., 1964c, Quality of surface waters of the United States, 1962; Parts 5 and 6, U.S. Geological Survey Water-Supply Paper 1943.

Love, S. K., 1964d, Quality of surface waters of the United States, 1962; Parts 7 and 8, U.S. Geological Survey Water-Supply Paper 1944.

Love, S. K., 1964e, Quality of surface waters of the United States, 1962; Parts 9–14, U.S. Geological Survey Water-Supply Paper 1945.

Love, S. K., 1965a, Quality of surface waters of the United States, 1959; Parts 1 and 2, U.S. Geological Survey Water-Supply Paper 1641.

Love, S. K., 1965b, Quality of surface waters of the United States, 1959; Parts 5 and 6, U.S. Geological Survey Water-Supply Paper 1643.

Love, S. K., 1965c, Quality of surface waters of the United States, 1959; Parts 7 and 8, U.S. Geological Survey Water-Supply Paper 1644.

Love, S. K., 1965d, Quality of surface waters of the United States, 1963; Parts 3 and 4, U.S. Geological Survey Water-Supply Paper 1948.

Love, S. K., 1966a, Quality of surface waters of the United States, 1959; Parts 9–14, U.S. Geological Survey Water-Supply Paper 1645.

Love, S. K., 1966b, Quality of surface waters of the United States, 1963; Parts 5 and 6, U.S. Geological Survey Water-Supply Paper 1949.

Love, S. K., 1966c, Quality of surface waters of the United States, 1963; Parts 7 and 8, U.S. Geological Survey Water-Supply Paper 1950.

Love, S. K., 1966d, Quality of surface waters of the United States, 1963; Parts 9–14, U.S. Geological Survey Water-Supply Paper 1951.

Love, S. K., 1966e, Quality of surface waters of the United States, 1961; Parts 5 and 6, U.S. Geological Survey Water-Supply Paper 1883.

Love, S. K., 1967a, Quality of surface waters of the United States, 1961; Parts 1 and 2, U.S. Geological Survey Water-Supply Paper 1881.

Love, S. K., 1967b, Quality of surface waters of the United States, 1961; Parts 7 and 8, U.S. Geological Survey Water-Supply Paper 1884.

Love, S. K., 1967c, Quality of surface waters of the United States, 1961; Parts 9–14, U.S. Geological Survey Water-Supply Paper 1885.

Love, S. K., 1967d, Quality of surface waters of the United States, 1963; Parts 1 and 2, U.S. Geological Survey Water-Supply Paper 1947.

Love, S. K., 1970, Quality of surface waters of the United States, 1965; Parts 3 and 4, U.S. Geological Survey Water-Supply Paper 1962.

Mairs, D. F., 1967, Surface chloride distribution in Maine lakes, *Water Resour. Res.* **3**, 1090–1092.

Mathews, W. H., 1975, Cenozoic erosion and erosion surfaces of eastern North America, *Am. J. Sci.* **275**, 818–824.

McCarthy, L. T., Jr., and Keighton, W. B., 1964, Quality of Delaware River water at Trenton, N.J., U.S. Geological Survey Water-Supply Paper 1779-X.

Moore, W. S., 1967, Amazon and Mississippi river concentrations of uranium, thorium, and radium isotopes, *Earth Planet. Sci. Letters* **2**, 231–234.

Nanz, R. H., Jr., 1953, Chemical composition of Precambrian slates with notes on the geochemical evolution of lutites, *J. Geol.* **61**, 51–64.

Pettijohn, F. J., 1957, *Sedimentary Rocks*, 2nd ed., Harper & Row, New York, 718 pp.

Poldervaart, A., 1955, Chemistry of the earth's crust, in *Crust of the Earth*, Geological Society of America Special Paper 62, pp. 119–144.

Reynolds, R. C., Jr., and Johnson, N. M., 1972, Chemical weathering in the temperate glacial environment of the northern Cascade Mountains, *Geochim. Cosmochim. Acta* **36**, 537–554.

Rona, E. and Urry, W. D., 1952, Radioactivity of ocean sediments. VIII. Radium and uranium content of ocean and river waters, *Am. J. Sci.* **250**, 241–262.

Ronov, A. B. and Yaroshevsky, A. A., 1967, Chemical structure of the earth's crust, *Geokhimiya*, 1285–1309.

Ronov, A. B. and Migdisov, A. A., 1971, Geochemical history of the crystalline basement and the sedimentary cover of the Russian and North American Platforms, *Sedimentology* **16**, 137–185.

Ronov, A. B., and Yaroshevsky, A. A., 1976, A new model for the chemical structure of the earth's crust, *Geokhimiya*, 1761–1795.

Sederholm, J. J., 1925, The average composition of the earth's crust in Finland, *Bull. Comm. Geol. Finl.*, No. 70.

Silker, W. B., 1964, Variations in elemental concentrations in the Columbia River, *Limnol. Oceanogr.* **9**, 540–545.

Strakhov, N. M., 1967, *Principles of Lithogenesis*, Vol. 1, Transl. J. P. Fitzsimmons, S. I. Tomkeieff, and J. E. Hemingway, Eds., Consultants Bureau, New York, 245 pp.

Sugawara, K., 1964, Migration of elements through phases of the hydrosphere and atmosphere, in *Khimiya Zemnoi Kory*, Vol. II, A. P. Vinogradov, Ed., Transl. 1967 by Israel Program for Scientific Translation.

Sylvester, R. O. and Rambow, C. A., 1967, Report No. 2, An Initial Study of the Water Resources of the State of Washington, Vol. 4, Water Quality of the State of Washington, State of Washington Water Research Center, Pullman, Wash.

Trimble, S. W., 1975, Denudation studies: can we assume stream steady state?, *Science* **188**, 1207–1208.

Turekian, K. K., 1969, The oceans, streams, and atmosphere, in *Handbook of Geochemistry*, Vol. I, Chap. 10, K. H. Wedepohl, Ed., Springer–Verlag, New York.

Turekian, K. K. and Johnson, D. G., 1966, The barium distribution in sea water, *Geochim. Cosmochim. Acta* **30**, 1153–1174.

Van Denburgh, A. S. and Feth, J. H., 1965, Solute erosion and chloride balance in selected river basins of the Western United States, *Water Resour. Res.* **1**, 537–541.

Van Moort, D. C., 1973, The magnesium and calcium contents of sediments, especially pelites, as a function of age and degree of metamorphism, *Chem. Geol.* **12**, 1–37.

Vinogradov, A. P. and Ronov, A. B., 1956, Evolution of the chemical composition of clays of the Russian Platform, *Geochemistry*, 123–139.

Visher, F. N. and Mink, J. F., 1964, Ground-water resources in Southern Oahu, Hawaii, U.S. Geological Survey Water-Supply Paper 1778.

Williams, P. M., 1968, Organic and inorganic constituents of the Amazon River, *Nature* **218**, 937–938.

White, D. E., 1965, Saline waters of sedimentary rocks, in *Fluids in Subsurface Environments—A Symposium*, Memoir No. 4 of the American Association of Petroleum Geologists, pp. 342–366.

White, D. E., Hem, J. D., and Waring, G. A., 1963, Chemical composition of subsurface waters, in *Data of Geochemistry*, 6th ed., Chap. F, U.S. Geological Survey Professional Paper 440-F.

CHAPTER V

THE CHEMISTRY OF THE
OCEANS

The previous chapter dealt with the composition of the present-day river fluxes into the oceans, and attempted to set reasonable limits on the variation of river inputs with time. This chapter explores the response of the oceans to inputs from rivers and other sources today, seeks to define the relationships between inputs to and outputs from the oceans, and attempts to develop criteria for using the geologic record, particularly the record in sedimentary rocks, to understand the geologic history of seawater.

THE COMPOSITION OF SEAWATER

Concentration of the Elements

Great numbers of analyses and compilations of analyses for the major components of seawater are now available. Table 5-1 follows the summary by Pytkowicz and Kester (1971) for the 11 most abundant constituents of seawater; the table includes all but 2 of the elements that are present in seawater in concentrations greater than 1 ppm. The mean concentration of many minor components is summarized in Table 5-2, which has been taken largely from Brewer (1975).

There are considerable similarities between the composition of ocean water and average river water. The same four cations and the same three anions dominate the composition of both types of water. However, average seawater is a much more concentrated solution than average river water, and the relative proportions of the major cations and the major anions are quite different. The composition of average river water is dominated by Ca^{2+} among the cations, by HCO_3^- among the anions, and by H_4SiO_4

153

Table 5-1

The Concentration and Mean Residence Time of the Major Constituents of Ocean Water

Constituent	Average Chlorinity Ratio[a]	Average Concentration in Ocean Water of Salinity 35 ‰		Concentration in Average River Water (mg/kg)[c]	Residence Time in Oceans[d]
		(mg/kg)[b]	(mmol/kg)		
Sodium	0.5561[e]	10,760	468.0	6.9	4.8×10^7
Magnesium	0.06679[e]	1,294	53.2	3.9	1.0×10^7
Calcium	0.02127[e]	412	10.2	15.0	8.5×10^5
Potassium	0.0206[e]	399	10.2	2.1	5.9×10^6
Strontium	0.00041[e]	7.9	0.090		4×10^6
Chloride	1.0000	19,350	545.0	8.1	7.3×10^7
Sulfate	0.1400	2,712	28.2	10.6	7.9×10^6
Bicarbonate	0.00749	145	2.38	55.9	8.0×10^4
Bromide	0.003473	67	0.84		1×10^8
Boron	0.000240	4.6	0.39		1×10^7
Fluoride	0.000067	1.3	0.068		5×10^5

[a] Data from Pytkowicz and Kester (1971).
[b] $(mg/kg)_i \equiv$ (chlorinity ratio)$_i \times 19,350$.
[c] Data from Gibbs (1972).
[d] See text for residence time computations.
[e] Mean of two averages.

among the neutral species. The composition of seawater is dominated by Na^+ among the cations and by Cl^- among the anions. The neutral species make up a much smaller fraction of the total dissolved matter in seawater than they do in average river water. Seawater is clearly not simply concentrated average river water. This is hardly surprising. If average river water were evaporatively concentrated, the cation and anion ratios would be altered by the precipitation of a variety of minerals; however, the composition of evaporated river water would resemble the composition of nonmarine evaporite brines much more than the composition of seawater. The conversion of average river water to seawater clearly involves processes other than evaporative concentration, unless the composition of present day river water is quite unrepresentative of the composition of average river water during the past few million years. The data in Chapter IV indicate that this is unlikely; much of Chapter V is devoted to a description of the processes that are at work removing the river inputs of dissolved salts from the oceans and of the manner in which they convert the cation and anion ratios of present-day river water into those of seawater.

Residence Times of the Elements

Some river inputs are removed on a much shorter time scale than others. The residence time $\tau_{R,O}$ of a substance in seawater is defined as the ratio of A_o, the total quantity of the substance in seawater, to $dA_{R,O}/dt$, the rate of river input

$$\tau_{R,O} = \frac{A_o}{dA_{R,O}/dt} \qquad (5\text{-}1)$$

A_o for a particular constituent is equal to its concentration in seawater times the mass of the oceans. $dA_{R,O}/dt$ is equal to the concentration of the constituent in average river water times the annual flux of river water to the oceans. For example, the concentration of Ca^{2+} in average seawater is 412 mg/kg. The total mass of Ca^{2+} in the oceans is approximately

$$(A_o)_{Ca^{2+}} \cong 412 \text{ mg/kg} \times \frac{1}{1000} \text{ g/mg} \times 1.4 \times 10^{21} \text{ kg}$$

$$\cong 5.8 \times 10^{20} \text{g}$$

Since the concentration of Ca^{2+} in average river water is approximately 15 mg/kg

$$(dA_{R,O}/dt)_{Ca^{2+}} \cong 15 \text{ mg/kg} \times \frac{1}{1000} \text{ g/mg} \times 4.6 \times 10^{16} \text{ kg/yr}$$

$$\cong 6.9 \times 10^{14} \text{ g/yr}$$

Therefore

$$(\tau_{R,O})_{Ca^{2+}} \approx \frac{5.8 \times 10^{20} \text{ g}}{6.9 \times 10^{14} \text{ g/yr}}$$

$$\approx 0.8 \times 10^6 \text{ yr}$$

The value of $\tau_{R,O}$ of the other major anions and cations in Table 5-1 ranges from somewhat less than 10^5 yr for HCO_3^- to nearly 10^8 yr for Cl^- and Br^-. Inclusion of the minor constituents of ocean water in Table 5-2 extends the range of residence times to values down to 10^2 yr, that is, to periods shorter than the mixing time of the oceans.

Although the function $\tau_{R,O}$ is mathematically well defined, its use for interpreting the removal rate of seawater constituents requires caution (see Chap. I). At steady state the rate of river input of dissolved salts is balanced by their output from the oceans

$$\frac{dA_{R,O}}{dt} = \frac{dA_{O,S}}{dt} \tag{5-2}$$

Since all the $\tau_{R,O}$'s are short compared to the age of the earth, equation 5-2 must be valid when the two rates are averaged over several residence times. However, most of the $\tau_{R,O}$ values are long compared to the time during which the system has been observed, and there is every reason to believe that both the rate of river input and the rate of removal of salts from the oceans have been influenced by the large climatic fluctuations during the past million years and, more recently, by human activity. The assumption that equation 5-2 holds precisely today is, therefore, almost certainly incorrect, and in any case requires support based on direct observation of both rates for all constituents of ocean water.

Values of $\tau_{R,O}$ for elements supplied exclusively to the oceans as dissolved constituents of river water and not recycled through the atmosphere are important parameters for gauging the response time of their concentration in the oceans to changes in their rate of supply. For ions such as Cl^- and Na^+, which are cycled extensively through the atmosphere and are returned via rivers, the interpretation of the $\tau_{R,O}$ values as calculated is more difficult. If the total quantity of Cl^- in the oceans were fixed, and if Cl^- were allowed to leave the oceans only into the atmosphere, then the calculated value of 7.3×10^7 yr for $(\tau_{R,O})_{Cl^-}$ would be a measure of the time required for all the Cl^- in the oceans to be cycled once via rivers. Fortunately, this is not entirely correct, but estimates of the rate of Cl^- input to the oceans from solution and weathering on land require that the river flux of Cl^- be corrected for atmospheric cycling. If this is done, characteristic times for Cl^- and Na^+ are obtained that are several times longer than the values of $\tau_{R,O}$ listed in Table 5-1.

Table 5-2
Minor Constituents of Ocean Water[a]

Constituent	Concentration[b] ($\mu g/kg$ in ocean water of salinity 35°/∞)	$\log \tau_{R,O}$ (yr)
Ag	0.008	5
Al	2	2
Ar	4	
As	3	5
Au	0.005	5
Ba	20 (4–28)[c]	4.5
Be	0.006	(2)
Bi	0.02	
Cd	0.1 (0.07–0.7)	4.7
Ce	0.001	
Co	0.05	4.5
Cr	0.3 (0.04–2.5)	3
Cs	0.4	5.8
Cu	0.3 (0.1–12.3)[d,e]	4
Dy	0.0009	
Er	0.0009	
Eu	0.0001	
F	1300	5.7
Fe	2 (0.1–60)	2
Ga	0.03	4
Gd	0.0007	
Ge	0.06	
He	0.0072	
Hf	<0.008	
Hg	0.02	5
Ho	0.0002	
I	60	6
In	$1. \times 10^{-4}$	
Kr	0.21	
La	0.0034	
Li	180	6.3
Lu	0.0001	
Mn	0.2 (0.05–0.88)[d]	4
Mo	10	5

Table 5.2 (continued)

Constituent	Concentration[b] ($\mu g / kg$ in ocean water of salinity $35°/_{oo}$)	$\log \tau_{R,0} (yr)$
N	1.5×10^4	6.3
Nb	0.015	
Nd	0.0028	
Ne	0.12	
Ni	0.6	4
P	60	4
Pa	2×10^{-10}	
Pb	0.03	(2.6)
Pr	0.0006	
Ra	1×10^{-3}	6.6
Rb	120	6.4
Sb	0.33	4
Sc	6×10^{-4}	4.6
Se	0.09	4
Si	2900 (100–5000)	3.8
Sm	0.0004	
Sn	0.01	
Ta	<0.0025	
Tb	0.0001	
Th	0.0015	(2)
Ti	1	4
Tl	0.01	
Tm	0.0002	
U	3.3	6.4
V	2.5 (0.2–4)	5
W	0.004	
Xe	0.047	
Y	0.001	
Yb	0.0008	
Zn	3 (0.2–15)[d]	4
Zr	0.03	

[a]After Turekian, 1969; Brewer, 1975; and others.
[b]Numbers in parentheses denote concentration ranges.
[c]Wolgemuth, 1970; Wolgemuth and Broecker, 1970.
[d]Slowey and Hood, 1971.
[e]Boyle et al., 1977.

For most constituents of seawater the river input of dissolved salts is by far the most important source. For these constituents $dA_{R,O}/dt$ is, therefore, an adequate measure of the total rate of supply. Some elements are, however, brought to the oceans in significant quantities adsorbed on clays and on other minerals from which they are easily removed. Some elements are apparently added to the oceans in significant quantities during the reaction of seawater with oceanic basalts both at temperatures close to $0°C$ and/or at much higher temperatures. For such elements the total rate of input is greater than the river input alone, and the true residence time of these elements in the oceans is smaller than $\tau_{R,O}$.

Despite the uncertainties in the data and in spite of all the questions and problems that surround the use of $\tau_{R,O}$ values as a measure of the residence time of dissolved constituents in the oceans, the values of $\tau_{R,O}$ in Tables 5-1, 5-2, and 5-3 are useful guides. The calculated range in $\tau_{R,O}$ values covers approximately 6 orders of magnitude, from 10^2 to 10^8 yr. Uncertainties in oceanic residence times due to the factors discussed above probably do not exceed an order of magnitude. The difference between the $\tau_{R,O}$ value of Be^{2+} and that of Na^+ is, therefore, certainly significant; the difference between $(\tau_{R,O})_{K^+}$ and $(\tau_{R,O})_{Mg^{2+}}$ is almost certainly not.

Table 5-3

Values of $\log_{10} \tau_{R,O}$ (yr)[a]

H 4.5																	He
Li 6.3	Be (2)											B 7.0	C 4.9	N 6.3	O 4.5	F 5.7	Ne
Na 7.7	Mg 7.0											Al 2	Si 3.8	P 4	S 6.9	Cl 7.9	Ar
K 6.7	Ca 5.9	Sc 4.6	Ti 4	V 5	Cr 3	Mn 4	Fe 2	Co 4.5	Ni 4	Cu 4	Zn 4	Ga 4	Ge	As 5	Se 4	Br 8	Kr
Rb 6.4	Sr 6.6	Y	Zr 5	Nb	Mo 5	Tc	Ru	Rh	Pd	Ag 5	Cd 4.7	In	Sn	Sb 4	Te	I 6	Xe
Cs 5.8	Ba 4.5	La 6.3	Hf	Ta	W	Re	Os	Ir	Pt	Au 5	Hg 5	Tl	Pb (2.6)	Bi	Po	At	Rn
Fr	Ra 6.6	Ac															

Ce	Pr	Nd	Pm	Sm	Eu	Gd	Tb	Dy	Ho	Er	Tm	Yb	Lu
Th (2)	Pa	U 6.4											

[a]For sources of data see Tables 5-1 and 5-2.

Since the residence time of all the dissolved constituents of seawater is a small fraction of the age of the earth, the oceans have clearly been a transient home rather than a constantly growing reservoir for the dissolved salts in seawater. This is also true for the particulate matter delivered to the ocean basins by rivers, wind, and volcanic injection. The volume of seawater today is 1.4×10^{24} cc. This volume would be replaced by sediments of density 1.8 g/cc delivered at the estimated current rate of 2×10^{16} g/yr (Holeman, 1968; Chap. IV) in a time Δt

$$\Delta t \approx \frac{2.5 \times 10^{24} \text{ g}}{2 \times 10^{16} \text{ g/yr}} = 120 \text{ million yr}$$

The embarrassment of such a short ocean filling time has been eliminated by the discovery of sea floor spreading, which provides for the renewal of the ocean basins on this time scale. The fate of the oceanic sediments deposited and subsequently removed during any given 200-million-yr period is obviously of importance for the later chemistry of the oceans.

The mixing time of the upper 100 m of ocean water is about 10 yr. Mixing of deep ocean water takes place on a time scale of ca. 1000 yr (see, for instance, Broecker et al., 1961; Broecker, 1963). Elements with residence times smaller than 1000 yr can, therefore, be expected to be heterogeneously distributed in the oceans. Elements with residence times well in excess of 1000 yr tend to be homogeneously distributed. This is certainly true of the major dissolved constituents in Table 5-1. However, a $\tau_{R,O}$ value in excess of 1000 yr is no guarantee of homogeneity. Phosphorus ($\tau_{R,O} = 10^{4.1}$ yr), silicon ($\tau_{R,O} = 10^{3.8}$ yr), nitrogen ($\tau_{R,O} = 10^{6.3}$ yr), and many of the other minor constituents of seawater are heterogeneously distributed due to the intense biological activity within the oceans. The concentration of these elements is reduced in surface waters where they are incorporated into organisms, enhanced in deeper waters where they are restored during the oxidative decay of organisms, and complicated by the physical transport of water masses within the ocean. The residence time of all these elements with respect to cycling through the marine biosphere is, of course, much shorter than the residence time with respect to final removal from the oceans.

Activity Coefficients and Complexing

The interpretation and prediction of the solubility of a large variety of solids in seawater demands a knowledge of the activity and of the activity coefficients of single ions in seawater. Garrels and Thompson (1962) constructed a self-consistent model for the major dissolved species in seawater based on the formation of ion pairs and on reasonable values of single ion activity coefficients. Although other models have been proposed (see for instance Whitfield, 1975) studies since 1962 by Berner (1965), Platford (1965a, b), Platford and Dafoe (1965), and Thompson (1966), have largely confirmed the usefulness of the Garrels–Thompson model; Kester and Pytkowicz (1967, 1975), Pytkowicz and Kester (1971), Hawley (1973), and Disteche (1974) have extended and refined the model. Single ion activity coefficients from these sources are summarized in Table 5-4, and several estimates of the distribution of the major dissolved species in seawater are listed in Table 5-5. The activity coefficient of neutral species is apparently close to unity, that of singly charged species between 0.6 and 0.7, and that of doubly charged species between 0.1 and 0.2. The major cations and chloride are largely unassociated, whereas bicarbonate, $H_2PO_4^-$, HPO_4^{2-}, and sulfate

Table 5-4

Single Ion Activity Coefficients in Seawater[a]

Dissolved Species	*Activity Coefficients*	*Source of Data*
$NaHCO_3^0$	1.0–1.13	(1)
$MgCO_3^0$	1.0–1.13	(1)
$CaCO_3^0$	1.0–1.13	(1)
$MgSO_4^0$	1.0–1.13	(1)
$CaSO_4^0$	1.0–1.13	(1)
HCO_3^-	0.50	(5)
$NaCO_3^-$	0.68	(1)
$NaSO_4^-$	0.68	(1)
KSO_4^-	0.68	(1)
Cl^-	0.62	(5)
	0.64	(2)
SO_4^{2-}	0.12	(5)
	0.115	(3)
CO_3^{2-}	0.20	(1)
	0.03	(5)
Na^+	0.69	(5)
	0.68 ± 0.01	(4)
K^+	0.62	(5)
$MgHCO_3^+$	0.68	(1)
$CaHCO_3^+$	0.68	(1)
Ca^{2+}	0.22	(5)
Mg^{2+}	0.25	(5)

[a] Ionic strength 0.7, chlorinity $19^0/_{00}$, 25°C. (1) Garrels and Christ (1965); (2) Platford (1965a,b); (3) Platford and Dafoe (1965); (4) Platford (1965b); (5) Pytkowicz et al., (1975).

Table 5-5

Speciation of the Major Ions in Seawater[a]

Ion	Source of Data	% Free	% SO_4^{2-} pair	% HCO_3^- pair	% CO_3^{2-} pair
Ca^{2+}	(1)	91	8	1	0.2
	(2)	88	11	0.6	0.1
Mg^{2+}	(1)	87	11	1	0.3
	(2)	89	10	0.6	0.1
Na^+	(1)	99	1.2	0.01	—
	(2)	97.7	2.2	0.03	—
K^+	(1)	99	1	—	—
	(2)	98.8	1.2	—	—

Ion	Source of Data	% Free	% Ca^{2+} pair	% Mg^{2+} pair	% Na^+ pair	% K^+ pair
Cl^-	(1)	100				
SO_4^{2-}	(1)	54	3	21.5	21	0.5
	(2)	39	4	19	37	—
HCO_3^-	(1)	69	4	19	8	—
	(2)	64	3	16	8	—
	(3)	81.3	1.5	6.5	10.7	—
	(4)	84.7	0.25	1.3	13.4	0.25
CO_3^{2-}	(1)	9	7	67	17	—
	(2)	10	6	72	12	—
	(3)	8.0	21.0	43.9	16.0	—
	(4)	18	20.2	46.8	14.5	0.2
F^-	(5)	48–50	2	47–49	1	—

[a]Salinity 35‰ at 25°C and 1 Atm. (1) Garrels and Thompson (1962); (2) Pytkowicz and Kester (1971); (3) Hawley (1973); (4) Disteche (1974); (5) Goldberg (1975).

are partially associated, and CO_3^{2-} and PO_4^{3-} are largely present as components of ion pairs. The stability of CO_3^{2-} and PO_4^{3-} complexes contributes heavily to the large apparent solubility products of carbonate and phosphate minerals in seawater. Considerable uncertainty is still attached to current estimates of the concentration of several of the complexes in Table 5-5, and a good deal more work is required, particularly at temperatures close to 0°C and at ocean floor pressures, to resolve contradictions between the several sets of recent calculations. However, the Garrels–Thompson (1962) model has withstood the rigors of extensive testing remarkably well, and promises to remain an important milestone in the history of chemical oceanography.

THE CARBONATE OUTPUT OF THE OCEANS

Calcium Carbonate Deposition

The current carbonate output from the oceans consists almost entirely of calcite, aragonite, and very small quantities of dolomite. Although siderite and ankerite are important constituents of sedimentary rocks associated with iron formations, the carbonates of calcium and magnesium have dominated the carbonate mineralogy of sediments for the past 2 billion yr, and only the small ratio of dolomite deposition to $CaCO_3$ deposition sets the present oceans apart from their precursors.

Organisms are responsible for nearly all of the calcium carbonate precipitation in the oceans today. The origin of oölites, grapestone, and certain aragonite needle muds has been debated extensively (see, for instance, Cloud, 1965; Bathurst, 1967). It seems likely that at least some of these and other, rather minor, carbonate sediments are inorganically precipitated (Ginsburg and James, 1976); their association with organic matter, and the lack of decisive criteria for distinguishing aragonite of biologic origin from aragonite of inorganic origin has, however, left the matter partly unresolved. Ocean water in areas of oölite, grapestone, and aragonite mud formation is supersaturated with respect to calcite and aragonite. Broecker and Takahashi (1966) have shown, for instance, that ocean water arrives on the Bahama Banks with an ion activity product $a_{Ca^{2+}} \cdot a_{CO_3^{2-}}$ of 1.68×10^{-8}. During residence on the Banks $CaCO_3$ is removed, and the ion activity product decreases to values just slightly larger than 0.8×10^{-8}, the solubility product of aragonite.

Most surface water in the oceans is somewhat supersaturated with respect to calcite and aragonite; deep ocean water is normally undersaturated with respect to these minerals (see Figure 5-1 and Lyakhin, 1968; Li, Takahashi, and Broecker, 1969; Gieskes, 1974; Broecker and Takahashi, 1976). This difference between shallow and deep water is due largely to the effects of decreasing temperature, increasing pressure (Pytkowicz and Connors, 1964; Pytkowicz, 1965; Hawley and Pytkowicz, 1969; Ingle, 1975), and the progressive oxidation of organic matter (Park, 1966, 1968). Undersaturation of seawater with respect to calcite results in the widespread dissolution of

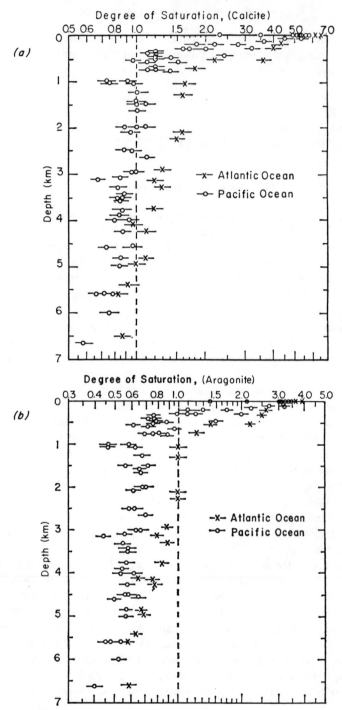

Figure 5-1. Degree of saturation of calcite (*a*) and aragonite (*b*) as a function of depth in the Atlantic and Pacific Oceans (after Li et al., 1969). Copyrighted by American Geophysical Union.

calcite and aragonite in the deep ocean. As a consequence, sediments in large portions of the floor of the Pacific Ocean are virtually carbonate free (see Figure 5-2). The level at which the rate of carbonate input from the surface is balanced by the rate of carbonate dissolution has been defined as the carbonate compensation depth (CCD). The dissolution of carbonate minerals in slightly undersaturated seawater at temperatures close to 2°C is quite slow. Carbonate remains can, therefore, accumulate at depths well below the surface below which seawater becomes undersaturated with calcite. The depth of the CCD in a given area depends on the rate of supply of carbonate remains, their chemical composition, their mineralogy, size, shape (Berger, 1970, 1971; Honjo and Erez, 1978), and the rate of bioturbation as well as on the depth where seawater becomes undersaturated with respect to calcite and argonite.

Roughly 80% of the calcium carbonate formed in the surface water of the Pacific dissolves in the deep ocean after sedimentation (Gieskes, 1974). Thus calcium ions entering the oceans as a constituent of river water are normally precipitated and redissolved several times before they are incorporated in a carbonate grain that is deposited in seawater supersaturated with respect to $CaCO_3$ or before they are deposited in an area where the rate of accumulation of $CaCO_3$ exceeds the rate of $CaCO_3$ dissolution.

The oceans as a whole are clearly not at thermodynamic equilibrium with calcite or other carbonate phases. However, the presence of a lysocline and of a carbonate compensation depth indicates that the oceans are close to steady state, that the assumption of near-saturation of the system as a whole is essentially correct, and that the oceans have probably always managed to adjust their carbonate metabolism sufficiently rapidly to keep up with the pulse of the earth.

The current calcium balance of the oceans is difficult to evaluate quantitatively. The uncertainty in the calcium input by rivers may be as large as 50%, and the uncertainty in the rate of $CaCO_3$ removal from the oceans is at least as large. Balance sheets such as those of Turekian (1965) certainly show that output on ocean floors can be made to balance current input, but the use of such data to make inferences regarding the calcium carbonate budget during glacial periods (Olausson, 1967) seems premature. It also seems quite dangerous to use such data to define the relative quantities of deep-water and shallow-water carbonate deposition. Broecker and Takahashi (1966) showed that on the Bahama Banks $CaCO_3$ is accumulating at a rate of about 50 mg/cm^2 yr. The total input rate of calcium into the oceans today is approximately 7×10^{14} g/yr (see above). This quantity of calcium yields 18×10^{14} g of $CaCO_3$. If spread uniformly over the floor of the oceans, the mean $CaCO_3$ sedimentation rate would be 0.5 mg/cm^2 yr or 0.5 g/cm^2 per 1000 yr. Only 1% of the total area of the oceans would, therefore, be required to remove the entire input of calcium at the current Bahamian

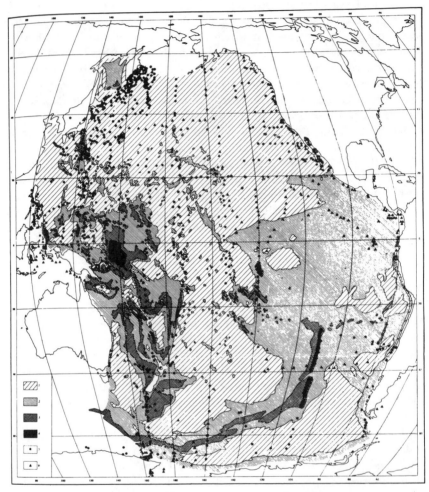

Figure 5-2. The distribution of planktonic foraminiferal tests on the floor of the Pacific Ocean, in specimens per grams of sediment; (1) absent, (2) 100 to 1000, (3) 1000 to 10,000, (4) >10,000, (5) VITYAZ stations for which there are quantitative data, (6) stations of non-Russian expeditions for which there are no quantitative data (Belyayeva, 1968). Copyrighted by American Geophysical Union.

rate; 100 regions like the Bahamas could remove the entire river input of calcium. The presence of areas of such intense calcium carbonate removal make very heavy demands on anyone contemplating a complete analysis of the calcium budget of the oceans. Attempts to sidestep this demand by using the distribution of strontium in marine carbonate sediments have been hampered by a lack of sufficiently definitive data for the mean strontium concentration of river water (Turekian, 1964).

Although the relative proportions of deep-water and shallow-water carbonates are still not known with certainty, it is clear that calcium is removed from ocean water today largely as carbonate. Gypsum and anhydrite are precipitated in evaporite areas in quantities that are of importance for the geochemical budget of sulfur but not for that of calcium. Calcium silicates are decidedly rare. Phillipsite is formed on the ocean floor, largely by the alteration of volcanic glass (see, for instance, Arrhenius, 1963). The calcium concentration of marine phillipsite is quite variable but is apparently always less than that of tholeiitic basalts (Sheppard, et al., 1970; Cronan, 1974). The formation of phillipsite, therefore, involves a loss of calcium to, rather than a gain of calcium from, seawater. Even if all of the calcium in such phillipsite were derived from ocean water, the rate of loss would be a small fraction of the present rate of calcium input to the oceans. If the rate of accumulation of sediment containing 50% phillipsite is 1 mm per 1000 yr, the rate of phillipsite accumulation is roughly 0.1 g/cm^2 per 1000 yr. On the other hand, if all of the river calcium were used to produce phillipsite containing 3% by weight calcium, the rate of accumulation of phillipsite over the entire ocean floor would be

$$\frac{7 \times 10^{14} \, g/yr}{3.6 \times 10^{18} \, cm^2} \times \frac{100}{3} \times 1000 \, g/cm^2 \text{ per } 1000 \text{ yr} = 6 \, g/cm^2 \text{ per } 1000 \text{ yr}$$

that is, about 60 times faster than the present rate of phillipsite accumulation in the Pacific.

A good deal of the montmorillonite brought to the oceans by rivers appears to be calcium montmorillonite. Keller (1963), Carroll and Starkey (1960), Russell (1970), and Sayles and Mangelsdorf (1976) have reported that calcium is replaced by sodium, potassium, and magnesium when such clays are treated with seawater. Calcium montmorillonite, the most likely candidate for the position of dominant calcium silicate mineral in ocean sediments, therefore seems to be unstable with respect to other types of montmorillonites in ocean water, and improbably large changes in the cation ratios of seawater would be required to reverse this situation (Holland, 1972).

The quantity of calcium released to ocean water by base exchange processes is approximately 10 to 20% of the flux of dissolved calcium in river water. Kennedy (1965) has determined the cation exchange capacity of sediments from 21 streams and has found that nearly all of the calcium exchange capacity values fell between 10 and 40 meq/100 g. Rivers containing large amounts of montmorillonite were found to have a proportionately large C.E.C. If we take the C.E.C. value of 26 ± 2 meq/100 g of sediment from the Mississippi River to be representative of river sediments as a whole, then such sediments carry approximately $26 \times 10^{-2} \times 20 \times 10^{15} = 5.2 \times 10^{15}$ meq of cations annually. A similar value is obtained from Russell's (1970) data for cation exchange of clays from the Ameca River, Mexico, as they enter the ocean in Banderas Bay (Drever, 1974).

In Chapter IV it was shown that about 20% of the calcium brought to the sea by rivers is derived from the alteration of silicate minerals. As virtually no calcium is removed from ocean water as a constituent of silicate minerals, enough CO_2 and SO_3 must be made available continuously for combining with river calcium to produce the calcium carbonate and calcium sulfate that are the dominant sinks for calcium in the marine environment. Limestones are much more abundant than gypsum and anhydrite in sedimentary rocks. It has, therefore, been the lot of CO_2 to remove the major part of the silicate calcium brought to the oceans by rivers. Conversely, the silicate calcium in rivers has been responsible in large part for removing CO_2 injected into the atmosphere–ocean system by volcanoes and hot springs. The rate of CO_2 injection has apparently been such that silicate calcium has been released in insufficient quantities to absorb the entire CO_2 input. Most of the excess CO_2 has been removed with magnesium, largely in the form of the double carbonate dolomite, $CaMg(CO_3)_2$. However, the CO_2 input into the atmosphere–ocean system has never been sufficiently large to demand the entire ocean input of magnesium. Although calcium is present overwhelmingly as a component of carbonate minerals in sedimentary rocks, magnesium has been present as a constituent both of carbonate and silicate phases in sediments formed during all of Phanerozoic time, and probably during all of the part of Precambrian time for which a sedimentary record is available (see Chap. IV). This suggests that the atmosphere–ocean system has normally been fairly close to the stability boundary between dolomite and one or more magnesium silicates; conversely, it suggests that dolomite and magnesium silicate minerals may have managed to act as rough CO_2 buffers during most of geologic time (Holland, 1965, 1968; Bartholomé, 1966).

Dolomite Deposition

Some skeletal carbonate remains consist of calcite containing large amounts of magnesium (see, for instance, Dodd, 1967). Nevertheless, average recent marine calcites and aragonites probably contain less than 5 mol% $MgCO_3$.

Since the mol ratio of Ca^{2+} to Mg^{2+} in average river water is approximately 2, magnesium removal as a constituent of calcite and aragonite sediments can account for less than ca. 10% of the river flux of magnesium, and is, therefore, not a major removal mechanism for magnesium from the oceans today (Drever, 1974).

The formation of dolomite has apparently been the major mechanism of magnesium carbonate output from oceans during most of geologic time. However, until surprisingly recently no modern dolomites were known. During the past 25 yr Holocene dolomite has been discovered in numerous saline lakes, marine evaporites, and supratidal marine environments (see, for instance, Friedman and Sanders, 1967, and Bathurst, 1972). During the same period the physical chemistry of dolomite has been explored in several laboratories, and a good deal of the chemistry involved in the formation of recent and ancient dolomites is now understood.

Studies of the chemical composition of cave waters (Holland et al., 1964) and of underground water in carbonate terrains (Hsu, 1963) suggested that the solubility product of dolomite at $25\,°C$ is near 10^{-17} (see Chap. II), and that an upward revision of earlier estimates of K_D was required. Langmuir's (1964) measurements of the solubility of dolomite confirmed this suggestion. The best current estimate for K_D at $25\,°C$ is $10^{-16.7}$. In solutions saturated with respect to both calcite and dolomite

$$a_{Ca^{2+}} \cdot a_{CO_3^{2-}} = K_c \tag{5-3}$$

and

$$a_{Ca^{2+}} \cdot a_{Mg^{2+}} \cdot a_{CO_3^{2-}}^2 = K_D \tag{5-4}$$

If equation 5-4 is divided by equation 5-3 squared, it follows that

$$\frac{a_{Mg^{2+}}}{a_{Ca^{2+}}} = \frac{K_D}{K_C^2} \tag{5-5}$$

At $25\,°C$ K_C^2 is nearly equal to $10^{-16.7}$. The ratio $a_{Mg^{2+}}/a_{Ca^{2+}}$ in solutions saturated with respect to both carbonates is, therefore, close to unity. In seawater approximately 10% of both the Ca^{2+} and Mg^{2+} ions are complexed (see Table 5-5). The activity coefficients of the free ions are comparable; thus the value of the ratio $a_{Mg^{2+}}/a_{Ca^{2+}}$ is essentially the same as the concentration ratio $m_{Mg^{2+}}/m_{Ca^{2+}}$, that is, 5.3. It follows that near-surface seawater saturated or supersaturated with respect to calcite is markedly supersaturated with respect to dolomite. This is true even of deep seawater that is undersaturated with respect to calcite, but not by as much as a factor of 5. The oceans as a whole are clearly out of equilibrium with respect to dolomite.

Many authors have experienced and commented on the difficulties of precipitating dolomite directly from solution at room temperature (see, for instance, Weyl, 1967; Berner, 1966). Experiments at temperatures above 150°C (Rosenberg and Holland, 1964; Sureau, 1974) have shown that dolomite is formed more easily by the replacement of calcite than by direct precipitation from solution; even at temperatures as high as 300°C, Ca–Mg–carbonate phases formed by direct precipitation are often metastable. It is, therefore, not surprising that the formation of dolomite from seawater requires a considerable degree of supersaturation, and that modern marine dolomites are largely formed by the replacement of aragonite and calcite.

The very slow growth rate of dolomite is almost certainly related to the problem of Mg^{2+}–Ca^{2+} ordering in the double carbonate. Calcite precipitated from seawater usually contains excess Mg^{2+} in solid solution, and protodolomites are usually nonstoichiometric. The solubility of disordered members of the $CaCO_3$–$MgCO_3$ series is considerably greater than that of the ordered phases (Berner, 1975). Direct precipitation of stable phases from seawater is, therefore, difficult, and ordering of the metastable phases is very slow.

Today dolomite is formed largely in evaporitic settings. During the early stages of evaporation the Mg^{2+}/Ca^{2+} ratio in seawater remains nearly constant. After saturation with respect to gypsum or anhydrite is reached, calcium can be removed quite rapidly. When this happens, the Mg^{2+}/Ca^{2+} ratio rises and moves seawater in the direction of the precipitation of magnesium carbonates. On the island of Bonaire, Deffeyes et al. (1965) found Mg/Ca ratios in excess of 20:1 in the hypersaline solutions of the Pekelmeer. In the Persian Gulf, Kinsman (1964, 1966, 1967) found similarly high Mg/Ca ratios in dolomitizing waters of the sabkhas near Abu Dhabi. At the highest Mg/Ca ratios encountered in this area magnesite ($MgCO_3$) and huntite ($CaMg_3(CO_3)_4$), rather than dolomite, are apparently forming. It is still not clear whether this is due largely to the abnormally high Mg^{2+}/Ca^{2+} ratio in solution or due to the effect of high salinity on the hydration state of Mg^{2+} (Sayles and Fyfe, 1973).

In the Abu Dhabi area downward movement of dense brines has almost certainly been taking place. It is very likely that some dolomitization of carbonate sediments occurs during refluxing, and that Fischer's interpretation (see Newell et al., 1953) of dolomitization by such a mechanism in the Permian Reef Complex of the Guadalupe Mountain Region of Texas and New Mexico is quite likely. A similar interpretation of recent dolomite in the "Solar Lake" of the Gulf of Elat has been proposed by Aharon et al. (1977).

The frequency of association of ancient dolomites with evaporite settings has been stressed repeatedly (see, for instance, Friedman and Sanders, 1967;

Strakhov, 1958), and it is likely that much of the dolomite formed in the past has involved the replacement of calcium carbonate in supratidal sediments and in nearby subsurface sediments of evaporitic areas. However, not all penecontemporaneous dolomites seem to be related to evaporites, and it is possible that a good deal of recent dolomite has formed by the replacement of $CaCO_3$ by seawater or by sabkha evaporite waters that have been diluted by fresh water (Land and Epstein, 1970; Folk and Land, 1975).

It is difficult to make a good estimate of the present-day rate of near-surface dolomite production, but the rate must be small compared to the rate of calcium carbonate formation, unless dolomitization is taking place on a large scale in quite unsuspected places. It was shown above that the rate of removal of magnesium as a component of calcitic tests is also currently a small fraction of the rate of magnesium input from rivers. Magnesium is, therefore, either accumulating in the oceans or is being removed as a constituent of one or more noncarbonate phases.

THE SILICATE OUTPUT OF THE OCEANS

Since the publication of Sillén's (1961) classic paper on the physical chemistry of seawater, the nature and quantity of silicate minerals formed in oceans have been at the center of controversy regarding controls on the composition of seawater. In Sillén's scheme the pH and cation ratios of seawater are controlled largely by reactions between seawater and silicate minerals. Mackenzie and Garrels (1966b) (see also Maynard, 1976) showed that this concept is consistent with a relatively simple scheme for the removal of river-borne cations from seawater. Their mass balance calculation for oceans is reproduced in Table 5-6. River-borne sodium is removed largely as NaCl, the remainder as a constituent of sodium montmorillonite, (reaction 9). Magnesium is removed largely as a constituent of chlorite (reaction 10), and potassium largely as a constituent of illite (reaction 11). Calcium is removed largely as a constituent of $CaCO_3$ (reactions 4 and 8) and of anhydrite or gypsum (reaction 2). The quantity of new montmorillonite, chlorite, and illite that would be produced annually in order to remove the current river flux of Mg^{2+}, and K^+ and Na^+ not balanced by Cl^- would be approximately 15×10^{14}, 5×10^{14}, and 14×10^{14} g/yr, respectively. Together this would amount to ca. 17% of the input of detrital sediments $(2 \times 10^{16}$ g/yr$)$. Since the new clays are assumed to form at the expense of detrital clays—largely kaolinite—in the Mackenzie–Garrels scheme, the proposed reverse weathering reactions increase the total mass of detrital sediments by much less than 17%. The calculated percentage of new clays is not overwhelmingly large, but they should be detectable if they were forming according to the Mackenzie–Garrels scheme. The search for a

Mackenzie and Garrels' (1966b) Mass Balance Calculation for

Step. No.	Reaction (*balanced in terms of mmol of constituents used*)	SO_4^{2-}	Ca^{2+}	Cl^-
	Amount of material to be removed from ocean in 10^3 yr ($\times 10^{21}$ mmol)	382	1220	715
1.	$95.5\ FeAl_6Si_6O_{20}(OH)_4 + 191\ SO_4^{2-} + 47.8\ CO_2 + 55.7\ C_6H_{12}O_6 + 238.8\ H_2O = 286.5\ Al_2Si_2O_5(OH)_4 + 95.5\ FeS_2 + 382\ HCO_3^-$	191	1220	715
2.	$191\ Ca^{2+} + 191\ SO_4^{2-} = 191\ CaSO_4$	0	1029	715
3.	$52\ Mg^{2+} + 104\ HCO_3^- = 52\ MgCO_3 + 52\ CO_2 + 52\ H_2O$	0	1029	715
4.	$1029\ Ca^{2+} + 2058\ HCO_3^- = 1029\ CaCO_3 + 1029\ CO_2 + 1029\ H_2O$	0	0	715
5.	$715\ Na^+ + 715\ Cl^- = 715\ NaCl$	0	0	0
6.	$71\ H_4SiO_4 = 71\ SiO_{2(s)} + 142\ H_2O$	0	0	0
7.	$138\ Ca_{0.17}Al_{2.33}Si_{3.67}O_{10}(OH)_2 + 46\ Na^+ = 138\ Na_{0.33}Al_{2.33}Si_{3.67}O_{10}(OH)_2 + 23.5\ Ca^{2+}$	0	24	0
8.	$24\ Ca^{2+} + 48\ HCO_3^- = 24\ CaCO_3 + 24\ CO_2 + 24\ H_2O$	0	0	0
9.	$486.5\ Al_2Si_{2.4}O_{5.8}(OH)_4 + 139\ Na^+ + 361.4\ SiO_2 + 139\ HCO_3^- = 417\ Na_{0.33}Al_{2.33}Si_{3.67}O_{10}(OH)_2 + 139\ CO_2 + 625.5\ H_2O$	0	0	0
10.	$100.4\ Al_2Si_{2.4}O_{5.8}(OH)_4 + 502\ Mg^{2+} + 60.2\ SiO_2 + 1004\ HCO_3^- = 100.4\ Mg_5Al_2Si_3O_{10}(OH)_8 + 1004\ CO_2 + 301.2\ H_2O$	0	0	0
11.	$472.5\ Al_2Si_{2.4}O_{5.8}(OH)_4 + 189\ K^+ + 189\ SiO_2 + 189\ HCO_3^- = 378\ K_{0.5}Al_{2.5}Si_{3.5}O_{10}(OH)_2 + 189\ CO_2 + 661.5\ H_2O$	0	0	0

Note: Constituent Balance ($\times 10^{21}$ mmol)

the Removal of River-Derived Constituents from the Ocean

Constituent Balance ($\times 10^{21}$ mmol)					HCO_3^- Consumed (−) Evolved (+)	CO_2 Consumed (−) Evolved (+)	Products ($\times 10^{21}$ mmol)	Percentage of Total Products Formed (mol basis)
Na^+	Mg^{2+}	K^+	SiO_2	HCO_3^-				
900	554	189	710	3118				
900	554	189	710	3500	+382	−48	96 Pyrite	3
							287 Kaolinite	8
900	554	189	710	3500			191 CaSO$_4$	5
900	502	189	710	3396	−104	+52	52 MgCO$_3$ in magnesian calcite	2
900	502	189	710	1338	−2058	+1029	1029 Calcite and/or aragonite	29
185	502	189	710	1338			715 NaCl	20
185	502	189	639	1338			71 Free silica	2
139	502	189	639	1338			138 Sodic montmorillonite	4
139	502	189	639	1200	−48	+24	24 Calcite and/or aragonite	1
0	502	189	278	1151	−139	+139	417 Sodic montmorillonite	12
0	0	189	218	147	−1004	+1004	100 Chlorite	3
0	0	0	29	−42	−189	+189	378 Illite	11

sufficiency of new clays has been discouraging. The distribution of clay minerals in the oceans (Biscaye, 1965), their isotopic composition (Dasch, 1969), their age (Hurley et al., 1963; Hurley, 1966), and their lack of reactivity have all tended to show that clay minerals are rather inert in the oceans. Cation exchange certainly takes place in near-shore areas and is probably important for the potassium budget of the oceans, but other mechanisms, perhaps reaction of seawater with basalt at elevated temperatures, may well be more important than reverse weathering for the marine metabolism of Mg^{2+} and Na^+.

Cation Exchange

The cation exchange capacity of sediments entering the oceans was estimated above to be ca. 5.2×10^{15} meq of cations/yr. Table 5-7 from Russell (1970) shows that during the reaction of seawater with clays from the Rio Ameca on the West Coast of Mexico, Ca^{2+} is removed and replaced by an approximately equal number of milliequivalents of K^+, Na^+, and Mg^{2+}. If the behavior of Rio Ameca clay is typical of river clays as a whole, then nearly all of the river flux of K^+, and ca. 10% of the total river flux of Na^+ and Mg^{2+} can be removed by near-shore cation exchange processes on clays. More recent measurements by Sayles and Mangelsdorf (1976) have shown that uptake of Na^+ and Mg^{2+} is less important than was suggested by the earlier work. Nevertheless, cation exchange may remove as much as 30% of the river sodium released during silicate weathering, and is, therefore probably of importance in the removal of noncyclic sodium from the oceans.

The cation exchange processes described above take place on a time scale of a few days to weeks. Drever (1971a, b) has proposed that on a longer time

Table 5-7

Comparison of Standard Rio Ameca Clay with Average Surface
Marine Clay from Banderas Bay, Mexico[a]

	−(1) River Clay		+(2) Marine Clay		=(3) Difference
Element	(wt %)	(meq/l)[b]	(wt %)	(meq/l)[b]	(meq/l)[b]
Na	0.43	0.191	0.61	0.271	+0.080
K	1.23	0.321	1.65	0.430	+0.109
Ca	1.02	0.519	0.44	0.224	−0.295
Mg	1.71	1.436	1.81	1.520	+0.084
Total	4.39	2.467	4.51	2.445	−0.022±0.073

[a]Russell (1970).
[b]1.02 g clay per l of river water.

scale iron is extracted from clay minerals in anoxic sediments to form sulfides, and that Mg^{2+} from the surrounding seawater replaces Fe^{2+} in the vacated cation sites. It is difficult to determine the rate at which this process supplies iron for the formation of pyrite in marine sediments. A reasonable maximum is set by the condition that all of the iron for iron sulfide formation is extracted from clay minerals (Drever, 1974). Approximately half of the river flux of SO_4^{2-} seems to have gone ultimately into pyrite formation during the past 100 million years (Holland, 1973).

Since 1 mol of pyrite contains 2 mol of sulfur, approximately

$$0.12/4 \text{ mmol/kg} \times 4.6 \times 10^{16} \text{kg/yr} = 1.4 \times 10^{12} \text{ mol/yr}$$

of iron are used annually to precipitate iron sulfides that are ultimately converted to pyrite in marine sediments. If the replaced iron is present as Fe^{2+}, 1 mole of Mg^{2+} is required per 1 mol of iron; if the replaced iron is present as Fe^{3+}, 1.5 mol of Mg^{2+} are required per 1 mol of iron.

Complete replacement of iron by magnesium from seawater would, therefore, require a flux of $(1.4 \text{ to } 2.1) \times 10^{12}$ mol Mg^{2+}/yr into marine sediments. This represents 20 to 30% of the river flux of Mg^{2+}, and is a strong upper limit to the removal rate of magnesium from seawater by the Mg–Fe exchange mechanism. Berner et al. (1970) found no evidence for a Mg^{2+}–Fe^{2+} replacement reaction in an intertidal mud from Long Island Sound and in an organic-rich sediment from Maine. The reaction is, therefore, by no means universal. Iron released during weathering accumulates in soils largely as ferric oxides and hydroxides, and it seems likely that at least some of this material becomes a source of iron for iron sulfides in marine sediments. Thus, since some iron clearly comes from clays and is replaced by Mg^{2+}, while just as clearly not all of the iron is derived in this fashion, Mg^{2+} exchange for iron probably removes about $10 \pm 5\%$ of the river flux of Mg^{2+} from the ocean. The removal mechanism for the major part of the river-borne flux of Mg^{2+}, therefore, remains to be identified.

Authigenic Clay Minerals

The Formation of Magnesium Silicates. Sepiolite, attapulgite (palygorskite), and talc have all been reported as authigenic minerals in marine sediments. Their position within the composition triangle $MgO–SiO_2–H_2O$ is shown in Figure 5-3, together with the probable tie lines at 25°C in this system. The composition and stability of sepiolite and attapulgite are not well defined. Two quite similar formulas have been proposed for sepiolite (see Grim, 1968, p. 119) $Mg_8Si_{12}O_{30}(OH)_4(OH_2)_4 \cdot 8H_2O$ and $Mg_9Si_{12}O_{30}(OH)_6(OH_2)_4 \cdot 6H_2O$. The formula given by Grim (1968, p. 115) for attapulgite and its

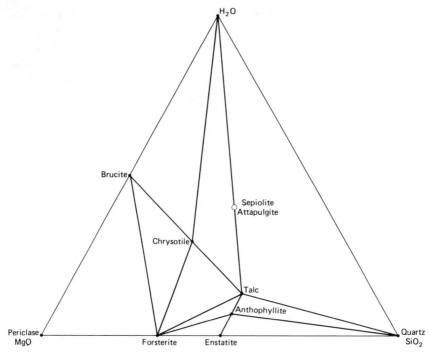

Figure 5-3. Mineral phases and tie lines in the system $MgO–SiO_2–H_2O$ at $25°C$. Forsterite $= Mg_2SiO_4$, Enstatite $= MgSiO_3$, Anthophyllite $= Mg_7Si_8O_{22}(OH)_2$, Talc $= Mg_3Si_4O_{10}(OH)_2$, Chrysotile $= Mg_3Si_2O_5(OH)_4$, Sepiolite $= Mg_9Si_{12}O_{30}(OH)_6(OH_2)_4 \cdot 6H_2O$ and $Mg_8Si_{12}O_{30}(OH)_4(OH_2)_4 \cdot 8H_2O$, Attapulgite $= Mg_5Si_8O_{20}(OH)_2 \cdot 4H_2O$.

synonym, palygorskite, is similar: $Mg_5Si_8O_{20}(OH)_2(OH_2)_4 \cdot 4H_2O$, although natural palygorskites frequently contain several percent of Al_2O_3; the three minerals have, therefore, been plotted together in the ternary diagram. Naturally occurring sepiolite seems to be thermodynamically unstable with respect to decomposition to quartz, talc, and water, but is present rather commonly in low-temperature hydrothermal veins and in sediments at least as old as Carboniferous (Millot, 1964, 1967).

The solubility of the magnesium silicates has been studied quite thoroughly, and the most recent data are summarized in Figures 2-2 and 5-4. Analyses of normal seawater fall within the area marked S; analyses of many interstitial waters fall within the area marked I. Seawater is supersaturated with respect to talc in nearly the entire extent of areas S and I (Hostetler et al., 1971). In the upper parts of the two areas seawater is also supersaturated with respect to chrysotile and natural, well-ordered sepiolite (C) (Christ, et al., 1973). Saturation of seawater with freshly precipitated

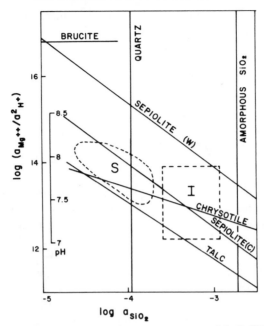

Figure 5-4. The solubility of some phases in the system MgO–SiO_2–H_2O at $25\,°C$ and 1 atm pressure (Drever, 1974). The area S covers the composition of normal seawater. The area I covers the composition of most interstitial waters.

sepiolite (W) (Wollast, et al., 1968) is achieved rarely, if ever. However, the precipitation of several magnesium silicates from seawater is thermodynamically possible, and kinetically not unexpected.

Sepiolite has been reported as a reasonably frequent constituent of sediments in highly saline alkaline lacustrine environments, as well as with dolomite and marl (Hathaway and Sachs, 1965). In recent sediments from the mid-Atlantic Ridge sepiolite and palygorskite have been found, possibly as alteration products of volcanic ash (Hathaway and Sachs, 1965; Bonatti and Joensuu, 1968; Siever and Kastner, 1967), or as the product of the interaction between seawater and hydrothermal solutions related to volcanism (Bowles et al., 1971; Drever, 1974); they have also been found in a saline lake, apparently as an alteration product of montmorillonite (Parry and Reeves, 1968). Talc seems to be more common in sediments than has been supposed (Isphording, 1972). It is generally detrital in origin, but does occur as an authigenic mineral. The mineral is present as a minor constituent of shallow water marine sediments near Hawaii (Moberly, 1963, and personal communication, 1968), and may be authigenic there. Other occurrences of talc, particularly in carbonates and evaporites, have been recorded by Friedman (1965).

All the reported occurrences of talc can account at best for a minute fraction of the current magnesium input into the ocean. If all of the river magnesium were precipitated as talc in the oceans, the mean talc sedimentation rate would be

$$\frac{3.9 \times 10^{-6} \text{g Mg/g river water} \times 4.6 \times 10^{19} \text{ g river water/yr}}{3.6 \times 10^{18} \text{ cm}^2}$$

$$\times \frac{379.3 \text{ g/mol talc}}{24.3 \text{ g/mol Mg}} \times 1000 = 0.8 \text{ g/cm}^2 \text{ per } 1000 \text{ yr} \qquad (5\text{-}6)$$

This rate exceeds the total present rate of sedimentation on the floor of the Pacific Ocean, and would account for about 10 to 15% of the total accumulation rate in shelf and slope environments. The reported occurrences of the authigenic magnesium silicates are trivial by comparison. However, if authigenic magnesium silicates were to form at a rate of just a few percent of the rate of input of terrigenous sediments, they could remove a nonnegligible fraction of the river input of magnesium from seawater (Drever, 1974).

The relatively small distance that separates the field of seawater and of interstitial waters from the line for sepiolite (W) is intriguing. Excursions of seawater composition beyond this line are virtually ruled out by the rapid precipitation of sepiolite when the line is overstepped; it is unlikely, therefore, that seawater has ever occupied a position well above the sepiolite (W) line.

The Formation of Chlorite. The concentration of aluminum and iron in ocean water is extremely small (see Table 5-2). Both cations are, however, brought to the oceans in part as constituents of unstable and reactive compounds: aluminum as a constituent of weakly crystalline to amorphous hydrated aluminum oxides and silicates, iron in hydrated ferric oxides (see for instance Mackenzie and Garrels, 1966a; Moberly, 1963; Drever, 1971a). In the presence of such materials, hydrated magnesium silicates are probably thermodynamically unstable with respect to chlorite. There are no low-temperature data for the system $MgO - Al_2O_3 - SiO_2 - H_2O$, but the data of Fawcett and Yoder (1966) in Figure 5-5 do suggest that near 25°C quartz can be in equilibrium with montmorillonite and talc or with higher hydrates such as sepiolite, and that at lower SiO_2 activities montmorillonite, serpentine, a serpentine–amesite solid solution, and kaolinite are stable phases in the presence of aqueous solutions. Phase relations among the montmorillonites and chlorites in oceanic settings are undoubtedly complicated by the presence of the additional components Fe_2O_3, FeO, CaO, Na_2O, and K_2O.

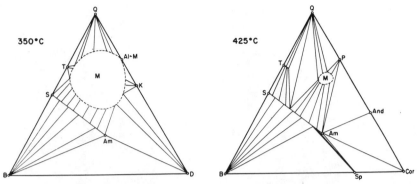

Figure 5-5. Isothermal projections in the system $MgO–Al_2O_3–SiO_2–H_2O$ at 2 kbar H_2O pressure (Fawcett and Yoder, 1966). Copyright by the Mineralogical Society of America. Al–M = Aluminum montmorillonite (beidellite), Am = Amesite, And = Andalusite, B = Brucite, Cor = Corundum, D = Diaspore, K = Kaolinite, M = Montmorillonite, P = Pyrophyllite, Q = Quartz, S = Serpentine, Sp = Spinel, T = Talc.

The abundance of detrital chlorite in oceanic sediments (see, for instance, Biscaye, 1965; Griffin et al., 1968) hinders the identification of authigenic chlorite, but Swindale and Fan (1967) have reported the presence of chloritic alteration of grains of gibbsite in Waimea Bay off the island of Kauai, Hawaii. Sediment brought into the bay by the Waimea River contains about 1 to 2% gibbsite. Occasional grains of gibbsite in the Bay were stained with iron oxide, but were otherwise unaltered. Most gibbsite grains, however, showed the extensive alteration to chlorite illustrated by Figure 5-6. Unfortunately, no chemical analyses of this chlorite are available, but the optical data ($n_\alpha = n_\beta = 1.57$; $n_\gamma = 1.59$, optically positive) suggest that it is of the highly aluminous, low-iron corundophilite variety (Hey, 1954).

Members of the high-iron septechlorite group of minerals were unknown as constituents of recent sediments until the discovery of chamosite in marine sediments from the Niger delta, the greater Orinoco delta, and the shelf of Sarawak by Porrenga (1965, 1966, 1967), its discovery near the coast of Guinea by Von Gärtner and Schellmann (1965), and near the coast of Gabon by Giresse (1965a, b). The relationship of chamosite and the high-iron septechlorite minerals to the chrysotile-amesite series is shown in Figure 5-7. The replacement of Mg^{2+} by Fe^{2+} in chrysotile produces members of the chrysotile-greenalite series, and the replacement of Mg^{2+} by Fe^{2+} in amesite produces members of the amesite-chamosite series. The replacement of Al^{3+} by Fe^{3+} in chamosite ultimately produces cronstedtite.

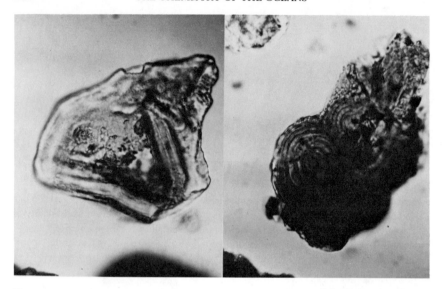

Figure 5-6. Photomicrographs showing gibbsite surrounded and penetrated by authigenic chlorite (Swindale and Fan, 1967). Copyright (1967) by the American Association for the Advancement of Science.

The appearance of chamosite in several modern shallow water marine sediments implies that sufficient Fe^{2+} is produced in these sediments by the reduction of Fe^{3+} to compete successfully with Mg^{2+} for inclusion in octahedral positions in the chlorite structure. The apparent absence of Fe^{2+} in the Hawaiian chlorite of Swindale and Fan (1967) speaks for its genesis in a more oxidizing environment, and the presence of both types of chlorite in modern sediments underscores the importance of detrital constituents and of the redox potential in sedimentary settings for the genesis of this group of minerals.

The few occurrences of authigenic chlorites in modern sediments together account for only a very small percentage of the current river input of magnesium into the ocean. Future discoveries may show that they represent a significant sink for oceanic magnesium; at present their role must still be considered rather minor.

The Formation of Montmorillonite. Authigenic marine montmorillonite is difficult to identify, because so much detrital montmorillonite is present in marine sediments. However, the development of montmorillonite during the alteration of volcanic ash in North and South Pacific sediments is well documented. Griffin et al. (1968) report that six montmorillonites from the

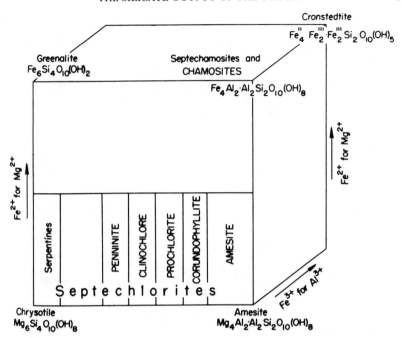

Figure 5-7. A revised "Winchell-type" scheme showing compositional relationships among members of the chlorite group (Nelson and Roy, 1958). Copyright by the Mineralogical Society of America.

Pacific Ocean contained from 7 to 15% Fe_2O_3. Rex (1967) discovered phillipsite and a yellow-buff montmorillonite with interlayer hydroxy-iron polymers suggestive of an iron chlorite-montmorillonite mixed layer clay that had been produced by the alteration of basaltic tuff breccia in seawater on Sylvania Guyot near Bikini Atoll in the Marshall Islands. Quite generally, the alteration of basaltic glass on the ocean floor yields zeolites, Fe-montmorillonite, Fe-hydroxides, and MnO_2 (Bonatti, 1975; see also Honnorez, 1972). The nature of the zeolites is discussed below. The alkali metals are usually enriched in altered basalts, but sodium is an exception. As shown in Table 5-8, sodium is enriched in some altered basalts, while it is depleted in others (Thompson, 1973). The mean direction of the transfer of sodium at the ocean floor as a whole is still in doubt, and there is currently little support for the Mackenzie–Garrels scheme of removing sodium from seawater by the formation of authigenic Na-montmorillonite.

Magnesium is, however, almost certainly lost from basalts during interaction with seawater on the ocean floor (Rivers, 1976). If such reactions take

Table 5-8

Change in Mass of Oxides (g/100 cm^3) in Some Basalts Weathered on the Seafloor Compared to the Fresh Precursor[a]

	1	2	3	4
SiO$_2$	−6.5	−10.7	−22.0	−18.4
Al$_2$O$_3$	+3.8	−1.8	−3.1	−8.7
FeO[b]	+4.4	−0.3	−3.1	−5.9
MgO	−7.8	−8.0	−8.4	−1.4
CaO	−3.6	−1.1	−24.8	−4.3
Na$_2$O	+1.4	+0.5	−1.7	−0.7
K$_2$O	+0.3	+0.4	+4.3	+1.9
H$_2$O	+2.2	+7.6	+24.0	+9.7
TiO$_2$	−0.77	+0.66	+0.20	−0.33
MnO	+0.02	−0.02	−0.11	−0.08
P$_2$O$_5$	+0.15	+0.01	+0.08	n.d.

[a]Thompson (1973). (1) Slightly altered interior of dredged pillow; age less than 10 million yr (Thompson, unpublished). (2) Altered interior of dredged pillow; age less than 10 million yr (Thompson, unpublished). (3) Very badly altered lava; age 80 million yr (Matthews, 1962). (4) Altered interior of lava flow, leg 2, site 10; age 16 million yr (Frey et al., 1972). n.d. = not determined; (+) = gained; (−) = lost
[b]Total Fe as FeO.

place extensively to a depth of several hundred meters below the ocean floor, the flux of magnesium into seawater from this source approaches the magnitude of the river flux of magnesium. At present there is little evidence in favor of such a large flux of Mg^{2+} up through the ocean floor (see for instance Kastner and Gieskes, 1976); there is, however, a rather substantial body of evidence for extensive reaction of seawater with basalt at elevated temperatures, where magnesium is transferred in the opposite direction, that is, from seawater to altered basalt (see text following).

Seawater appears to be in equilibrium or in near equilibrium with iron-rich montmorillonite. This is also suggested by the somewhat uncertain extrapolation of the data of Hemley et al. (1961) for the system Na$_2$O–Al$_2$O$_3$–SiO$_2$–H$_2$O to lower temperatures. The range of the ratio a_{Na^+}/a_{H^+} and the activity of dissolved silica in seawater almost certainly straddle the montmorillonite–kaolinite boundary, and are presumably not far from the Na-montmorillonite–phillipsite boundary. Analcite and albite typically form in environments with considerably higher values of a_{Na^+}/a_{H^+} and $a_{H_4SiO_4}$ (see, for instance, Hay, 1966; Iijima and Utada, 1966; Gregor, 1967).

If the pH of seawater were raised to 8.5 or above, Mg^{2+} would apparently be extracted rapidly by montmorillonite (Deffeyes, 1965, and quoted by Russell, 1970), probably by the formation of Mg(OH)$_2$ interlayers. This suggests that magnesium removal from seawater by such a mechanism becomes rapid close to the pH at which sepiolite can begin to precipitate from seawater (see Fig. 5-4). The two mechanisms together could remove the river flux of Mg^{2+} and SiO_2 quite efficiently.

Other mechanisms of Mg^{2+} and SiO_2 removal are almost certainly dominant today, but it is likely that an increase in the pH of seawater by only a few tenths of a pH unit would produce a drastic increase in the importance of the sepiolite and the $Mg(OH)_2$ interlayer removal mechanisms of Mg^{2+} and SiO_2.

The Formation of Illites. There is abundant evidence in support of the proposal that illitic clays have been and are currently forming in marine environments. These illites are largely iron-rich and are normally classed as glauconites. Their composition falls in the range $(K,Na,Ca)_{1.2-2.0}(Fe^{3+}, Al, Fe^{2+}, Mg)_{4.0}(Si_{7-7.6}Al_{1-0.4}O_{20})(OH)_4 \cdot n(H_2O)$ (Deer et al., 1962, Vol. 3, p. 35). The origin and distribution of glauconite and the history of confusion in glauconite nomenclature have been reviewed by Millot (1964, pp. 238–245). More recently Bell and Goodell (1967) have presented a comparative study of glauconite and associated clay fractions in modern marine sediments from six environments; they demonstrate that glauconite pellets in marine sediments often contain the same clays as nonpelletized clay minerals in the surrounding sediments. Despite this demonstration, and despite the fact that well-crystallized glauconite in recent marine sediments is often detrital rather than authigenic, the evidence for the formation of glauconite in the oceans is convincing. Porrenga (1967) has shown that off the Niger delta glauconite forms from fecal pellets composed of mixed detrital clays. During the transformation the MgO content of the pellets increases from 1.0 to 2.2%.

Figure 5-8 illustrates the occurrence of glauconite in near-shore sediments along many coastlines today. Rather curiously, glauconite containing largely trivalent iron appears to form at greater depths than chamosite, in which iron is present largely in the divalent state. The mechanism of formation of glauconite pellets is still so poorly understood that it is difficult to go beyond the rather trite proposal that redox conditions in marine sediments must be of importance in determining whether a ferric illite or a ferrous chlorite develops within a particular near-shore sediment.

Although glauconite is reasonably abundant in modern sediments, it is very difficult to estimate its present rate of production and to make a convincing mass balance calculation for the input and output of magnesium and potassium from oceans today. Drever (1974) estimates that the glauconite content of recent sediments is no greater than 3%. If this is correct, not more than ca. 3% of the river flux of magnesium and not more than ca. 20% of the river flux of potassium are removed from seawater via the formation of glauconite.

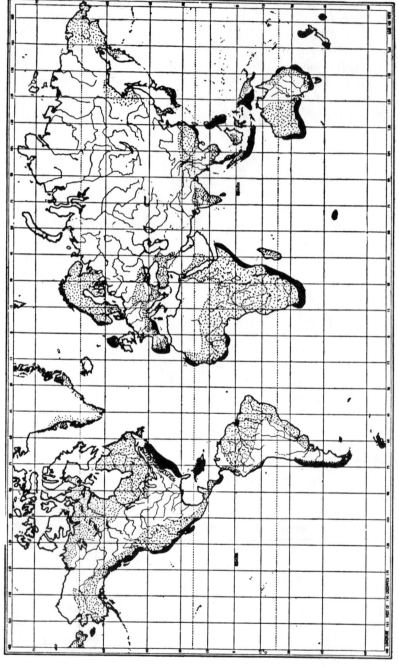

Figure 5-8. Glauconite occurrences in recent sediments are shown in solid black (Galliher, 1935).

The Formation of Zeolites

Phillipsite and clinoptilolite are the most important authigenic zeolites in modern marine sediments. Phillipsite was discovered by Murray and Renard (1891); its distribution and properties have since been described by many investigators, including Goldberg (1961), Arrhenius (1963), Bonatti (1963), Sheppard et al. (1970), and Bernat et al. (1970); their findings have been summarized by Cronan (1974). Phillipsite is common in slowly depositing deep-sea sediments. In sediments from the Pacific Ocean phillipsite is associated with iron and manganese oxides, clay minerals of the smectite group, palagonite, and other volcanic debris. The common association of phillipsite with volcanic debris on the ocean floor indicates that the mineral is generally formed during the reaction of these debris with seawater.

Table 5-9 lists some averages of the composition of marine phillipsites. If Rex's (1967) average is recast so that the sum of $Al + Si = 8.0$ atoms per formula unit, one obtains the composition

$$[Na_{1.6}K_{0.8}Mg_{0.04}Ca_{0.03}][Al_{2.2}Si_{5.8}]O_{16.1} \cdot 5.7\ H_2O.$$

Similar formulas would be obtained by recasting the other averages. Marine phillipsites, therefore, tend to be somewhat more Si-rich and somewhat more Al- and alkali-poor than ideal phillipsites of composition

$$[Na, K, \tfrac{1}{2}Ca]_3[Al_3Si_5O_{16}] \cdot 6H_2O.$$

Although most phillipsite is very rich in the alkali metals and is frequently a major component of deep-sea sediments, it does not seem to be a major oceanic sink for the alkalies. If all of the river potassium were removed as a

Table 5-9

Average Composition of Marine Phillipsites (Wt %)

	Goldberg (1961)	Rex (1967)	Sheppard et al. (1970)
SiO_2	53.47	53.20	54.06
Al_2O_3	16.62	16.80	17.71
MgO	0.06	0.30	0.49
CaO	0.29	0.30	1.29
BaO			0.24
Na_2O	7.41	7.40	3.82
K_2O	6.14	6.0	6.77
H_2O		15.70	15.58

constituent of phillipsite containing 6% K_2O, the sedimentation rate of phillipsite in the entire ocean would be approximately

$$\frac{2.1 \times 10^{-6}(\text{g K/g river water}) \times 4.6 \times 10^{19}(\text{g river water/yr})}{(0.06 \text{ g K/g phillipsite}) \times 3.6 \times 10^{18} \text{ cm}^2}$$

$$= 4.5 \times 10^{-4} \text{ g/cm}^2 \text{ yr}$$

or 0.45 g/cm^2 1000 yr. The mean sedimentation rate in the Pacific Ocean is approximately 0.1 g/cm^2 1000 yr, and the average K_2O content of Pacific pelagic sediments is approximately 2.5% (Goldberg and Arrhenius, 1958; McMurtry, 1975). If all of the potassium in these sediments were derived from seawater, approximately

$$\frac{2.5}{6.0} \times \frac{0.1}{0.45} \times 100 = 9\%$$

of the river flux of potassium would be removed as a constituent of phillipsite. This figure, like most others of its kind, is uncertain. Some of the potassium in marine phillipsites is almost certainly derived from altering volcanics; if the volcanic contribution of K^+ is important, proportionately less river potassium is removed as a constituent of phillipsite. On the other hand, the mean sedimentation rate of pelagic sediments in the Pacific Ocean may be somewhat greater than 0.1 g/cm^2 1000 yr. It is likely, then, that phillipsite formation removes a nonnegligible, but not a major fraction of the river flux of K^+ and of Na^+ today.

Clinoptilolite is present in many marine sediments, but apparently in smaller quantities than phillipsite. The mineral tends to form as an alteration product of silicic volcanic glass, and its relative scarcity may reflect the small proportion of silicic compared to basaltic glass in marine sediments. It seems likely that the formation of clinoptilolite in marine sediments contributes less than the growth of phillipsite to the removal of the alkalies from seawater.

The Deposition of Silica

The annual flow of dissolved silica into the oceans as a constituent of river water is equal to

$$\left(13 \times 10^{-6} \text{ g SiO}_2/\text{g river water}\right) \times \left(4.6 \times 10^{19} \text{ g river water/yr}\right)$$

$$= 6 \times 10^{14} \text{ g/yr}$$

The controls of the concentration of dissolved silica in ocean water have been discussed extensively. In 1957 Siever reviewed the silica budget of the oceans, and concluded that siliceous organisms are probably responsible for keeping the oceans undersaturated with respect to amorphous silica. On the other hand Mackenzie and Garrels (1965), Mackenzie et al. (1967) Siever (1968) and Siever and Woodford (1973) have shown that clays carried in the suspended load of streams release SiO_2 rapidly to silica-deficient seawater, and that they take up dissolved SiO_2 rapidly from silica-enriched seawater. These authors concluded that the interaction of seawater with silicates exerts a major control on the concentration of dissolved SiO_2 in the oceans. Several lines of evidence speak in favor of the earlier view, that siliceous organisms rather than clays are acting as the major controls on the distribution of dissolved SiO_2 in the oceans, and that most of the river flux of dissolved silica leaves the oceans as siliceous tests.

Wollast (1974) has recently described the marine geochemistry of silica. The concentration of dissolved SiO_2 in seawater ranges from undetectably low values at many surface stations and in shallow seas, to about 8 mg of SiO_2/kg in the deep waters of the Pacific and Indian Oceans. The mean value of dissolved silica in these oceans falls between 4 and 6 mg of SiO_2/kg. In the Atlantic Ocean the concentration of silica is roughly half that in the Pacific and Indian Oceans. The very low concentration of dissolved SiO_2 in surface waters is due to the extensive biologic uptake of SiO_2. The higher concentrations of dissolved SiO_2 at depth are determined in large part by the dissolution of siliceous tests. Lisitsyn and Bogdanov (1968) and Hurd (1973) have shown that the rate of formation of siliceous tests must exceed their rate of burial by more than a factor of 100. The dissolution of siliceous tests is, therefore, a process of major importance for the geochemistry of SiO_2 in the oceans (Grill, 1970; Heath, 1974). The actual rate of burial of siliceous tests is hard to determine precisely since siliceous tests are X-ray amorphous and are difficult to distinguish from other forms of silica in bulk chemical analyses. Wollast's (1974) estimate of 3.6×10^{14} g/yr for the rate of free silica removal is sufficiently imprecise that clay–sea water reactions are not ruled out as a potentially important mechanism for controlling the SiO_2 content of bottom waters and as sinks for the river flux of dissolved SiO_2. Figure 5-9 shows the results of the experiments of Mackenzie et al. (1967) on the effect of clays on the SiO_2 concentration of silica-deficient and silica-enriched seawater. The increase of SiO_2 in SiO_2-deficient seawater, and the decrease of SiO_2 in SiO_2-enriched seawater is quite striking, and led the authors to conclude that silicates exert a major control on the silica concentration of the ocean. This is not necessarily true. In Figure 5-9 the maximum SiO_2 uptake is shown by glauconite. One gram of glauconite was shown to have removed 17 ppm of SiO_2 from 200 ml of seawater. This corresponds to an increase of 0.34% in the SiO_2 content of the glauconite sample. An average uptake of this quantity of SiO_2 on all detrital sediments

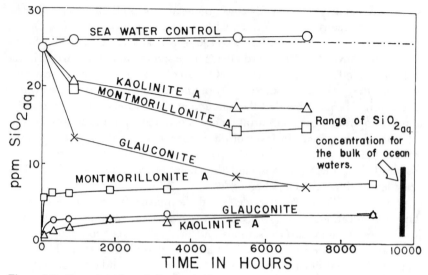

Figure 5-9. Concentration of dissolved silica as a function of time for suspensions of silicate minerals in seawater. Curves are for 1 g ($< 62\ \mu$) mineral samples in 200 ml of silica-deficient (SiO_2 in water was initially 0.03 ppm) and silica-enriched (SiO_2 was initially 25 ppm) seawater at room temperature. Size of symbols indicates precision of SiO_2 determinations. Dash-dot line shows minimum SiO_2 concentration of seawater in equilibrium with a hydroxylated magnesium silicate at the pH's of the experiments. (Mackenzie et al., 1967.) Copyright (1967) by the American Association for the Advancement of Science.

would account for the removal of

$$0.34 \times 10^{-2} \times 20 \times 10^{15} = 0.7 \times 10^{14}\ \text{g}\ SiO_2/\text{yr}$$

that is, of roughly 12% of the river input of dissolved silica. Laboratory experiments to determine the total silica capacity of clays have not been carried out, but the corresponding field observations indicate that their capacity is not large enough. Measurements of the concentration of dissolved silica in the interstitial water of sediment cores by Siever et al. (1965), Fanning and Pilson (1971), Hurd (1973) and by numerous other investigators have shown that these concentrations are usually between 20 and 50 ppm, and may reach values as high as 85 ppm. It is hard to understand how clays can control the silica concentration of the oceans, when they apparently fail to control the SiO_2 content of interstitial waters at the levels suggested in Figure 5-9. It is essentially certain that the dissolution of siliceous tests is responsible for the high dissolved SiO_2 concentrations in interstitial waters, and that the capacity of clays for silica uptake is insufficient to remove more than a small fraction of the available supply of silica. Wollast (1974) and Schink et al. (1975) have modeled the dissolution of siliceous tests and the diffusion of silica in interstitial waters through the sediment–water interface.

If this interpretation is correct, the deposition of free silica from the oceans is a consequence of the overall chemistry of the transformation of average igneous rock into sediments, and siliceous organisms merely act as catalysts for the inevitable precipitation of free SiO_2. This is in agreement with the data for the quartz content of sedimentary rocks. The quartz content of shales is ca. 30% (Shaw and Weaver, 1965). Sandstones contain on the average 65% quartz (Pettijohn et al., 1972) and carbonate rocks contain very little quartz. Average sedimentary rock consists of approximately 60% shale, 20% sandstone, and 20% carbonate rocks. The mean free quartz content of sedimentary rocks must, therefore, be between 30 and 40% by weight. On the other hand average igneous rock contains much less free quartz. Granites contain approximately 30% quartz, basalts and gabbros virtually none; average igneous rock consisting of a $1:1$ mixture of granite and basalt contains ca. 15% of quartz. Since the total SiO_2 content of average sedimentary rocks is nearly equal to that of a $1:1$ mixture of granite and basalt, the much greater percentage of free quartz in sedimentary rocks than in a mechanical mixture of 1 part granite and 1 part basalt is consistent with the release of SiO_2 during weathering followed by its precipitation as free SiO_2 in the ocean.

In Chapter IV Holeman's (1968) figure of 2×10^{16} g/yr was adopted for the input rate of particulate matter to the oceans. If one-quarter of this quantity of sediment is of igneous parentage, and if 20% of free silica must be added to this detritus to complete its conversion to average sediment, then approximately

$$\frac{2 \times 10^{16}}{4} \times 0.20 = 10 \times 10^{14} \text{ g } SiO_2/\text{yr}$$

ultimately appear as free SiO_2 in sedimentary rocks. This figure is similar to the estimated total river flux of dissolved silica (see Chap. IV). The approximate agreement between the two figures does not, however, rule out completely the formation of new clays in the oceans. Hower et al. (1976) have pointed out that the breakdown of expanded clays to illite and quartz during diagenesis may be an important source of free silica in sedimentary rocks. At least some river-borne silica may, therefore, be used to form new expanded clays, only to be re-released later during the diagenetic conversion of expanded clays to illite and quartz.

The river flux of dissolved silica to the oceans is augmented by a flux due to the flow of seawater through midocean ridge basalts. It is shown below that this flux of SiO_2 is almost certainly small compared to the river flux, but that the process may be of importance for the marine geochemistry of several dissolved species, notably Mg^{2+} and SO_4^{2-}.

Interaction of Seawater with Basalt at Elevated Temperatures

The small quantity of authigenic marine clay and the deposition of a large fraction of the river flux of dissolved SiO_2 in the form of siliceous tests both suggest that only a minor fraction of the river flux of Mg leaves the ocean as a constituent of silicates on the ocean floor. The removal of Mg below the ocean floor is an attractive alternative disposal mechanism. Metamorphic rocks have been dredged from the ocean; there is excellent evidence that seawater was the metasomatic fluid, and that during the reaction of seawater with midocean ridge basalts Mg is removed almost quantitatively from seawater. There is also reasonably compelling evidence that the quantity of seawater that is currently cycling through midocean ridges is large enough to remove much, if not all, of the river flux of Mg to the oceans.

Numerous authors have described serpentinites and greenstones from the Mid-Atlantic Ridge (see, for instance, Melson and van Andel, 1966; Melson et al., 1968; Bonatti et al., 1970; Aumento et al., 1971; Miyashiro et al., 1971; Bonatti et al., 1975; Humphris and Thompson, 1978). Zeolite facies minerals from abyssal metabasalts include analcime, natrolite, thomsonite, chabazite, laumontite, stilbite, montmorillonite, and mixed-layer chlorite-montmorillonite. Typical greenschist assemblages in hydrothermally altered basalts dredged from the ocean floor consist of chlorite plus albite plus or minus actinolite plus or minus epidote accompanied by montmorillonite, quartz, pyrite, sphene, and relict plagioclase and pyroxene.

A similar alteration assemblage has been described by Tómasson and Kristmannsdóttir (1972) in basalt from the Reykjanes Peninsula geothermal areas in Iceland. In these geothermal wells montmorillonite is the dominant sheet silicate at temperatures below 200°C. Mixed layer chlorite-montmorillonite dominates between 200 and 230°C. Chlorite first appears between 230 and 280°C and replaces the other sheet silicates toward higher temperatures. The zeolite assemblage, including mordenite, stilbite, mesolite, analcime, and wairakite, is confined to temperatures below 230°C. Prehnite is present toward higher temperatures from just below the zeolite zone. Epidote forms at temperatures as low as 200°C and becomes a major constituent above 260–270°C. K-feldspar, quartz, calcite, anhydrite, and pyrite are present in varying proportions throughout the system. Hematite only occurs in the upper 150 m of the geothermal systems.

The most notable differences between the mineralogy of the rocks in the Reykjanes area and the ocean floor metabasalts is the presence of anhydrite, calcite and K-feldspar, and the absence of actinolite and talc at Reykjanes. The absence of actinolite and talc is almost certainly due to the relatively low metamorphic temperatures ($T \leqslant 280°C$) at Reykjanes compared to those experienced by most metabasalts (Spooner and Fyfe, 1973).

The source of the geothermal solutions at Reykjanes is certainly seawater. The chloride concentration of brines from Reykjanes drill holes 2, 4, and 8 in brines from Njardvikurheidi drill hole 1 (see Table 5-10) is nearly identical to that of seawater. The somewhat higher chloride concentration in the Reykjanes spring water is almost certainly due to evaporation, and the lower salinity of brines from Reykjanes drill hole 7 is probably due to dilution of seawater with meteoric water.

Hydrothermal springs issuing on the ocean floor have been inferred in the Transatlantic Geotraverse area (Scott et al., 1974; Scott et al., 1976) and have been sampled near the Galapagos spreading center (Edmond, personal communication 1977). Seawater is also almost certainly the major fluid during metamorphism of ocean floor basalts. The few metabasalts studied to date have $^{87}Sr/^{86}Sr$ ratios higher than that of normal, unaltered ocean floor basalts and lower than that of present-day seawater. Hart (1972) has pointed out that a minimum seawater-rock ratio of about 2 would be required to explain the observed enrichment of ^{87}Sr in the metabasalts which he analyzed. Bonatti et al. (1975) report two elevated $^{87}Sr/^{86}Sr$ ratios that would require seawater/rock ratios of between 3 and 7 to account for their difference from those of average unaltered midocean ridge basalts. Such water/rock ratios are consistent with heat flow data in the vicinity of midoceanic ridges (see below). Jehl's (1975) studies of fluid inclusions in a variety of metabasalts from the Mid-Atlantic Ridge have shown that the salinity of the inclusion fluids is typically somewhat higher than that of seawater, and can reach values as high as 17% weight equivalent NaCl. These results indicate that either solutions other than seawater can also be involved in ocean floor metamorphism, or that in the systems from which Jehl's samples were taken the seawater/rock ratios were sufficiently low, so that the removal of water by hydration of initially anhydrous basalts and ultramafic rocks has increased the salinity of the circulating fluids. It is perhaps significant that the highest salinities were found in inclusion fluids in serpentinites where hydration reactions may have abstracted particularly large quantities of water.

Although the salinity of the Reykjanes brines is very similar to that of seawater, the concentration of several of the major and minor constituents of the brines is very different from their concentration in seawater. As shown in Table 5-10 Mg^{2+} and SO_4^{2-} are present in much smaller concentrations in the Reykjanes brines than in seawater, whereas Ca^{2+}, K^+, and SiO_2 are

Table 5-10

The Chemical Composition of Geothermal Brines from the Reykjanes Peninsula, Iceland[a]

Locality	Temperature (°C)	Source of Data	Na+	K+	Ca²⁺	Mg²⁺	SO₄²⁻	S²⁻	ΣCO₂	Cl⁻	SiO₂	B	Sr²⁺	Ba²⁺	Fe	Mn
Reykjanes drillhole 7	10	(1)	3,150	200	66	"low"	276	—	—	4,170	3	—	—	—	—	—
Njardvikurheidi drillhole 1	48	(1)	9,170	359	3,776	24	1535	—	—	20,070	76	—	—	—	—	—
Reykjanes spring	99	(1)	14,325	1,670	2,260	123	206	0.2	5.0	29,100	544	12	—	—	0.2	—
Reykjanes spring 1918	99	(2)	13,550	1,931	2,223	50	—	—	—	27,600	612	12	12	7.1	0.3	4.5
Reykjanes small spring	99	(2)	13,425	1,950	2,320	18	—	—	—	27,660	—	—	12	10	—	5
Reykjanes drillhole 2	221	(1)	10,440	1,382	1,812	8	72	51	2,650	20,745	374	11.6	—	—	0.48	—
Svartsengi well 3	232	(2)	7,090	1,042	1,029	0.7	—	—	—	13,390	407	—	7.4	2.4	<0.24	0.48
Svartsengi well 3		(1)	—	—	—	—	30.8	—	174	—	—	—	—	—	—	—
Reykjanes drillhole 4	240	(1)	10,320	1,287	1,555	30	75	—	—	18,950	520	—	—	—	—	—
Reykjanes drillhole 8	277	(1)	9,610	1,348	1,530	16	30.8	45.3	1,926	19,260	636	—	—	—	—	—
Reykjanes drillhole 8	269	(2)	9,520	1,397	1,590	0.67	22.7	—	1,900	19,280	600	7.4	9.1	10.3	0.5	2.0

[a]Concentrations in mg/kg of solution. Sources of data (1) Björnsson et al. (1972); Arnórsson (1974). (2) Mottl (1976).

present in much larger concentrations. Na^+ seems to be slightly depleted in the brines. It could be argued that these differences between the composition of seawater and the Reykjanes brines speaks against a seawater origin for the brines. However, several recent sets of experiments have shown that seawater reacts rapidly with oceanic basalts, and assumes a composition nearly identical to that of the Reykjanes brines within a matter of months. Bischoff and Dickson (1975) have studied the reaction of seawater with basalt at 200°C and 500 bars during a 6-month period. Hajash (1975) has reacted basaltic glass with seawater between 200 and 500°C. Mottl, Corr, and Holland (1974); Mottl (1976); and Mottl and Holland (1978) have reported on somewhat more extensive experiments with glass and crystalline rocks in the same temperature range. The three sets of results agree on all important points. Figures 5-10 and 5-11 record the changes in the concentration of the major ions during the Bischoff–Dickson (1975) experiments. Mg^{2+} is removed rapidly and nearly quantitatively as a constituent of montmorillonite.

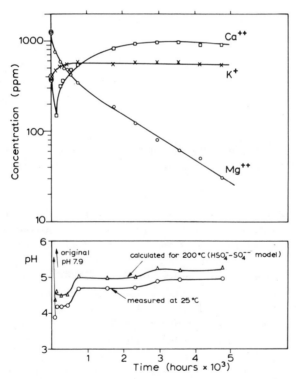

Figure 5-10. The pH and the concentration of Ca, Mg, and K in Copenhagen seawater during interaction with basalt powder (BCR-1) at 200°C, 500 bar; Bischoff and Dickson (1975).

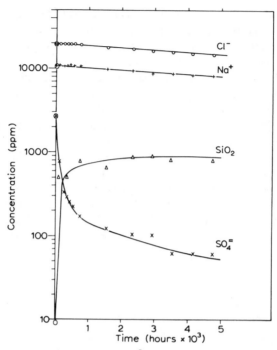

Figure 5-11. The concentration of SO_4^{2-}, dissolved SiO_2, Na^+, and Cl^- in Copenhagen seawater during interaction with basalt powder (BCR-1) at 200°C and 500 bar; Bischoff and Dickson (1975).

Mottl (1976) found that this was also true at temperatures up to 500°C, even though chlorite is almost certainly the stable phyllosilicate at such high temperatures.

In the Bischoff-Dickson experiments the concentration of K^+ increased rapidly to a steady state final value. Mottl and Holland (1978) have shown that a large fraction of the potassium in starting basalts was removed into the solutions, and that the increase in the K^+ concentration in any given experiment was determined largely by the potassium content of the basalt sample and by the ratio of seawater to basalt in the particular experiment.

The concentration of Ca^{2+} first decreased rapidly in both sets of experiments, then recovered, and finally reached values several times greater than the concentration of Ca^{2+} in seawater. The initial drop was due to the precipitation of anhydrite, which has a retrograde solubility. The recovery was due to the release of calcium from the calcium silicate minerals in the basalt. The SO_4^{2-} concentration decreased rapidly during all of the experiments. The initial decrease was due to the precipitation of anhydrite in heated seawater (Bischoff and Seyfried, 1977). Later, anhydrite was precipitated following Ca^{2+} release from calcium silicates. The sulfur content of average midocean ridge basalt is approximately equal to that of an equal weight of seawater. The total sulfur budget of submarine hydrothermal systems is therefore dominated by seawater SO_4^{2-} only when the seawater/basalt ratio is greater than 1. Mottl (1976) has discussed the redox state and redox balance in such systems in some detail, and has shown that their oxidation state is influenced strongly by the FeO content of basalts and by the seawater/rock ratio.

Changes in the Na^+ concentration of the hydrothermal solutions are rather slight compared to the quantity present. Although the data are still somewhat equivocal, it appears that up to about 10% of seawater Na^+ are transferred to the rock. This is particularly true at temperatures between 300 and 500°C. The SiO_2 concentration in the solutions generally approaches the solubility of quartz, although in some experiments at 200°C supersaturation was maintained with respect to quartz.

Figures 5-12 and 5-13 summarize Mottl's (1976) and Bischoff and Dickson's (1975) data for the Fe and Mn content of their solutions. At 200°C nonequilibrium seems to prevail for rather long periods of time. The equilibrium, or at least the steady state, concentration of both elements, as well as the Fe/Mn ratio increase rapidly with increasing temperature.

A comparison of the composition of the run products of the three sets of experiments with the composition of the Reykjanes brines shows that they are remarkably similar even though the mineralogy of the solid residues from the experiments is not identical with that of the altered basalts at Reykjanes. Apparently, the composition of seawater changes rapidly until only small differences remain between its composition and that of a solution in equilibrium with the appropriate equilibrium mineral assemblage. The driving force for mineral reactions is then so small that solution–mineral reactions are slow on a time scale of months.

The similarity between the composition of the Reykjanes brines and of seawater reacted with basalts between 200 and 500°C removes the objection raised earlier to a seawater parentage for the brines at Reykjanes. It also shows that the reaction of seawater with basalt at elevated temperatures can have a major effect on the chemistry of the oceans if the flux of seawater through marine hydrothermal systems is sufficiently large. The best evidence for the magnitude of the seawater flux through such systems is derived from the magnitude of heat flow anomalies in the vicinity of active spreading ridges. For some time now it has been known that the conductive heat flow measured near spreading ridges is much lower than the expected heat flow from cooling lithospheric plates. It seems likely that the heat flow deficit is mainly due to the removal of heat by hydrothermal solutions that exit from

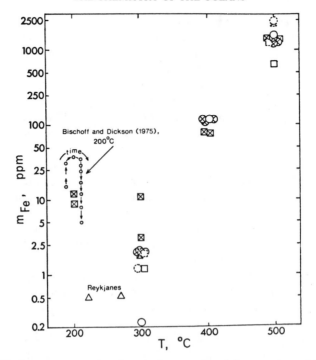

Figure 5-12. Iron concentration in solutions produced by reaction with basalt at Reykjanes (triangles) and in experiments (circles and squares) (Mottl, 1976). Dashed symbols are for experiments using a Na–K–Ca–Cl solution; all others used seawater. Small circles at 200°C are for Bischoff and Dickson's (1975) 6-month experiment.

the crust in hot springs along active ridges (Deffeyes, 1970; Spooner and Fyfe, 1973). Several estimates of the heat deficit have been made recently. Wolery and Sleep (1976) have proposed that the loss of heat by hydrothermal advection amounts to about 40×10^{18} cal/yr. Williams and Von Herzen (1974) estimate that the heat loss by this mechanism is $\geqslant 60 \times 10^{18}$ cal/yr. Lister's (1973) calculation translates into a heat loss of $(14–18) \times 10^{18}$ cal/yr. The differences between these figures are a fair measure of the uncertainty in the starting data.

The mean exit temperature of seawater from seafloor hydrothermal systems is not known. Preliminary calculations (Cathles and Fehn, personal communication, 1976) have shown that exit temperatures probably range from ca. 200 to 20°C. Wolery and Sleep have taken 50 and 250°C to be the most likely minimum and maximum values for the mean exit temperature and, therefore, obtain an annual flow rate of $(2–9) \times 10^{17}$ g/yr for the flow rate of seawater through midocean ridges. If Lister's (1973) values for the

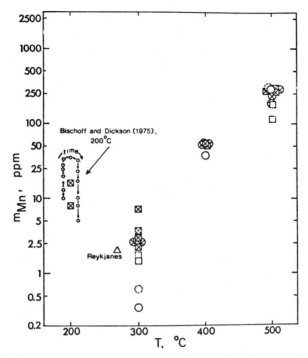

Figure 5-13. Manganese concentration in solutions produced by reaction with basalt at Reykjanes (triangle) and in experiments (circles and squares) (Mottl, 1976). Dashed symbols are for experiments using a Na–K–Ca–Cl solution; all others used seawater. Small circles at 200°C are for Bischoff and Dickson's (1975) 6-month experiment.

heat flow anomaly are used, the estimated flow rate is $(0.6–3.6) \times 10^{17}$ g/yr.

Fehn and Cathles (1978) arrive at a figure of ca. 1×10^{17} g/yr. These fluxes are probably sufficiently large to remove much, if not all, of the river flux of Mg^{2+}. If virtually all of the Mg^{2+} in a volume of seawater cycling through ocean floor basalt is removed, then the loss of Mg is ca. 1.3 mg/g of cycled seawater, and the total river flux of Mg is removed by cycling

$$\frac{1.9 \times 10^{14} \text{ g/yr}}{1.3 \times 10^{-3} \text{ g/g cycled seawater}} = 1.5 \times 10^{17} \text{ g cycled seawater/yr}$$

Such a flux is somewhat greater than Deffeyes' (1970) estimate, is within Lister's range, and is slightly lower than the minimum annual flow estimated by Wolery and Sleep. The interpretation of the heat flow data is still uncertain, but it is hard to avoid the conclusion that the cycling of seawater through midocean ridges is a major mechanism for the removal of the river flux of Mg from the oceans unless the interpretation of the heat flow data turns out to be wrong.

A rough, independent estimate of the seawater flux can be obtained by determining the rate of accumulation of the metalliferous sediments so commonly associated with spreading ridges. Unfortunately the data for the mean rate of accumulation of these sediments are rather poor. However, Mottl (1976) has shown that there is currently no discrepancy between the flow rates estimated in this manner and the rates based on heat flow calculations.

The annual cycling of such large quantities of seawater through midocean ridges can have a sizable effect on the cycle of the other major cations and anions in seawater. If we choose a cycling rate of 1.0×10^{17} g seawater/yr and assume that nearly all of the Mg in the cycled seawater is removed by reaction with submarine basalts, then we can calculate, at least roughly, the annual gain or loss of other constituents. The main uncertainties in such calculations involve the maximum temperature reached by the solutions and the seawater/rock ratios. The figures in Table 5-11 have been derived from Table 5-10; they, therefore, assume that the chemistry of the Icelandic brines is truly representative. Mg removal is essentially complete at all temperatures between 50 and 250°C. However, the change in the concentration of the other constituents depends on the maximum temperatures. This is particularly true for K^+. At low temperatures seawater cycling probably involves very little exchange of K^+ between basalts and seawater. At high temperatures the K^+ flux from basalts to seawater could be so large that it equals the river flux of K^+ if the seawater/basalt ratio in average submarine hydrothermal systems is as small as it is in the Reykjanes system. There are reasons to believe, however, that the mean seawater/basalt ratio in submarine geothermal systems is considerably greater than at Reykjanes. The best current estimate for the creation of new ocean floor is ca. 3×10^{10} cm^2/yr (Deffeyes, 1970). If seawater cycling involves the entire upper 5 km of ocean crust, then the overall seawater/rock ratio is ca.

$$\frac{1.0 \times 10^{17} \text{ g/yr}}{3 \times 10^{10} \text{ cm}^2/\text{yr} \times 5 \times 10^5 \text{ cm} \times 3.0 \text{ g/cc}} \approx 2$$

However, it seems likely that only a rather small fraction of the oceanic crust reacts with seawater, since basalt in the few Deep Sea Drilling Project holes that have penetrated the basement by more than 100 m is largely unaltered. If 20% of the upper 5 km were altered, the effective seawater/rock ratio would be 10; in that case the increase in the K^+ concentration of cycled

Table 5-11

The Flux of Dissolved Substances Between the Oceans and Midocean Ridges Due to Seawater Cycling at a Rate of 1×10^{17} g/yr

	Approximate Mean Change in Concentration (mg/g seawater)		Total Annual Flux		Percentage of Annual River Flux	
	Max T = 50°C	Max T = 250°C	Max T = 50°C	Max T = 250°C	Max T = 50°C	Max T = 250°C
Mg^{2+}	-1.2	-1.3	-1.2×10^{14}	-1.3×10^{14}	-63%	-68%
Ca^{2+}	+3.4	+1.2	$+3.4 \times 10^{14}$	$+1.2 \times 10^{14}$	+48%	+17%
Na^+	-1.4	-1.0	-1.4×10^{14}	-1.0×10^{14}	-23%	-16%
K^+	~0	+0.9	0	$+0.9 \times 10^{14}$	~0	+100%
SO_4^{2-}	-1.1	-2.5	-1.1×10^{14}	-2.5×10^{14}	-24%	-50%
SiO_2	+0.1	+0.6	$+0.1 \times 10^{14}$	$+0.6 \times 10^{14}$	+26%	+15%

seawater would only be about 0.2 mg/g seawater, and the K^+ flux to the oceans with cycled seawater would be about 20% of the river flux of this element.

The apparently rather minor amount of Na removal in Table 5-11 is somewhat misleading. The critical figure is really not the percentage of the total flux of river sodium removed, but the percentage removal of the river flux of the sodium derived from silicate weathering. Since only about one-third of the total river flux has this origin (see Chap. IV), the removal of sodium during seawater cycling through midocean ridges is potentially of major importance.

The removal of seawater sulfate may also be important. At 50°C the loss of SO_4^{2-} due to the precipitation of gypsum or anhydrite is minor. At 250°C sulfate removal is much more complete, because the solubility of anhydrite decreases rapidly with increasing temperature; anhydrite precipitation and sulfate reduction both probably play a major part in sulfate removal. However, the apparent absence of anhydrite from greenschist-facies rocks from the ocean floor is rather curious. Sulfate reduction followed by sulfide precipitation may be the dominant sulfate-removal mechanism in submarine hydrothermal systems, but the modest increases in the Fe_2O_3/FeO ratio that accompany greenschist formation rule out large amounts of SO_4^{2-} reduction.

The silica budget of the oceans is apparently affected only slightly by seawater cycling through midocean ridges. The maximum effect is probably less than the uncertainty in the estimated river flux of dissolved silica, and has no influence on the arguments advanced in the previous section of this chapter.

The most easily detectable chemical effect of seawater cycling through midocean ridges is the removal of Mg from seawater, and it seems rather likely that Mg removal in this fashion is the solution to the present-day magnesium problem (Drever, 1974). The difference between the Li content of sediments (50 ± 5 ppm) and the Li content of average igneous rocks (30 ± 5 ppm) is probably due to seawater cycling through midocean ridges as well (Holland et al., 1976; Styrt, 1977). Before the Jurassic period much more Mg seems to have been removed from seawater as a constituent of dolomite than is being removed today. The greatly diminished importance of dolomite in the geologic record since the end of Triassic time may, therefore, be due to an increase in the cycling of seawater through ridges, the exchange of Mg for Ca during this process, and consequently an increase in the rate of limestone formation approximately equal to the decrease in the rate of dolomite formation. The implication of this hypothesis for the chemical evolution of seawater is discussed in the next section. An examination of the Phanerozoic sedimentary record bears out the reasonableness of the hypothesis.

THE FORMATION OF EVAPORITES

Marine evaporites form in basins with restricted access to the oceans when loss of water by evaporation exceeds rainfall and the influx of river water. Evaporite basins are commonly found in areas of extensive carbonate precipitation. Calcite and dolomite frequently precede the evaporite minerals, and have often been lumped together with the true components of marine evaporites. In the laboratory the evaporation of seawater first yields a minor precipitate of calcite and aragonite. After the quantity of water has been reduced by a factor of 4 to 5, gypsum or anhydrite precipitates, and these are joined by halite when the volume of water has been reduced by a factor of 10.5 (Holser and Kaplan, 1966). When the volume of the solution is less than 5% of the original volume, polyhalite ($K_2Ca_2Mg(SO_4)_4 \cdot 2H_2O$), or at higher temperatures glauberite, $Na_2Ca(SO_4)_2$, take the place of previously precipitated anhydrite or gypsum. Polyhalite separates together with halite until the solution is saturated with magnesium-bearing sulfates free of calcium and potassium (Stewart, 1963). On continued evaporation a complicated series of potassium and magnesium chlorides and sulfates precipitate. Their precipitation sequence depends rather critically on temperature and on the degree of equilibrium maintained between the solution and the precipitated phases. The large body of experimental data bearing on the formation of these sequences has been summarized by Lotze (1957), Braitsch (1962), Stewart (1963), Borchert and Muir (1964), and Borchert (1965). Table 5-12 contains a list of minerals found in marine evaporites. In most marine evaporites, langbeinite, $K_2Mg_2(SO_4)_3$, kainite, $KMg(SO_4)Cl \cdot 3H_2O$, and kieserite, $MgSO_4 \cdot H_2O$, are the only common sulfates of the late precipitates. Sylvite, KCl, and carnallite, $KMgCl_3 \cdot 6H_2O$, are the only chlorides other than halite, NaCl, that occur in significant quantities in marine evaporite deposits. Carnallite is in part a primary, in part a replacement mineral. Sylvite may be of primary origin but is largely derived from earlier carnallite. Kinsman (1966, 1974a) has shown that the sequence of minerals observed in the recent evaporites of the Persian Gulf follows quite closely the sequence predicted on the basis of laboratory experiments. Along the Trucial Coast marine sediments consisting largely of aragonite form the extensive salt flats (sabkhas) shown in Figure 5-14. Gulf water enters these salt flats laterally and during infrequent periods of flooding. As evaporation normally exceeds precipitation, the salinity of the interstitial solutions in the sediments increases away from the coast line. The first mineral to precipitate is usually either gypsum, anhydrite, or celestite, $SrSO_4$. Table 5-13 shows that the sulfate content of the interstitial waters decreases across the entire sabkha; this indicates that the precipitation of sulfate minerals continues throughout the entire extent of evaporation in the area.

Table 5-12

Minerals of Marine Evaporties[a]

Chlorides:

Halite	NaCl
Sylvite	KCl
Bischofite	$MgCl_2 \cdot 6H_2O$
Koenenite	$Mg_9Al_4Cl_8(OH)_{22} \cdot 7H_2O$
Zirklerite	Basic chloride of Al and Fe^{2+}, with minor Ca and Mg
Chlorocalcite (= hydrophilite).	$KCaCl_3$
Carnallite	$KMgCl_3 \cdot 6H_2O$
Tachyhydrite	$CaMg_2Cl_6 \cdot 12H_2O$
Douglasite	$K_2FeCl_4 \cdot 2H_2O$?
Erythrosiderite	$K_2FeCl_5 \cdot H_2O$
Rinneite	NaK_3FeCl_6

Fluorides:

Fluorite	CaF_2
Sellaite	MgF_2

Sulfates:

Aphthitalite (glaserite)	$(K,Na)_3Na(SO_4)_2$
Thenardite	Na_2SO_4
Barite	$BaSO_4$
Celestite	$SrSO_4$

Carbonates:

Calcite	$CaCO_3$
Magnesite	$MgCO_3$
Siderite	$FeCO_3$
Aragonite	$CaCO_3$
Strontianite	$SrCO_3$
Dolomite	$CaMg(CO_3)_2$
Ankerite	$Ca(Fe, Mg)CO_3)_2$

Borates:

Pinnoite	$Mg(BO_2)_2 \cdot 3H_2O$
Kurgantaite	$(Sr,Ca)_2B_4O_8 \cdot H_2O$
Priceite (Pandermite)	$Ca_4B_{10}O_{19} \cdot 7H_2O$
Ulexite	$NaCaB_5O_9 \cdot 8H_2O$
p-Veatchite	$SrB_6O_{10} \cdot 2H_2O$
Colemanite	$Ca_2B_6O_{11} \cdot 5H_2O$
Hydroboracite	$CaMgB_6O_{11} \cdot 6H_2O$
Inderborite	$CaMgB_6O_{11} \cdot 11H_2O$
Inyoite	$Ca_2B_6O_{11} \cdot 13H_2O$
Kurnakovite	$Mg_2B_6O_{11} \cdot 15H_2O$
Inderite	$Mg_2B_6O_{11} \cdot 15H_2O$

Anhydrite $CaSO_4$
Vanthoffite $Na_6Mg(SO_4)_4$
Glauberite $Na_2Ca(SO_4)_2$
Langbeinite $K_2Mg_2(SO_4)_3$
Mirabilite $Na_2SO_4 \cdot 10H_2O$
Syngenite $K_2Ca(SO_4)_2 \cdot H_2O$
Loeweite $Na_4Mg_2(SO_4)_4 \cdot 5H_2O$
Blödite (Astrakanite) $Na_2Mg(SO_4)_2 \cdot 4H_2O$
Leonite $K_2Mg(SO_4)_2 \cdot 4H_2O$
Picromerite (Schoenite) $K_2Mg(SO_4)_2 \cdot 6H_2O$
Polyhalite $K_2Ca_2Mg(SO_4)_4 \cdot 2H_2O$
Görgeyite $K_2Ca_5(SO_4)_6 \cdot H_2O$
Bassanite $2CaSO_4 \cdot H_2O$
Kieserite $MgSO_4 \cdot H_2O$
Sanderite $MgSO_4 \cdot 2H_2O$
Gypsum $CaSO_4 \cdot 2H_2O$
Starkeyite (Leonhardtite) $MgSO_4 \cdot 4H_2O$
Pentahydrite (Allenite) $MgSO_4 \cdot 5H_2O$
Hexahydrite $MgSO_4 \cdot 6H_2O$
Epsomite (Reichardtite) $MgSO_4 \cdot 7H_2O$
Kainite $KMg(SO_4)Cl \cdot 3H_2O$
Anhydrokainite $KMg(SO_4)Cl$
D'Ansite $MgNa_{21}(Cl_3SO_4)(SO_4)_9$

Howlite $Ca_2SiB_5O_9(OH)_5$
Paternoite $MgB_8O_{13} \cdot 4H_2O$
Ginorite $Ca_2B_{14}O_{23} \cdot 8H_2O$
  (Cryptomorphite).
Kaliborite $KMg_2B_{11}O_{19} \cdot 9H_2O$
Volkovite Hydrous borate of Sr and K
Ivanovite Hydrous chloroborate of Ca
  (and K?)
Szaibelvite (Ascharite) $(Mg)(BO_2)(OH)$
Boracite $Mg_3B_7O_{13}Cl$
Ericaite $(Fe,Mg,Mn)_3B_7O_{13}Cl$
Hilgardite $Ca_8(B_6O_{11})_3Cl_4 \cdot 4H_2O$
Parahilgardite $Ca_8(B_6O_{11})_3Cl_4 \cdot 4H_2O$
Strontiohilgardite $(Ca,Sr)_2[B_5O_8(OH)_2Cl]$
Heidornite $Na_2Ca_3Cl(SO_4)_2 \cdot B_2O_3(OH)_2$
Lueneburgite $Mg_3B_2(OH)_6(PO_4)_2$
Sulphoborite $Mg_6H_4(BO_3)_4(SO_4)_2 \cdot 7H_2O$
Danburite $CaSi_2B_2O_8$

Elements, sulfides, oxides, silicates, phosphates:

Sulfur, pyrite, hauerite, hematite, goethite (limonite), magnetite, quartz, opal, talc, illite, kaolinite, goyazite.

[a] Stewart (1963).

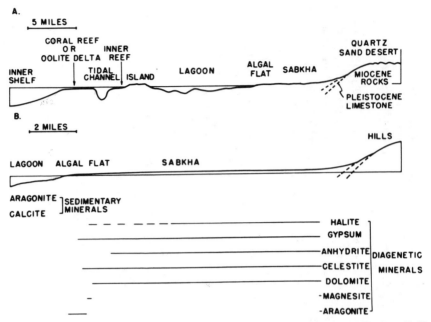

Figure 5-14. Topography and mineralogy along the Trucial Coast, Persian Gulf (Kinsman, 1966). (A) Diagrammatic profile across the inner shelf, lagoon barrier, lagoon, and supratidal sediment surface (sabkha). (B) Sabkha profile indicating approximate relative positions at which the early diagenetic minerals developed. Printed by Permission of the Northern Ohio Geological Society.

Table 5-13

Analyses of Interstitial Ground Waters from the Intertidal Algal Flats and from the Sabkha Environment, Trucial Coast, Persian Gulf[a]

	Algal Flat Ground Waters			*Sabkha Ground Waters*				
Water table depth (cm)	17.8	15.2	30.5	45.7	81.3	76.2	76.2	61.0
Field temperature (°C)	26.7	—	27.5	—	28.4	31.4	30.5	31.2
Field pH	—	7.45	7.10	7.55	6.20	6.10	—	—
Density at 22°C (g/cc)	1.085	1.087	1.125	1.128	1.183	1.192	1.203	1.211
Cl^-	63.8	63.9	90.9	96.3	134.5	151.6	159.1	165.9
SO_4^{2-}	7.66	8.53	11.68	8.57	2.51	1.98	0.44	0.39
Ca^{2+}	0.92	1.12	0.93	1.22	2.48	2.56	9.00	10.43
Mg^{2+}	4.07	4.34	6.49	6.00	5.53	6.81	9.51	15.53
Sr^{2+}	0.02	0.02	0.02	0.03	0.05	0.06	0.21	0.24
K^+	1.24	1.18	1.64	1.80	2.74	3.84	3.13	4.15
Na^+	35.9	35.5	50.3	53.3	75.1	81.4	75.1	65.1

[a]Kinsman (1966); concentrations in units of g/kg ground water.

204

In seawater the molar concentration of calcium is less than that of sulfate. Virtually calcium-free brines are, therefore, produced during the precipitation of gypsum or anhydrite from seawater unless additional calcium is supplied from an external source. The dolomitization of $CaCO_3$, which was discussed earlier in this chapter, can produce such a supply. It can be shown that the reaction

$$2CaCO_3 + Mg^{2+} + SO_4^{2-} + 2H_2O \rightarrow CaMg(CO_3)_2 + CaSO_4 \cdot 2H_2O \quad (5\text{-}7)$$

Calcite   Dolomite   Gypsum
Aragonite

explains the variation of the magnesium, calcium, and sulfate concentration in the interstitial sabkha waters. The same mechanism can explain the deficiency of magnesium and sulfate observed in the late precipitates of many ancient marine evaporite deposits (see, for instance, Braitsch, 1963). Borchert and Muir (1964), however, believe that bacterial sulfate reduction is more important than dolomitization in reducing the sulfate content of late-stage evaporite assemblages, and it is likely that some of the Mg^{2+} and SO_4^{2-} deficits are due to the loss of late-stage brines during periods of mixing with seawater.

The question of the relative stability of gypsum and anhydrite as a function of temperature and salinity seems finally to have been settled by Hardie (1967). Figure 5-15 shows that in pure water the two minerals are in equilibrium at $58 \pm 2°C$. In seawater saturated with halite, gypsum and anhydrite are in equilibrium at $18°C$. These data differ quite markedly from the previously accepted equilibrium relations, but agree very well with the temperature and salinity at which gypsum gives way to anhydrite in the sabkha sediments of the Trucial Coast.

Halite was the only chloride observed in the sabkha sediments. Saturation with carnallite or sylvite was apparently not reached. Polyhalite was not found. Kinsman (1966) suggests that the lack of polyhalite could be due to the removal of magnesium by dolomitization and of sulfate by the subsequent precipitation of gypsum or anhydrite. The degree of concentration of brines in this area may, however, simply not have been sufficiently great, perhaps because the relative humidity on the sabkha is generally too high (Kinsman, 1976). Holser (1966) has discovered recent polyhalite at the Laguna Ojo de Liebre in Baja California, an area of noncarbonate sediments that is climatically similar to the Trucial Coast. It seems likely that the inability of solutions in this area to lose magnesium and sulfate via reaction 5-7 contributes to the appearance of polyhalite as a replacement of gypsum, but the very much greater concentration of these brines (60 × seawater) may be a more important difference between the mineralogy of the evaporites in this area and those of the Trucial Coast.

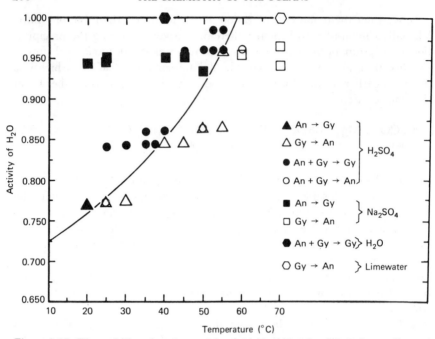

Figure 5-15. The stability of gypsum and anhydrite determined experimentally as a function of temperature and activity of H_2O at atmospheric pressure. Only runs in which a conversion was achieved are plotted, Hardie (1967).

Marine evaporites have been found in sedimentary rocks of all geologic periods from the recent to the late Precambrian (see, for instance, Kozary et al., 1968). They typically consist of a basal carbonate facies followed by a gypsum (or anhydrite) facies and a gypsum (or anhydrite)-halite facies. In some cases these facies are followed by a polyhalite-halite or by a glauberite-halite facies and by an economically important Mg–K–SO$_4$–Cl facies. Major differences between marine evaporites generally involve the magnesium and sulfate content of the later facies (Braitsch, 1963). Large evaporites normally contain more than one sequence of evaporite formation. The stratigraphy of the German Zechstein evaporites has been studied very thoroughly, and the complex succession of mineralogic zones in these deposits is well demonstrated in Figure 5-16. Periods of ready access of seawater must have alternated with periods of extensive evaporation. The relative proportion of the minerals in the Zechstein successions is probably not a measure of the relative proportion of the constituent ions in Permian seawater. As shown in Figure 5-17, carbonates and sulfates are typically overrepresented, whereas chlorides are typically underrepresented in

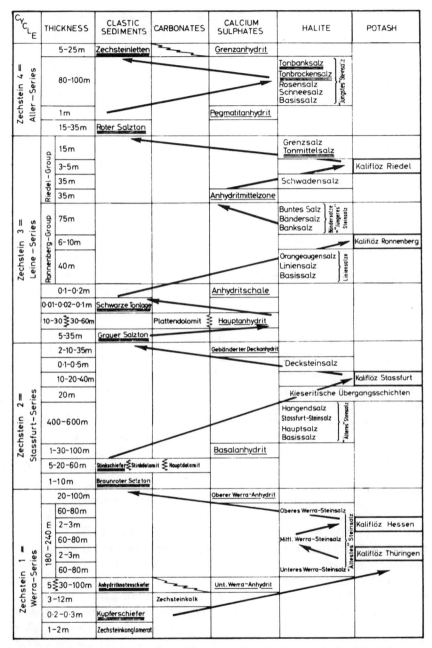

Figure 5-16. The standard Zechstein succession showing variations in brine concentration during the formation of successive cycles, after Richter-Bernburg (1955). Diagram from Borchert and Muir (1964).

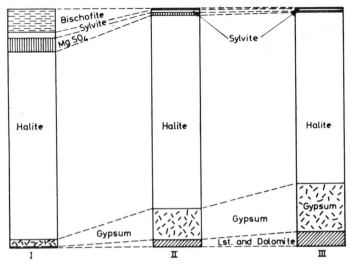

Figure 5-17. Comparative precipitation profiles from I, the experimental evaporation of seawater; II, Zechstein evaporites; and III, the average of numerous other marine salt deposits (Borchert and Muir, 1964).

evaporites, possibly because downward reflux of dense evaporite brines often prevents precipitation of the more soluble evaporite minerals.

The sequence of mineral precipitation in marine evaporites has been essentially invariant. This does not mean that the composition of seawater has remained completely constant since the formation of the oldest preserved marine evaporites in late Precambrian time, but the constancy of the precipitation sequence does set fairly severe limits on possible excursions in the composition of seawater during the last 600 million yr (Holland, 1972, 1974). Changes by more than a factor of 3 in the concentration of any of the major ions in seawater during this period are distinctly unlikely.

Evaporite basins can form both on single lithospheric plates and between lithospheric plates (Kinsman, 1974b, 1975a, 1975b; Burke, 1975). Evaporites on single lithospheric plates tend to be areally extensive, but their thickness rarely exceeds 1 km. Cratonic basins containing such evaporites are typically long lived; they frequently develop during a period of several hundred million yr and subside rather irregularly at a mean rate averaging about 10 mm/1000 yr. Evaporites formed between lithospheric plates can develop during periods of plate convergence as well as during periods of plate divergence. The Cenozoic evaporites of the Mediterranean can be considered to have been produced during a restriction of seawater access accompanying plate convergence. It is likely, however, that such evaporites are less widespread than those developed in juvenile, actively spreading oceans following the rupture of a continental plate. Such evaporites typically occur in linear belts along continental margins or within basins where an earlier phase of continental rifting subsequently aborted. Evaporite masses formed in protooceans are commonly 500–4000 km long, 100–600 km wide, and 1–7 km thick, and are formed in periods on the order of 10–20 million yr (Kinsman, 1975b). Until recently it was thought that marine evaporites in cratonic basins were quantitatively dominant. The discovery of intraplate evaporites has cast doubt on this conclusion, and is forcing a considerable upward revision of the inventory of marine evaporites. Braitsch's opinion (1962, p. 156) that the salt content of the oceans is very large compared to that of marine evaporites is almost certainly wrong.

Marine evaporites can extract dissolved salts from seawater at an alarmingly rapid rate. Consider a shallow basin that annually receives 1 m of seawater, which subsequently evaporates completely to dryness. Although an evaporation rate of 1 m/yr is about average, complete evaporation is uncommon; to this extent the basin is not entirely typical. Due to evaporation in such a basin the oceans would annually lose 0.27 g of SO_4^{2-} and 1.9 g of Cl^- per cm^2 of basin area. The area of such a basin or the combined area of several such basins that would be required to remove the SO_4^{2-} and Cl^- brought to the oceans annually by rivers and derived from the dissolution of earlier evaporites (see Table 4-13) is approximately

$$A_{SO_4^{2-}} \approx \frac{0.07 \text{ meq/kg} \times 48 \times 10^{-3} \text{ g/meq} \times 4.6 \times 10^{16} \text{ kg/yr}}{0.27 \text{ g/cm}^2 \text{ yr}}$$

$$\approx 6 \times 10^{14} \text{ cm}^2$$

$$\approx 6 \times 10^4 \text{ km}^2$$

and

$$A_{Cl^-} \approx \frac{0.16 \text{ meq/kg} \times 35 \times 10^{-3} \text{ g/meq} \times 4.6 \times 10^{16} \text{ kg/yr}}{1.9 \text{ g/cm}^2 \text{ yr}}$$

$$\approx 1.3 \times 10^{14} \text{ cm}^2$$

$$\approx 1.3 \times 10^4 \text{ km}^2$$

The difference between the value of $A_{SO_4^{2-}}$ and A_{Cl^-} is due in part to the greatly oversimplified model for evaporite basins, and in part to the uncertainties in the figures in Table 4-13. Both figures show that only a small fraction of the oceans has to behave like the proposed evaporite basin in order to remove the entire river input of SO_4^{2-} and Cl^-. $A_{SO_4^{2-}}$ is approximately 0.02%, A_{Cl^-} approximately 0.004%. of the total area of the oceans.

The area of evaporite basins active during Phanerozoic time is not well known; Kozary et al., (1968) estimate that approximately one-fourth of the area of the continents is underlain by evaporites, and that about 60% of the areas underlain by evaporites contain chloride salts. No oceanic areas were included by these authors; thus at least 400×10^{15} cm^2 of the earth's surface are underlain by evaporites. In some areas, such as the Persian Gulf, salt deposition has recurred several times since late Precambrian time. Two million years is, therefore, probably a reasonable minimum value for the mean period of evaporite formation in any given area. Since virtually all known evaporites are Phanerozoic, the probable minimum area of evaporite basins during the last 600 million yr is roughly

$$A \geqslant \frac{2 \text{ million yr}}{600 \text{ million yr}} \times 400 \times 10^{15} \text{ cm}^2$$

$$\geqslant 13 \times 10^{14} \text{ cm}^2$$

The rough agreement between A, $A_{SO_4^{2-}}$ and A_{Cl^-} is reassuring.

The actual area of evaporite formation has probably varied considerably during geologic history. The Permian period, for instance, seems to have been a time of particularly intense evaporite formation. The response of the oceans to changes in the area of evaporite formation is simple in principle, but undoubtedly complex in reality. If the area of evaporite formation were suddenly to increase, the rate of SO_4^{2-} and Cl^- removal would probably exceed the rate of supply. The concentration of SO_4^{2-} and Cl^- in seawater would, therefore, decrease until the precipitation of sulfates and chlorides in the enlarged evaporite area again balances the rate of river input. Conversely, a decrease in the area of evaporite formation would lead to an increase in the SO_4^{2-} and Cl^- content of seawater until dynamic equilibrium is reestablished. The sedimentologic consequences of an increase in the area of evaporites, that is, in the intensity of evaporite formation, are not simply the inverse of the consequences of a decrease in the intensity of evaporite formation. A drastic decrease in the area of evaporite deposition could ultimately lead to saturation of the entire oceans with respect to gypsum and/or anhydrite unless sulfate reduction were increased considerably. Gypsum would then tend to appear not only in evaporitic settings but also in cold water sediments, since its solubility decreases with decreasing temperature below 30°C. This has not been observed, and it seems likely that the oceans have always responded to a decrease in the area of intense evaporite formation by the precipitation of gypsum and/or anhydrite in areas of less intense evaporation and by an increase in the rate of pyrite formation (see text following).

The removal of NaCl has probably never been a problem; seawater is now so far undersaturated with respect to halite that even the dissolution of the entire inventory of halite in evaporite deposits would probably not bring the NaCl concentration of seawater up to halite saturation. A rather simple calculation supports this contention. Virtually no pre-Phanerozoic evaporites are preserved. The mean half-life of evaporites with respect to dissolution is, therefore, about $(1-2) \times 10^8$ yr (Garrels and Mackenzie, 1971, p. 272). The corresponding decay constant, $k = 0.7/\tau$, is $(7-3) \times 10^{-9}$ yr^{-1}. The present rate of NaCl solution and river transport (see Table 4-13) is ca. 4×10^{14} g/yr. If the rate of solution is related to the total quantity, N, of halides in evaporites by the expression $dN/dt \cong -kN$ then

$$N \approx \frac{4 \times 10^{14} \text{ g/yr}}{(7-3) \times 10^{-9} \text{ yr}^{-1}} = (6-12) \times 10^{22} \text{ g}$$

The total quantity of NaCl in the oceans is 3×10^{22} g. The quantity of halite in evaporites may, therefore, be as much as two to four times the NaCl content of the oceans.

ORGANIC CARBON, NITROGEN, PHOSPHORUS AND SULFIDES IN MARINE SEDIMENTS

The Primary Productivity of the Oceans

The abundance and distribution of elemental carbon, nitrogen, phosphorus, and sulfides in marine sediments are closely related to the photosynthetic fixation of carbon in near-surface waters, the oxidative and fermentative destruction of organic matter, and the burial of organic matter at sea. Fogg (1975) has summarized the several methods commonly used to measure the rate of primary productivity in the oceans. Thousands of determinations of primary productivity have shown that the rate of photosynthesis varies in time and space by several orders of magnitude. The most important factors controlling primary productivity at sea are the availability of nutrients and the availability of light. Light is probably a limiting factor only in the polar regions. In all other areas nutrients limit the local photosynthetic rate; among these nitrate and phosphate are almost certainly the most important

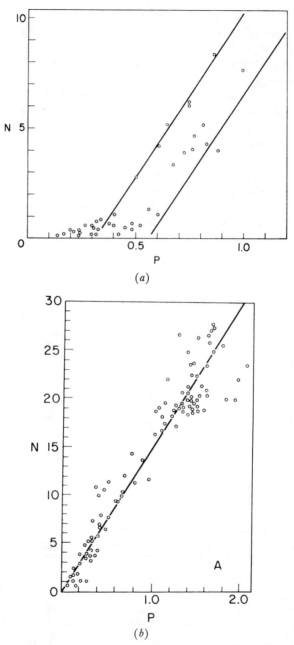

Figure 5-18. Correlation between phosphate phosphorus, and nitrate nitrogen (mg atoms/m^3) in (*a*) coastal waters south of Long Island and (*b*) waters of the western Atlantic (Redfield, Ketchum, and Richards, 1963). (*c*) Correlation between (NO_3^- + NO_2^-) and (PO_4^{3-}–P) in waters from the Costa Rica Dome (Strickland, 1965).

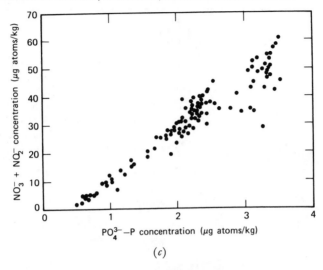

(c)

Figure 5-18. (*Continued*)

(Spencer, 1975). Silicon is frequently exhausted in surface waters during the growth of siliceous phytoplankton, but the element is probably not important for the growth of nonsiliceous marine algae. Various other elements and several classes of organic compounds have been proposed as limiting nutrients in the oceans, but although there are probably regions where neither nitrate nor phosphate is the limiting nutrient, their areal extent appears small. In most parts of the oceans nitrate and phosphate disappear virtually simultaneously during the height of photosynthetic activity (see Fig. 5-18); such behavior suggests that the availability of one or both of these nutrients limits the growth of phytoplankton; were this not so, photosynthesis would be observed to cease while both nitrate and phosphate are still present in more than trace amounts.

Nitrate and phosphate are both supplied to the oceans mainly by rivers. However, the intensity of primary productivity is largely independent of the geographic distribution of the river input. As shown in Figure 5-19, primary productivity in the open ocean is heavily concentrated in areas of upwelling. On the continental shelves and slopes primary productivity is enhanced by rapid vertical mixing that tends to speed the recycling of nutrients.

Numerous estimates of the mean level of primary production in the oceans have been published (see Fogg, 1975). These have tended to converge gradually on values between 85 g of carbon/m^2 yr (Platt and Subba Rao, 1973) and 120 g of carbon/m^2 yr (Bruyevich and Ivanenkov, 1971). The total net rate of oceanic carbon fixation is apparently close to 35×10^{15} g of C/yr, the value proposed by Koblentz-Mishke, et al. (1970).

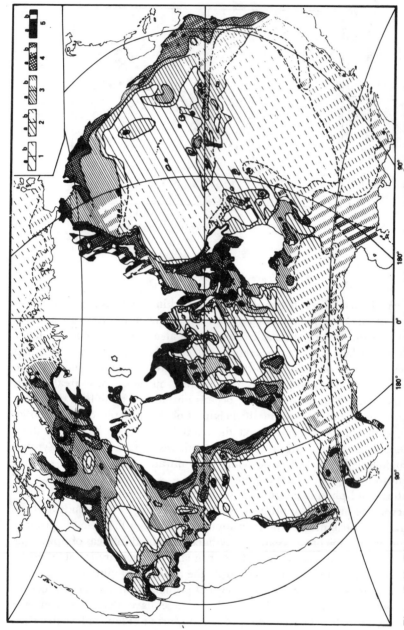

Figure 5-19. Distribution of plankton primary production in the world ocean (mg of C/m^2 day). (1) Less than 100, (2) 100 to 150, (3) 150 to 250, (4) 250 to 500, (5) more than 500. Koblentz-Mishke et al. (1970). Reproduced with the permission of the National Academy of Sciences.

The Rate of Burial of Organic Carbon

All but a very small fraction of the organic carbon produced in the oceans is recycled rather than buried. The total rate of sediment input from rivers to the oceans is approximately 2×10^{16} g/yr (see Chap. IV), that is, roughly half of the net rate of carbon fixation in the oceans. Nevertheless the reduced carbon content of marine sediments is quite small. The concentration of organic carbon in near-shore marine sediments is frequently more than 0.3% and less than 3% of the weight of dry sediment. Bordovskiy (1965b) has summarized a good deal of the Russian data, and has shown that an organic carbon content of 0.3 to 3% is typical of recent sediments in the Caspian Sea, the western half of the Bering Sea, and in sediments of the northern Pacific. Emery (1960) has shown that organic carbon is present in similar amounts in recent sediments off Southern California, although the concentration of organic carbon in some basin sediments exceeds 3%. Gross et al. (1967) record organic carbon concentrations of up to 2.5% in continental shelf sediments off the northwestern part of the United States. On the other hand, Arrhenius (1963) has recorded organic carbon concentrations of between 0.1 and 1.0% in deep-sea sediments from the eastern Pacific, and Chester (1965) has proposed a mean value of 0.25% for the organic carbon content of pelagic sediments. Gehman's (1962) summary of data for the content of total organic matter (organic $C \times 1.22$) of lime muds from the Gulf of Batabano, Florida Bay, and the Persian Gulf, and of clay sediments from the Orinoco Delta, the Cariaco Trench, the Gulf of Mexico, and Lake Maracaibo is shown in Figure 5-20. The range and mean of the concentration of organic matter in these modern clay sediments is very similar to that of shales, whereas modern lime muds appear to contain a larger complement of organic matter than their lithified counterparts. Van Andel (1964) has summarized data pertaining to the organic carbon content of a number of other areas. These data do not permit a precise calculation of the mean average organic carbon content of modern sediments, but they suggest strongly that the mean value is roughly 0.6%, a figure that is surprisingly close to Trask and Patnode's (1942, p. 25) estimate of 0.8 to 1.1% for the average quantity of organic carbon in ancient sediments of the North American continent. Ronov (1958) found an average organic carbon content of 0.40% for the sedimentary rocks of the Russian Platform. He ascribes the difference between his data and those of Trask and Patnode (1942) to a preponderance of samples from petroliferous provinces in the North American average, and points out that the carbon content of an equivalent selection of Russian samples is 0.85%. The quantity of organic carbon that is buried annually as a constituent of marine sediments is clearly very small compared to the annual rate of carbon fixation.

This fact is demanded in part by limitations on the supply of phosphorus to the oceans. Garrels et al, (1975, p. 109) and Lerman, et al. (1975) propose

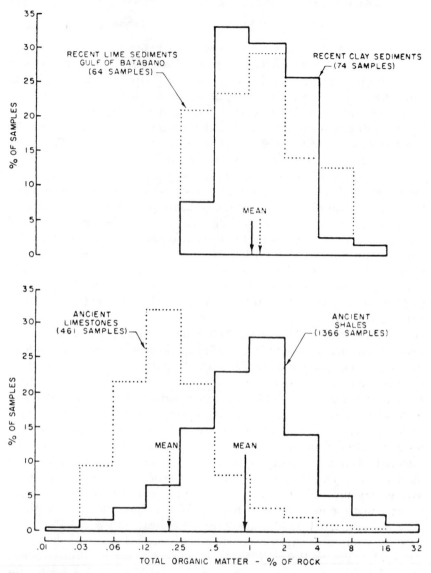

Figure 5-20. The total organic content of recent and ancient limestones and shales (Gehman, 1962).

216

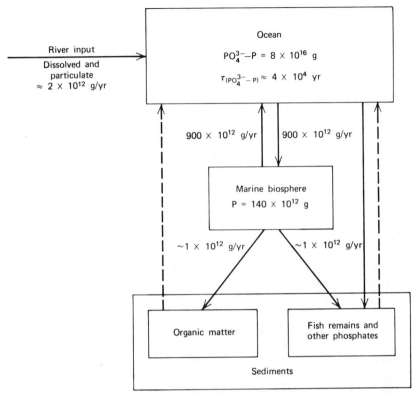

Figure 5-21. The marine geochemistry of phosphorous; river input after Kazakov (1950), Silker (1964), and Williams (1968); remaining figures based largely on Armstrong (1965). See also Garrels et al., (1975, p. 105).

that 0.06×10^{12} mol (1.8×10^{12} g) of dissolved phosphorus are added to the ocean annually as a component of river water. Kazakov (1950) has proposed an annual river influx of 4.4×10^{11} g of phosphorus (see Figure 5-21). If the data of Silker (1964) for the Columbia River are representative of average river water, the annual phosphorus input is 19×10^{-9} g/cc $\times 4.6 \times 10^{19}$ cc/yr $= 9 \times 10^{11}$ g/yr. Williams (1968) has studied the dissolved and organic particulate phosphorus content at six stations along the length of the Amazon River. At each station particulate organic phosphorus was present in larger concentrations than was dissolved phosphorus. Total phosphorus ranged from 6.5 to 94 μg/kg. If we take the mean of 50 μg/kg to be representative of world river water, the annual input of dissolved phosphorus is about 23×10^{11} g. Many additional phosphate determinations are obviously needed before much beyond an order of magnitude precision can be claimed for our knowledge of phosphorus inputs to the oceans. It would also be helpful to know how much phosphorus is carried to the oceans adsorbed on sediment grains, how much of this phosphorus becomes available for participation in the marine biological cycle, how much of the river phosphorus is recycled by rain, and whether detrital phosphate minerals are dissolved to a significant extent on the ocean floor. Nevertheless, the available data are sufficiently consistent to warrant their use, in a cautious manner.

The ratio of carbon to phosphorus atoms in phytoplankton is variable but is frequently close to $106:1$. The world input of phosphorus computed on the basis of Williams' (1968) data for the Amazon River would be sufficient to account for the annual removal of 23×10^{11} g/yr $\times (106$ mol of C/mol of P$) \times (12$ g of C mol$^{-1}/31$ g of P mol$^{-1}) = 1 \times 10^{14}$ gm C/yr as a constituent of organic matter with a C/P ratio typical of phytoplankton. This quantity of carbon represents about 0.3% of the annual net rate of carbon fixation, and is surprisingly close to the present estimated elemental carbon burial rate of 1.2×10^{14} gm/yr; just how close both numbers are to the true figure is somewhat difficult to say with certainty.

The C:N:P ratio of organic matter in marine sediments is usually not the same as that of average living plankton (see, for instance, Bader, 1955; Emery, 1960; Emery and Rittenberg, 1952; and Gordon, 1971). The several constituents of phytoplankton apparently have rather different degrees of resistance toward degradation and oxidation (Duursma, 1965); nitrogen and phosphorus are generally released preferentially during the decomposition of organic matter (Sholkovitz, 1973), so that buried organic matter generally contains roughly half as much nitrogen and phosphorus as average phytoplankton.

Some of the phosphate released during the decomposition of phytoplankton leaves the oceans as adsorbed phosphate (Berner, 1973), some as a constituent of phosphate minerals. Arrhenius (1963) pointed out the importance of phosphatic fish remains in pelagic sediments. D'Anglejan (1967) pointed out the large extent of the marine phosphorites within recent continental shelf sediments off the west coast of Baja California, and it seems likely that phosphorites have been a significant sink of marine phosphorus at least during much of Phanerozoic time. Unfortunately, quantitative data regarding accumulation rates of phosphorus in phosphorites are lacking, and it is difficult to assess the relative importance of phosphorus removal as a constituent of organic matter and as a constituent of phosphorites.

Despite the present uncertainties in the rate of phosphorus input into the oceans and in its removal from seawater, it seems likely that a fair fraction, perhaps as much as half, of the marine phosphate is removed from the oceans as a constituent of organic matter. An increase by more than a factor of 2 in the burial rate of organic matter is probably ruled out by the limits on current phosphorus availability in the oceans. This conclusion has important consequences for the mechanisms that control the oxygen content of the atmosphere (see Chap. VI).

The Distribution of Organic Matter in Marine Sediments

Figure 5-22 summarizes a large number of analyses of the organic carbon content of modern marine sediments. The similarities between this map and the map of Figure 5-19 detailing the distribution of primary productivity in the world oceans are striking. The organic carbon content of sediments clearly reflects the productivity of the overlying water column. Nevertheless, the ratio of the two parameters is not constant. This must be due in part to variations in the total sedimentation rate and water depth, but a large number of environmental parameters are surely of importance as well.

Various authors have suggested that there is a correlation between the oxygen content of bottom waters and the degree of oxidation of organic

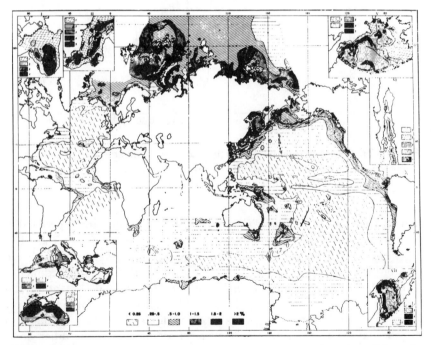

Figure 5-22. The distribution of organic carbon in the surface layer of marine sediments (Lisitzin, 1972). Insets: (I) Riga Bay, (II) Baltic Sea, (III) Mediterranean Sea, (IV) Black Sea, (V) Bering Sea, (VI) Gulf of California, (VII) Sea of Okhotsk.

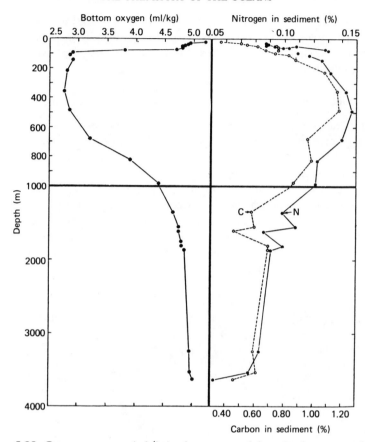

Figure 5-23. Oxygen content (ml/kg) of water overlying the bottom, and non-carbonate carbon content (%) and nitrogen content (%) of sediments along a traverse in the Gulf of Mexico (Richards and Redfield, 1954).

matter. Richards and Redfield's (1954) data for a line of stations perpendicular to the coast of the Gulf of Mexico are reproduced in Figure 5-23. The region of low oxygen content in the bottom water correlates well with the area of high organic carbon and nitrogen content of the bottom sediments. The authors suggest that the maximum in the organic content of the sediments is due to the presence of water of low oxygen content immediately above the bottom, and to the smaller oxygen diffusion gradient available to provide oxygen to the subsurface layers of sediment. On the other hand Figure 5-24 shows the rather poor correlation found by Gross (1967) between the dissolved oxygen content of seawater and the organic carbon content of surface sediments in the Northeast Pacific Ocean. The correlation

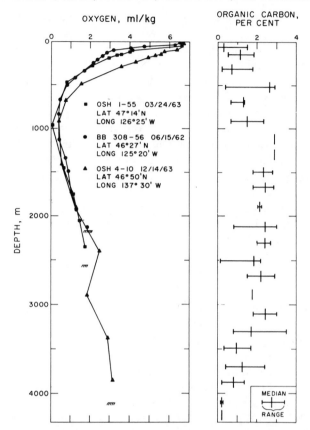

Figure 5-24. Variation with depth of the dissolved oxygen content of seawater from the Northeast Pacific Ocean, and the organic carbon content of surface sediments, grouped by 200-m depth intervals; Gross (1967).

observed by Richards and Redfield (1954) may, therefore, be noncausal, and the distribution of organic carbon in the Gulf of Mexico may well be determined more by differences in the ratio of the sedimentation rate of organic debris to that of detrital sediments than by the oxygen content of the bottom waters. Emery (1960) has suggested that such differences in the rate of sedimentation control the organic carbon content of marine sediments off southern California. A thorough study of the relationship between the rate of burial of organic matter and local physical, chemical, and biological processes is badly needed.

The river flux of terrestrial organic matter is by no means negligible; its survival and distribution in the oceans are still a matter of some debate. The

$\delta^{13}C$ value of terrestrial organic carbon is frequently somewhat more negative than that of marine organic carbon. In the Gulf of Mexico the difference between the mean $\delta^{13}C$ value of organic carbon in the Mississippi River system ($\sim -26 \,°/\text{oo}$) and in marine organic carbon ($-18 \,°/\text{oo}$) is sufficiently large, so that the proportion of terrestrial organic carbon in sediments of the Gulf of Mexico can be estimated on the basis of their isotopic composition (Sackett and Thompson, 1963). Shultz and Calder (1976) have shown that little terrestrial organic carbon can be detected in surface sediments more than a few tens of kilometers from the mouth of the Mississippi, and that none seems to be present in sediments more than 70 km from the river mouth. The data of Newman et al. (1973) suggest that Pleistocene sediments in the Gulf of Mexico contain a much larger proportion of terrestrial organic carbon, but other interpretations of their data are possible (Shultz and Calder, 1976); it seems likely that most of the organic matter buried at sea is of marine origin.

The Role of Nitrogen

The quantity of molecular nitrogen in the atmosphere, which is enormous compared to the nutrient demands of the oceans, is quite out of thermodynamic equilibrium with respect to the oxygen content of the atmosphere and the nitrate content of the oceans (see Chap. VI). Nevertheless nitrate is certainly one of the limiting nutrients in the oceans. Apparently, the rate of formation of nitrate in the atmosphere and its transfer to seawater is so slow, and the denitrification of combined nitrogen in the oceans so rapid, that marine NO_3^-, NO_2^-, and NH_4^+ are perennially in short supply. The shortness of this supply does not, however, seem to be sufficiently severe to give blue-green algae and other phytoplankton that are capable of fixing molecular nitrogen a strong competitive advantage today over photosynthetic organisms that cannot fix molecular nitrogen.

The major features of the marine geochemical cycle of nitrogen are shown in Figure 5-25. Neither the marine inputs nor the marine outputs are well known. It is likely, however, that the river flux of dissolved and particulate combined nitrogen is of roughly the same magnitude as the atmospheric input of combined nitrogen, and it is possible that nitrogen fixation in the oceans is of relatively minor importance. It is also likely that marine organic matter and denitrification to N_2 and N_2O are the major sinks of combined marine nitrogen. Arrhenius (1963) has pointed out that the C/N ratio of organic matter increases with depth from a value near 5 to a value near 15 in Pacific north equatorial clay sediments. Bordovskiy (1965a) has summarized a great deal of the literature bearing on the C/N ratio in recent sediments, and finds that the value of this ratio is frequently between 7 and

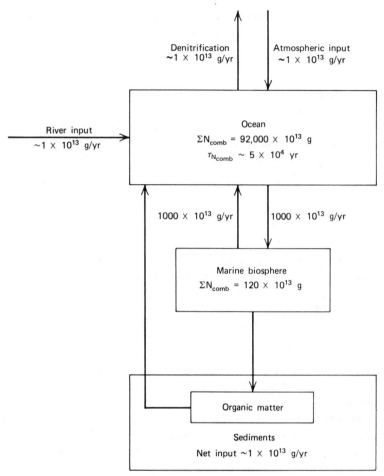

Figure 5-25. The marine geochemistry of combined nitrogen based largely on Vaccaro (1965); river input after Livingstone (1963); atmospheric input after Junge (1958), Gambell and Fisher (1964) and Tarrant et al. (1968); see also Figure 6-19.

10, that is, about $1\frac{1}{2}$ times the C/N ratio of planktonic organisms. Since organic carbon is currently being buried with sediments at roughly a rate of 1.2×10^{14} g/yr, nitrogen is removed from the oceans at the rate of roughly 1×10^{13} g/yr.

The oxidation of organic matter in most marine water columns takes place in the presence of free oxygen, and the conversion of organic nitrogen to nitrates is nearly complete. Free oxygen is the major oxidizing agent even at and below the water–sediment interface in well-aerated basins where bottom sediments are constantly turned over by benthic organisms. However, in areas where the oxygen concentration is $\leqslant 0.2$ ml/kg, combined nitrogen species are frequently lost, and molecular N_2 and N_2O are formed (Spencer, 1975). Although the rate of denitrification processes for the oceans as a whole is not well known, it is likely that they represent a major sink for combined nitrogen in oceans.

The data in Figure 5-25 indicate that the residence time of combined nitrogen in oceans is about 50,000 yr. This is essentially equal to the marine residence time of phosphorus and about half the residence time of carbon. The relatively short residence time of combined nitrogen and of phosphorus has important implications for the stability of atmospheric oxygen (see Chap. VI).

The Formation of Sulfides

Molecular oxygen is the major oxidant of organic material in most parts of the oceans. However, in some areas, particularly in stagnant basins, the biological oxygen demand exceeds the O_2 supply, and anoxic conditions develop. Below the sediment–water interface anoxia is the rule rather than the exception. There, oxidation reactions first reduce the available nitrate, and then turn to sulfate as a source of oxygen. Sulfate is reduced bacterially via several intermediate states to sulfide, which is then usually removed as a constituent of one or more iron sulfide minerals (Berner, 1971, 1972, 1974).

The generation of dissolved sulfide species and of iron sulfide precipitates depends in large part on the quantity and nature of the organic matter present in sediments, on the quantity and availability of iron, and on the diffusion of sulfate from the overlying water mass into the sediments. It is, therefore, not surprising that the vertical distribution of dissolved sulfate and sulfide, and the quantity of preciptated iron sulfide are quite variable, even in a rather restricted area such as the basins of the central part of the Gulf of California (Berner, 1964a). Figure 5-26 illustrates the distribution of dissolved sulfate and sulfide in interstitial waters from cores in this area. Curves describing the variation of the sulfate concentration with depth may be either convex, linear, or concave in the upper 100 cm of sediment. A depth distribution of sulfate that is amenable to simple mathematical analysis (Berner, 1964b) is not particularly common, but a number of conclusions have been established quite firmly regarding the process of sulfate reduction. It is clear that most sulfate is reduced within 100 cm of the water–sediment interface (Kaplan et al., 1963), that diffusion of sulfate into sediments and its introduction by bioturbation are sufficiently fast so that the amount of sulfur present in sulfides frequently exceeds the quantity initially present in the sulfate of interstitial waters. Nevertheless the oxidation of organic matter by anaerobic bacteria rarely, if ever, goes to completion.

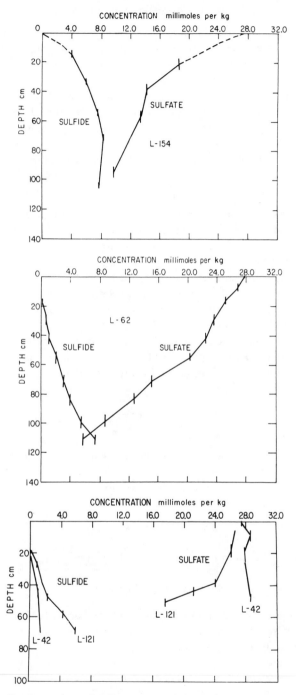

Figure 5-26. Concentration of sulfate and sulfide in interstitial water from cores taken in the central part of the Gulf of California (Berner, 1964a).

225

Bordovskiy (1965b) reports that approximately 20% of the organic carbon in the top 40 cm of three sediment cores from the Bering Sea were consumed in the reduction of iron compounds, and that an additional 20 to 25% were expended in this fashion after burial to depths greater than 40 cm. Loss of carbon becomes negligible at depths in excess of about 300 to 500 cm. The total initial carbon content of these sediments was between 1 and 2%. Berner (1969) found approximately 1% of sulfide sulfur in sediments from the central part of the Gulf of California. The precipitation of this quantity of sulfur required the loss of roughly 0.7% of organic carbon by the reaction

$$7CH_2O + 4SO_4^{2-} \rightarrow 2S_2^{2-} + 7CO_2 + 4OH^- + 5H_2O$$

The organic carbon content in the upper 20 cm of the sediments in these cores was approximately 4%. Loss of carbon associated with sulfate reduction leading to pyrite precipitation in these sediments was, therefore, roughly 15%. Berner's (1964a) analysis of Kaplan et al.'s data (1952) for a core from the sea off southern California suggested a loss of about 33% of the initial content of organic carbon. If these data are representative, they suggest that about 15–45% of the organic matter present in recent sediments are lost by bacterial reduction of sulfate to sulfide.

Berner (1971) has shown that the concentration $c(x,t)$ of an ion or compound produced during bacterial decay of organic matter in marine sediments not undergoing compaction is given by the equation

$$\frac{\partial c(x,t)}{\partial t} = D_s \frac{\partial^2 c(x,t)}{\partial x^2} - \omega \frac{\partial c(x,t)}{\partial x} + \left. \frac{dc(x,t)}{dt} \right|_{\text{org}} \tag{5-8}$$

where x is the depth below the sediment–water interface (where $x=0$), D_s is the diffusion constant of the ion or compound in the interstitial waters, and ω is the rate of sedimentation. If the decomposition of organic matter is a first-order process so that

$$dN(x,t)/dt = -kN(x,t) \tag{5-9}$$

where $N(x,t)$ is the quantity of organic carbon, then

$$dc(x,t)/dt\big|_{\text{org}} = \beta LkN(x,t) \tag{5-10}$$

where L is the ratio of the number of moles of the particular ion produced or consumed per mole of organic carbon oxidized; and β is the conversion factor for the units of N to the units of c.

At steady state

$$\partial c(x,t)/\partial t = 0 \tag{5-11}$$

and equation 5-8 becomes

$$0 = D_s \frac{\partial^2 c}{\partial x^2} - \omega \frac{\partial c}{\partial x} + \beta L k N_o e^{-k(x/\omega)} \tag{5-12}$$

Berner (1971) has shown that for ions such as SO_4^{2-} with a boundary condition

$$c(0,t) = c_o$$

the solution of equation 5-12 is

$$c(x) = c_o + \frac{\beta L N_o \omega^2}{D_s k + \omega^2} \left[1 - e^{-(k/\omega)x} \right] \qquad (L < 0) \tag{5-13}$$

when D_s, ω, and k are independent of x and t and when

$$c_o \geqslant \frac{\beta N_o \omega^2 |L|}{D_s k + \omega^2} \tag{5-14}$$

It is unlikely that steady state conditions generally obtain during sedimentation. Fortunately, non-steady state solutions of equation (5-8) are similar to steady state solutions except where sedimentation rates are in excess of about 200 cm/1000 yr, because diffusion within the interstitial water column smoothes out the effects of nonuniform distributions of decomposable organic matter (Lasaga and Holland, 1976).

The rate of sulfate reduction does not seem to be a strong function of the sulfate concentration of interstitial waters until $m_{SO^{2-}}$ is close to zero. Lasaga and Holland (1976) have shown that when the rate of SO_4^{2-} reduction is independent of the SO_4^{2-} concentration and when the concentration of SO_4^{2-} at depth in interstitial waters in carbonaceous marine sediments goes essentially to 0, the rate of reduction, dM_s/dt, of sulfate in the oceans as a whole is

$$\frac{dM_s}{dt} \cong c_o \times 10^{-6} \sum_i \phi_i \omega_i \left(1 + \frac{k_i D_{s_i}}{\omega_i^2} \right) \cdot A_i \text{ mol/yr} \tag{5-15}$$

where c_o is the concentration of SO_4^{2-} in seawater in mol/kg; ω_i is the rate of sedimentation in area i in cm/1000 yr; k_i is the rate constant of organic matter oxidation in area i in yr^{-1}; D_{s_i} is the rate of diffusion of sulfate in the sediment of area i in $cm^2/1000$ yr; and A_i is the extent of area i in cm^2.

It is interesting to note that dM_s/dt is proportional to the concentration of sulfate in seawater even though the rate of bacterial reduction of SO_4^{2-} in interstitial waters is essentially independent of their SO_4^{2-} content. It also follows from equation 5-15 that dM_s/dt depends on the ω_i's and that in a given area the minimum rate of sulfate reduction occurs when

$$\frac{k_i D_{s_i}}{\omega_i^2} = 1 \tag{5-16}$$

This analysis is clearly oversimplified, but is probably sufficiently detailed to demonstrate that the removal of sulfate from the oceans by reduction to sulfide and precipitation as iron sulfides depends on the SO_4^{2-} concentration in seawater, on the area covered by organic-rich sediments, on the bacterial rate of decay of organic matter, and on the rate of sedimentation of carbonaceous sediments. Even if the sulfate content of seawater and the total burial rate of organic carbon were constant, the rate of sulfate removal might be quite variable, since it depends on the distribution of the buried organic matter and on its dilution by noncarbonaceous material. The three major mechanisms for the removal of sulfate from seawater—evaporite formation, seawater cycling through midocean ridges, and pyrite precipitation—are, therefore all dependent on events and physical configurations in areally minor parts of the world oceans.

Large temporal variations in the isotopic composition of sulfur in seawater SO_4^{2-} are well documented by variations in the isotopic composition of sulfur in the gypsum and anhydrite of Phanerozoic evaporites (Holser and Kaplan, 1966). It is likely that these variations are due largely to fluctuations in the ratio of the rate of sulfate removal as a constituent of sulfate minerals to the rate of sulfate removal by reduction followed by the precipitation of sulfides, and that these fluctuations have occurred largely in response to changes in the geography of the ocean basins (Holland, 1973).

THE OUTPUT OF INTERSTITIAL SEAWATER

Marine sediments near the water–sediment interface frequently contain nearly equal weights of seawater and sediment grains. During burial the porosity of sediments decreases rapidly, and large quantities of water are lost from the sediment column. The removal of interstitial seawater and its dissolved salts from the oceans is difficult to estimate because the magnitude of return flow of formation water and the increase in the salinity of formation waters in marine sediments with increasing depth of burial are not well known.

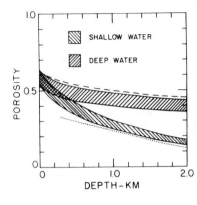

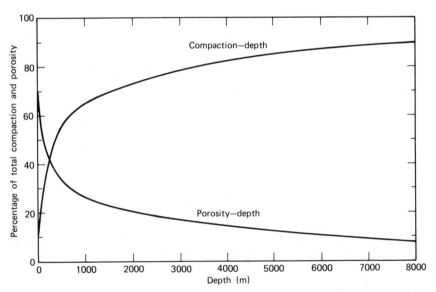

Figure 5-27. Porosity-depth dependence for average deep- and shallow-water sediments. The dashed curve was obtained from artificial compaction data. The dotted curve is a combination of the Venezuelan and Kansan bore hole data of Hedberg (1936). Figure from Nafe and Drake (1957).

A very strong maximum for the output of ocean water and of its salts can be set by calculating the output of ocean water with uncompacted marine sediments. Figure 5-27 shows Nafe and Drake's (1957) data for the porosity–depth dependence of average deep- and shallow-water sediments. These data are in satisfactory agreement with those of Hedberg (1936). Dickinson's (1953) data for the sediments in the Gulf of Mexico are summarized in Figure 5-28. They agree well with the shallow-water curves of Nafe and

Figure 5-28. Relation of porosity and compaction of shales to depth of burial in the Gulf Coast area; figure adapted from Dickinson (1953).

Drake (1957) and McDuff and Gieskes (1976), and extend their results to depths in excess of 7 km. Rieke and Chilingarian's (1974, p. 42) compilation of porosity–depth relationships in shales and argillaceous sediments shows a good deal of variability and scatter around the curves in Figures 5-27 and 5-28.

The porosity of marine sediments within a few meters of the water–sediment interface is frequently between 60 and 80% (see, for instance, Broecker et al., 1958; Goldberg and Arrhenius, 1958). If 70% is taken as an average value for the porosity of such sediments, 1.02 g/cc for the density of seawater, and 2.67 g/cc for the grain density, then 100 cc of surface sediment contain 72 g of seawater and 80 g of sediment grains, and the ratio of the weight of seawater to that of sediment grains is 0.90. The present-day rate of sedimentation is approximately 2×10^{16} g/yr (Holeman, 1968); the absolute maximum of sodium burial rate is, therefore,

$$1.07 \times 10^{-2} \text{ g Na/g seawater} \times 0.90 \text{ g seawater/g sediment}$$

$$\times 2 \times 10^{16} \text{ g sediment/yr} = 2 \times 10^{14} \text{ g Na}^+/\text{yr}$$

and the absolute maximum of chloride burial rate is

$$1.94 \times 10^{-2} \text{ g Cl}^-/\text{g seawater} \times 0.90 \text{ g seawater/g sediment}$$

$$\times 2 \times 10^{16} \text{ g sediment/yr} = 3.5 \times 10^{14} \text{ g Cl}^-/\text{yr}$$

The estimated annual river inputs of sodium and chloride corrected for atmospherically recycled salts (see Table 4-13) are

$$0.21 \times 10^{-3} \text{ mol/liter} \times 23 \text{ g/mol Na} \times 4.6 \times 10^{16} \text{ liters/yr}$$

$$= 2.2 \times 10^{14} \text{ g Na/yr}$$

$$0.16 \times 10^{-3} \text{ mol Cl}^-/\text{liter} \times 35.5 \text{ g/mol Cl}^- \times 4.6 \times 10^{16} \text{ liters/yr}$$

$$= 2.6 \times 10^{14} \text{ g Cl}^-/\text{yr}$$

The maximum rate of burial of sodium and chloride as a component of interstitial water in marine sediments is, therefore, essentially equal to their annual rate of river input.

The actual rate of burial must be very much smaller than the maximum rate calculated above. The bulk of marine sediments is deposited in relatively shallow water, and the porosity of these sediments decreases rapidly with increasing depth of burial. At a depth of 1 km their porosity is about 30%, and at a depth of 2 km their porosity is about 20%. Loss of water on compaction is accompanied by a somewhat irregular increase in salinity of

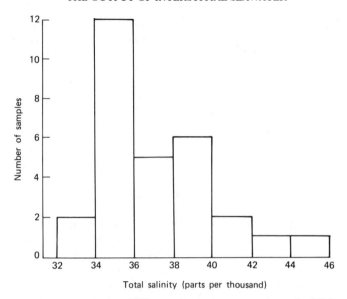

Figure 5-29. The total salinity of Pliocene formation waters from the Midway-Sunset area, Kern County, California; depths between 800 and 1350 m; Chave (1960).

the remaining pore water. In formation waters from the California basins the increase in salinity is slight in the upper kilometer of sediments. Figure 5-29 demonstrates this quite well for 29 samples of formation waters from the Pliocene of Kern County, California. On the other hand, Timm and Maricelli (1953) have shown that salinities up to $4\frac{1}{2}$ times that of normal seawater are found in Gulf Coast sediments, even at depths only slightly in excess of 1 km, and that there is no simple relationship between depth of burial and the salinity of formation waters in this area. The difference between the two areas is probably due, at least in part, to the presence of salt deposits in the Gulf Coast (Manheim and Sayles, 1970).

A reasonable maximum value for the rate of more or less permanent burial of sodium and chloride with interstitial water can be based on the assumption that seawater is lost from sediments until a burial depth of 1 km is reached, and that at greater depths the loss of water accompanying the decrease in porosity is offset by a progressive increase of salinity. A sediment porosity of 30% at a depth of 1 km corresponds to a value of 0.16 for the ratio of the weight of seawater to the weight of sediment grains. The annual rate of burial of sodium and chloride with interstitial water calculated on this basis is 0.3×10^{14} and 0.6×10^{14} g/yr, and corresponds to 15 and 25%, respectively, of the river input of sodium and chloride corrected, as above, for atmospherically recycled salts. These figures are obviously somewhat unsatisfactory; they indicate, however, that removal of sodium and chloride from the oceans as a constituent of interstitial water in marine sediments is probably a minor, but not a trivial mechanism for removing these ions from the oceans. Removal with interstitial water is of most importance for constituents of seawater that have a long residence time. Bromine (see Table 5-3) is the only constituent of seawater other than Na^+ and Cl^- for which removal with interstitial water is apt to be of some importance.

SUMMATION

The composition of seawater and its variation in space and time during the last few decades is now well known. The river inputs to the oceans are less well known, but it is clear that the residence time of all dissolved substances in the oceans is much smaller than the age of the earth. The range of residence times is very large. For some elements it is less than the mixing time of the oceans; for others it approaches 10^8 yr. For all but a few elements output from the oceans must be balancing inputs on a geologically short time scale. It is, therefore, a fundamental theorem of ocean chemistry that the concentration of nearly all constituents of seawater adjusts itself so that chemical outputs balance chemical inputs on a time scale of a few multiples of their residence times.

Numerous disposal mechanisms are available to the oceans. The bulk of this chapter has been devoted to identifying the dominant removal mechanisms for the major cations, anions, and neutral species in seawater. The results are summarized in Table 5-14. They have had to be cast in terms of major, minor, and trace processes, because nearly all of the individual output rates are still only poorly known. Even this classification is somewhat suspect in several instances, and much more work is required before a definitive table of outputs can be assembled. In a sense Table 5-14 is a retrograde step from Mackenzie and Garrels' scheme in Table 5-6: the rather specific numbers in the earlier table have been replaced by a much vaguer notation; but the experience of the past 10 yr has shown that the apparent certainty of Table 5-6 was somewhat illusory, and that even today a rather cautious attitude toward proposed schemes of ocean outputs is advisable.

Sodium is a case in point. Much of the river input of Na^+ is certainly recycled as sea spray, and a good deal must be removed as a constituent of halite, but the removal rate via cation exchange on marine clays and by the formation of albite during the cycling of seawater through basalt near midocean ridges is probably not negligible.

Potassium removal seems to take place largely via cation exchange on marine clays. However, K^+ uptake during the weathering of basalt at the seafloor has been demonstrated, and may turn out to be an important K^+ sink if weathering extends to depths of several hundred meters below the ocean floor.

Calcium is certainly removed largely as a constituent of calcite and aragonite; a small fraction of the marine Ca^{2+} budget leaves as a constituent of gypsum and anhydrite.

Magnesium is probably removed in part as a constituent of montmorillonite and chlorite formed during the cycling of seawater through midocean ridges, and in part on cation exchange sites of marine clays. Authigenic silicates on the ocean floor are probably not of great importance. Dissolved iron and iron hydroxides are probably removed largely as marine precipitates of $Fe(OH)_3$ and as a constituent of pyrite and iron-rich authigenic silicates such as chamosite and glauconite.

Chloride removal takes place largely as a constituent of sea spray, halite, and other chlorides.

Bicarbonate mainly disproportionates into CO_3^{2-} and H_2CO_3. The former is precipitated largely as a constituent of calcite and aragonite, the latter is largely returned to the atmosphere. A minor proportion of the carbon in HCO_3^- and CO_3^{2-} is removed as a constituent of organic matter together with a portion of the river input of dissolved and particulate organic matter. The net oxygen released during photosynthesis and by the burial of organic matter escapes to the atmosphere and is used predominantly in weathering reactions on land (see Chap. VI).

Marine sulfate is removed largely as a constituent of pyrite, gypsum, and anhydrite. All three minerals are deposited in marine sediments, and it is likely that some SO_4^{2-} is removed from seawater as a constituent of pyrite and anhydrite in hydrothermal settings near midocean ridges.

Nitrate leaves the oceans as a constituent of organic matter in marine sediments and as N_2 and nitrogen oxides to the atmosphere after denitrification. The relative importance of the two outputs is still not well known.

Phosphate is also removed in part as a constituent of organic matter, but a good deal of phosphorus may be removed as a constituent of apatite, other phosphates, or as adsorbed PO_4^{3-} on carbonates, iron, and manganese oxides and hydroxides.

Silica apparently leaves the oceans largely as free SiO_2. Estimates of the sedimentation rate of SiO_2 in diatoms, radiolarians, and other siliceous organisms agree quite well with estimates of the river input of SiO_2, and it is likely that authigenic silicates play only a small part in the marine geochemistry of silica. The large percentage of quartz in sedimentary rocks is also best explained if the role of reverse weathering reactions in the removal of dissolved SiO_2 from seawater is minor.

Table 5-14
Summary of Output Mechanisms of the Major and of some Minor Constituents of Seawater[a]

Input Rate [10^{14} g/yr]	Na^+	K^+	Ca^{2+}	Mg^{2+}	Fe^{3+}	Cl^-	CO_3^{2-} + HCO_3^-	SO_4^{2-}	NO_3^-	PO_4^{3-}	SiO_2	Organic Compounds	Amorphous hydroxides and silicates	Volcanic glass	Sub-sea floor Intrusives	Particulate organic matter
							Dissolved							**Particulate**		
Inputs to the Oceans																
1. Dissolved	3.2	1.0	6.9	1.8	0.3	3.7	26	4.9	0.1	0.02	6.0	1.4	—	—	—	—
2. Net Desorbed	—	2.1	—	—	—	—	—	—	—	—	—	—	—	—	—	—
3. Released by high-T reaction with basalt	—	0.4	2	—	—	—	—	—	—	—	—	—	—	—	—	—
Outputs from the Oceans																
Carbonates																
Calcite + aragonite	—	—	M	—	—	—	M	—	—	—	—	—	—	—	—	M
Dolomite	—	tr	—	tr	—	—	m	—	—	—	—	—	—	—	—	m
Silicates, oxides and hydroxides (sea floor)																
Sepiolite, attapulgite, and talc	—	—	—	tr	—	—	—	—	—	—	tr	—	—	—	—	—
Chlorite	—	tr	—	tr	m	—	—	—	—	—	tr	—	m	—	—	—
Montmorillonite	m	M	—	m	m	—	—	—	—	—	tr	—	m	M	—	—
Illite	—	m	—	tr	m	—	—	—	—	—	tr	—	m	m	—	—
Phillipsite and clinoptilolite	tr	—	—	—	—	—	—	—	—	—	—	—	—	M	—	—
Silica	tr	tr	—	—	—	—	—	—	—	—	M	—	—	M	—	—
Hydrous Fe and Mn oxides	—	—	—	—	m	—	—	—	—	m?	—	—	m	—	—	—

234

Silicates, Sulfides, and Sulfates (sub sea floor)											
Zeolites	—	—	—	—	—	—	—	—	—	—	m
Montmorillonite	—	—	—	M	—	—	—	—	—	—	m
Chlorite	—	—	—	⎰	—	—	—	—	—	—	m
Actinolite	—	—	—	⎱	—	—	—	—	—	—	m
K-feldspar	—	—	—	—	—	—	—	—	—	—	tr
Albite	m/M	—	—	—	—	—	—	m	—	—	M
Pyrite	—	—	—	—	m	—	—	m	—	—	tr
Anhydrite	—	—	—	—	m	—	—	m	—	—	tr
Evaporites											
Gypsum and anhydrite	M	—	m	—	M	—	—	—	—	—	—
Halite	—	m	M	—	—	—	—	—	—	—	—
Sylvite	m	—	m	—	—	—	—	—	—	—	—
Other halides and sulfates	tr	tr	m	—	—	—	—	—	—	—	—
Outputs related to organic matter deposition											
Organic carbon	—	—	—	—	m	—	—	m	—	m	m
Organic nitrogen	—	—	—	—	—	M	—	—	—	tr	tr
Organic phosphorus	—	—	—	—	—	M	M	—	—	tr	tr
Apatite and other phosphates	—	—	—	—	—	—	M	—	—	—	—
Pyrite and other sulfides	—	m	—	—	M	—	—	M	m	—	—
Sea spray	M	m	M	m	m	—	—	M	—	M	—
Interstitial seawater	m	—	m	—	—	—	—	m	—	m	—
Cases											
CO₂	—	—	—	—	M	—	—	M	—	M	M
O₂	—	—	—	—	m	—	—	m	—	m	m
N₂ + NOₓ	—	—	—	—	m	—	—	M	—	m	m

ᵃM = major (> 25% of input); m = minor (~10–25% of input); tr = trace (~5–10% of input).

235

In many respects Table 5-14 is similar to Table 5-6. However, the importance of the formation of authigenic silicates via reactions 9–11 in Table 5-6 has been reduced markedly, and the reaction of seawater with basalt at elevated temperatures is proposed as a major alternative. These changes are not trivial. They suggest that the chemistry of the oceans is linked to tectonic cycles and plate motions, and they imply that changes in the pulse of the earth are reflected in the chemistry of the oceans. They also suggest that the composition of seawater today is not such that the formation of authigenic silicates at the seafloor is sufficiently rapid or widespread to exert a dominant influence on the chemistry of the oceans; authigenesis of silicates apparently becomes important only when the ion activity product of silicates in the oceans is significantly greater than it is today.

A much better understanding of the kinetics of dolomite and of silicate mineral authigenesis is required for a more quantitative formulation of the response of seawater chemistry to changes in chemical inputs and geographic variables. Fortunately, the chemistry of the oceans does not seem to be a very sensitive function of these variables; the observed changes in the nature of sedimentary rocks can be understood in terms of the proposed model for the oceans and by the coupling of the oceans to the atmosphere and to the endogenic cycle.

REFERENCES

Aharon, P., Kolodny, Y., and Sass, E., 1977, Recent hot brine dolomitization in the "Solar Lake," Gulf of Elat, isotopic, chemical, and mineralogical study, *J. Geol.* **85**, 27–48.

Armstrong, F. A. J., 1965, Silicon, *Chemical Oceanography*, Vol. 1, Chap. 10, J. P. Riley and G. Skirrow, Ed., Academic Press, New York.

Arnórsson, S., 1974, The composition of thermal fluids in Iceland and geological features related to the thermal activity, in *Geodynamics of Iceland and the North Atlantic Area*, L. Kristjansson, Ed., D. Reidel Publishing, Dordrecht-Holland, Boston, pp. 307–323.

Arrhenius, G., 1963, Pelagic sediments, in *The Sea*, Vol. 3, Chap. 25 M. N. Hill, Ed., Interscience, New York.

Aumento, F., Loncarevic, B. D., and Ross, D. I., 1971, Hudson geotraverse: geology of the Mid-Atlantic Ridge at 45°N, *Philos. Trans. R. Soc. Lond. A* **268**, 623–650.

Bader, R. G., 1955, Carbon and nitrogen relations in surface and subsurface marine sediments, *Geochim. Cosmochim. Acta* **7**, 205–211.

Bartholomé, P., 1966, The abundance of dolomite and sepiolite in the sedimentary record, *Chem. Geol.* **1**, 33–48.

Bathurst, R. G. C., 1967, Depth indicators in sedimentary carbonates, *Marine Geol.* **5**, 447–471.

Bathurst, R. G. C., 1972, *Carbonate Sediments and Their Diagenesis*, Elsevier, New York, 620 pp.

Bell, D. L. and Goodell, H. G., 1967, A comparative study of glauconite and the associated clay fraction in modern marine sediments, *Sedimentology* 9, 169–202.

Belyayeva, N. V., 1968, Quantitative distribution of planktonic foraminiferal tests in recent Pacific sediments, *Oceanology* (transl.) 8, 85–89.

Berger, W. H., 1970, Planktonic foraminifera: selective solution and the lysocline, *Marine Geol.* 8, 111–138.

Berger, W. H., 1971, Sedimentation of planktonic foraminifera, *Marine Geol.* 11, 325–358.

Bernat, M., Bieri, R. H., Koide, M., Griffin, J. J., and Goldberg, E. D., 1970, Uranium, thorium, potassium, and argon in marine phillipsites, *Geochim. Cosmochim. Acta* 34, 1053–1071.

Berner, R. A., 1964a, Distribution of diagenesis of sulfur in some sediments from the Gulf of California, *Marine Geol.* 1, 117–140.

Berner, R. A., 1964b, An idealized model of dissolved sulfate distribution in recent sediments, *Geochim. Cosmochim. Acta* 28, 1497–1503.

Berner, R. A. 1965, Activity coefficients of bicarbonate, carbonate, and calcium ions in sea water, *Geochim. Cosmochim. Acta* 29, 947–965.

Berner, R. A., 1966, Diagenesis of carbonate sediments: interaction of magnesium in sea water with mineral grains, *Science* 153, 188–191.

Berner, R. A., 1969, Migration of iron and sulfur within anaerobic sediments during early diagenesis, *Am. J. Sci.* 267, 19–42.

Berner, R. A., 1971, *Principles of Chemical Sedimentology*, McGraw-Hill, New York, 240 pp.

Berner, R. A., 1972, Sulfate reduction, pyrite formation, and the oceanic sulfur budget, in *The Changing Chemistry of the Oceans*, D. Dyrssen and D. Jagner, Ed., Wiley, New York, pp. 347–361.

Berner, R. A., 1973, Phosphate removal from sea water by adsorption on volcanogenic ferric oxides, *Earth Planet. Sci. Lett.,* 18, 77–86.

Berner, R. A., 1974, Kinetic models for the early diagenesis of nitrogen, sulfur, phosphorus, and silicon in anoxic marine sediments, *The Sea*, Vol. 5, Chap. 13, E. D. Goldberg, Ed., Wiley-Interscience, New York.

Berner, R. A., 1975, The role of magnesium in the crystal growth of calcite and aragonite from sea water, *Geochim. Cosmochim. Acta* 39, 489–504.

Berner, R. A., Scott, M. R., and Thomlinson, C., 1970, Carbonate alkalinity in the pore waters of anoxic marine sediments, *Limnol. Oceanogr.* 15, 544–549.

Biscaye, P. E., 1965, Mineralogy and sedimentation of recent deep-sea clay in the Atlantic Ocean and adjacent seas and oceans, *Bull. Geol. Soc. Am.* 76, 803-832.

Bischoff, J. L. and Dickson, F. W., 1975, Seawater-basalt interaction at 200°C and 500 bars: implications for origin of sea-floor heavy-metal deposits and regulation of seawater chemistry, *Earth Planet. Sci. Lett.* 25, 385–397.

Bischoff, J. L. and Seyfried, W. E., 1977, Seawater as a geothermal fluid: chemical behavior from 25°C to 350°C, Proceedings of the Second International Symposium on Water-Rock Interaction I.A.G.C., Strasbourg, France, pp. IV 166–IV 172.

Björnsson, S., Arnórsson, S., and Tómasson, J., 1972, Economic evaluation of Reykjanes thermal brine area, Iceland, *Am. Assoc. Pet. Geol. Bull.* **56**, 2380–2391.

Bonatti, E., 1963, Zeolites in Pacific pelagic sediments, *N. Y. Acad. Sci. Trans. II*, **25**, 938–948.

Bonatti, E., 1975, Metallogenesis at oceanic spreading centers, in *Annual Review of Earth and Planetary Sciences*, Vol. 3, F. A. Donath, et al., Eds. Annual Reviews, Palo Alto, California, pp. 401–431.

Bonatti, E. and Joensuu, O., 1968, Palygorskite from Atlantic deep sea sediments, *Am. Miner.* **53**, 975–983.

Bonatti, E., Honnorez, J., and Ferrara, G., 1970, Equatorial Mid-Atlantic Ridge: Petrologic and Sr isotopic evidence for an Alpine-type rock assemblage, *Earth Planet. Sci. Lett.* **9**, 247–256.

Bonatti, E., Honnorez, J., Kirst, P., and Radicati, F., 1975, Metagabbros from the Mid-Atlantic Ridge at 6°N; contact-hydrothermal-dynamic metamorphism beneath the axial valley, *J. Geol.* **83**, 61–78.

Borchert, H., 1965, Principles of oceanic salt deposition and metamorphism, in *Chemical Oceanography*, Vol. 2, Chap. 19, J. P. Riley and G. Skirrow, Ed., Academic Press, New York.

Borchert, H. and Muir, R. O., 1964, *Salt Deposits; The Origin, Metamorphism and Deformation of Evaporites*, Van Nostrand, New York, 338 pp.

Bordovskiy, O. K., 1965a, Sources of organic matter in marine basins, *Marine Geol.* **3**, 5–31.

Bordovskiy, O. K., 1965b, Accumulation of organic matter in bottom sediments, *Marine Geol.* **3**, 33–82.

Bowles, F. A., Angino, E. A., Hosterman, J. W., and Galle, O. K., 1971, Precipitation of deep-sea palygorskite and sepiolite, *Earth Planet. Sci. Lett.* **11**, 324–332.

Boyle, E. A., Sclater, F. R., and Edmond, J. M., 1977, The distribution of dissolved copper in the Pacific, *Earth Planet. Sci. Letters* **37**, 38–54.

Braitsch, O., 1962, *Entstehung und Stoffbestand der Salzlagerstätten*, in *Mineralogie und Petrographie in Einzeldarstellungen*, Vol. 3 Springer-Verlag, Berlin, 282 pp.

Braitsch, O., 1963, Evaporite aus normalem und veränderten Meerwasser, *Fortschr. Geol. Rheinl. Westfall.* **10**, 151–172.

Brewer, P. G., 1975, Minor elements in sea water, in *Chemical Oceanography*, 2nd ed. Vol. 1, Chap. 7, J. P. Riley and G. Skirrow, Eds., Academic Press, New York.

Broecker, W. S., 1963, Radioisotopes and large-scale oceanic mixing, in *The Sea*, Vol. 2, Chap. 4, M. N. Hill, Ed. Interscience, New York.

Broecker, W. S., Turekian, K. K., and Heezen, B. C., 1958, The relation of deep sea Atlantic Ocean sedimentation rates to variations in climate, *Am. J. Sci.* **256**, 503–517.

Broecker, W. S., Gerard, R. D., Ewing, M., and Heezen, B. C., 1961, Geochemistry and physics of ocean circulation, in *Oceanography*, American Association for the Advancement of Science, Publication 67, pp. 301–322.

Broecker, W. S. and Takahashi, T., 1966, Calcium carbonate precipitation on the Bahama Banks, *J. Geophys. Res.* **71**, 1575–1602.

Broecker, W. S. and Takahashi, T., 1976, The relationship between lysocline depth and in situ carbonate ion concentration, ONR Conference, Honolulu, January 1976.

Bruyevich, S. V. and Ivanenkov, V. N., 1971, Chemical balance of the world ocean, *Oceanology* **11**, 694–699.

Burke, K., 1975, Atlantic salt deposits formed by evaporation of water spilt from the Pacific, Tethyan and Southern Oceans, *Am. Geophys. Union Trans.* **56**, 457.

Carroll, D. and Starkey, H. C., 1960, Effect of seawater on clay minerals in *Clays and Clay Minerals*, Proceedings of the National Conference on Clays and Clay Minerals, Vol. 5, Pergamon Press, New York, pp. 80–101.

Cathles, L. and Fehn, U., 1976, personal communication.

Chave, K. E., 1960, Evidence on history of sea water from chemistry of deeper subsurface waters of ancient basins, *Bull. Am. Assoc. Pet. Geol.* **44**, 357–370.

Chester, R., 1965, Elemental geochemistry of marine sediments, in *Chemical Oceanography*, Vol. 2, Chap. 15, J. P. Riley and G. Skirrow, Eds. Academic Press, New York.

Christ, C. L., Hostetler, P. B., and Siebert, R. M., 1973, Studies in the system $MgO–SiO_2–CO_2–H_2O$(III); the activity-product constant of sepiolite, *Am. J. Sci.* **273**, 65–83.

Cloud, P. E., Jr., 1965, Carbonate precipitation and dissolution in the marine environment, in *Chemical Oceanography*, Vol. 2, Chap. 17, J. P. Riley and G. Skirrow, Ed., Academic Press, New York.

Cronan, D. S., 1974, Authigenic minerals in deep-sea sediments, in *The Sea*, Vol. 5, Chap. 15, E. D. Goldberg, Ed., Wiley-Interscience, New York.

D'Anglejan, B. F., 1967, Origin of marine phosphorites off Baja California, Mexico, *Marine Geol.* **5**, 15–44.

Dasch, E. J., 1969, Strontium isotopes in weathering profiles, deep-sea sediments, and sedimentary rocks, *Geochim. Cosmochim. Acta* **33**, 1521–1552.

Deer, W. A., Howie, R. A., and Zussman, J., 1962, *Sheet Silicates*, Vol. 3 of *Rock-Forming Minerals*, Wiley, New York, 270 pp.

Deffeyes, K. S., 1965, The Columbia River flood and the history of the oceans, Pacific NW Oceanographers' Annual Meeting, Corvallis, Oregon.

Deffeyes, K. S., 1970, The axial valley: a steady-state feature of the terrain, in *The Megatectonics of Continents and Oceans*, Chap. 9, H. Johnson and B. L. Smith, Eds., Rutgers University Press, New Brunswick, New Jersey.

Deffeyes, K. S., Lucia, F. J., and Weyl, P. K., 1965, Dolomitization of recent and Plio-Pleistocene sediments by marine evaporite waters on Bonaire, Netherlands Antilles, in *Dolomitization and Limestone Diagenesis*, L. C. Pray and R. C. Murray, Eds., Society of Economic, Paleontologists and Mineralogists pp. 71–88.

Dickinson, G., 1953, Geological aspects of abnormal reservoir pressures in Gulf Coast Louisiana, *Bull. Am. Assoc. Pet. Geol.* **37**, 410–432.

Disteche, A., 1974, The effect of pressure on dissociation constants and its temperature dependence, in *The Sea*, Vol. 5, Chap. 2, E. D. Goldberg, Ed., Wiley-Interscience, New York.

Dodd, J. R., 1967, Magnesium and strontium in calcareous skeletons: A review, *J. Paleontol.* **41**, 1313–1329.

Drever, J. I., 1971a, Early diagenesis of clay minerals, Rio Ameca Basin, Mexico, *J. Sediment. Pet.* **41**, 982–994.

Drever, J. I., 1971b, Magnesium-iron replacement in clay minerals in anoxic marine sediments, *Science* **172**, 1334–1336.

Drever, J. I., 1974, The magnesium problem, in *The Sea*, Vol. 5, Chap. 10, E. D. Goldberg, Ed. Wiley-Interscience, New York.

Duursma, E. K., 1965, The dissolved organic constituents of sea water, in *Chemical Oceanography*, Vol. 1, Chap. 11, J. P. Riley and G. Skirrow, Eds., Academic Press, New York.

Edmond, J., 1977, personal communication.

Emery, K. O., 1960, *The Sea off Southern California, A Modern Habitat of Petroleum*, Wiley, New York, 366 pp.

Emery, K. O. and Rittenberg, S. C., 1952, Early diagenesis of California Basin sediments in relation to origin of oil, *Bull. Am. Assoc. Pet. Geol.* **36**, 735–806.

Fanning, K. A. and Pilson, M. E. Q., 1971, Interstitial silica and pH in marine sediments: some effects of sampling procedures, *Science* **173**, 1228–1231.

Fawcett, J. J. and Yoder, H. S., Jr., 1966, Phase relationships of chlorites in the system $MgO-Al_2O_3-SiO_2-H_2O$, *Am. Miner.* **51**, 353–380.

Fehn, U., and Cathles, L. M., 1978, Hydrothermal convection at slow-spreading mid-ocean ridges, *Tectonophysics*, to be published.

Fogg, G. E., 1975, Primary productivity, in *Chemical Oceanography*, Vol. 2, Chap. 14, 2nd Ed. J. P. Riley and G. Skirrow, Eds. Academic Press, New York.

Folk, R. L. and Land, L. S., 1975, Mg/Ca ratio and salinity: two controls over crystallization of dolomite, *Bull. Am. Assoc. Pet. Geol.* **59**, 60–68.

Frey, F. A., Bryan, W. B., and Thompson, G., 1972, Petrological and geochemical results for basalts from DSDP Legs 2 and 3, *Geol. Soc. Am. Abstr. Programs* **4**, 511.

Friedman, G. M., 1965, Occurrence of talc as a clay mineral in sedimentary rocks, *Nature* **207**, 283–284.

Friedman, G. M. and Sanders, J. E., 1967, Origin and occurrence of dolostones, in *Carbonate Rocks*, Chap, 6, G. V. Chilingar, H. J. Bissell, and R. W. Fairbridge, Eds., Elsevier, New York.

Galliher, E. W., 1935, Geology of glauconite, *Bull. Am. Assoc. Pet. Geol.* **19**, 1569–1601.

Gambell, A. W. and Fisher, D. W., 1964, Occurrence of sulfate and nitrate in rainfall, *J. Geophys. Res.* **69**, 4203–4210.

Garrels, R. M. and Thompson, M. E., 1962, A chemical model for seawater at 25°C and one atmosphere total pressure, *Am. J. Sci.* **260**, 57–66.

Garrels, R. M. and Christ, C. L., 1965, *Solutions, Minerals, and Equilibria*, Harper & Row, New York, 460 pp.

Garrels, R. M. and Mackenzie, F. T., 1971, *Evolution of Sedimentary Rocks*, Norton, New York, 397 pp.

Garrels, R. M., Mackenzie, F. T., and Hunt, C., 1975, *Chemical Cycles and the Global Environment*, William Kaufmann, Los Altos, California, 206 pp.

Gehman, H. M., Jr. 1962, Organic matter in limestones, *Geochim. Cosmochim. Acta* **26**, 885–897.

Gibbs, R. J., 1972, Water chemistry of the Amazon River, *Geochim. Cosmochim. Acta* **36**, 1061–1066.

Gieskes, J. M., 1974, The alkalinity-total carbon dioxide system in sea water, in *The Sea*, Vol. 5, Chap. 3, E. D. Goldberg, Ed. Wiley-Interscience, New York.

Ginsburg, R. N. and James, N. P., 1976, Submarine botryoidal aragonite in Holocene reef limestones, Belize, *Geology* **4**, 431–436.

Giresse, P., 1965a, Oolithes ferrugineuses en voie de formation au large du Cap Lopez (Gabon), *Comptes Rendus Acad. Sci. Paris* **260**, 2550–2552.

Giresse, P., 1965b, Observations sur la présence de "Glauconie" actuelle dans les sédiments ferrugineux peu profonds du bassin gabonais, *Comptes Rendus Acad. Sci. Paris* **260**, 5597–5600.

Goldberg, E. D., 1961, Chemical and mineralogical aspects of deep-sea sediments, in *Physics and Chemistry of the Earth*, Vol. 4, L. H. Ahrens, F. Press, K. Rankama, and S. K. Runcorn, Eds., Pergamon Press, New York, pp. 281–302.

Goldberg, E. D., Ed., 1975, *The Nature of Sea Water*, Dahlem Workshop Report, Abakon Verlagsgesellschaft, Berlin, 719 pp.

Goldberg, E. D. and Arrhenius, G. O. S., 1958, Chemistry of Pacific pelagic sediments, *Geochim. Cosmochim. Acta* **13**, 153–212.

Gordon, D. C., Jr., 1971, Distribution of particulate organic carbon and nitrogen at an oceanic station in the Central Pacific, *Deep Sea Res.* **18**, 1127–1134.

Gregor, B., 1967, *The Geochemical Behavior of Sodium*, Drukkerij Holland, N. V., Amsterdam, 66 pp.

Griffin, J. J., Windom, H., and Goldberg, E. D., 1968, The distribution of clay minerals in the World Ocean, *Deep Sea Res.* **15**, 433–459.

Grill, E. V., 1970, Mathematical model for the marine dissolved silicate cycle, *Deep Sea Res.* **17**, 245–266.

Grim, R. E., 1968, *Clay Mineralogy*, 2nd Ed., McGraw-Hill, New York, 596 pp.

Gross, M. G., 1967, Organic carbon in surface sediment from the Northeast Pacific Ocean, *Int. J. Oceanol. Limmol.* **1**, 46–54.

Gross, M. G., McManus, D. A., and Ling, H-Y., 1967, Continental shelf sediment, Northwestern United States, *J. Sediment. Pet.* **37**, 790–795.

Hajash, A., 1975, Hydrothermal processes along mid-ocean ridges: an experimental investigation, *Contrib. Miner. Pet.* **53**, 205–226.

Hardie, L. A., 1967, The gypsum-anhydrite equilibrium at one atmosphere pressure, *Am. Miner.* **52**, 171–200.

Hart, S. R., 1972, Strontium isotopic composition of the oceanic crust, *Carnegie Inst. Yearb.* **71**, 288–290.

Hathaway, J. C. and Sachs, P. L., 1965, Sepiolite and clinoptilolite from the mid-Atlantic ridge, *Am. Miner.* **50**, 852–867.

Hawley, J. E., 1973, Bicarbonate and carbonate ion association with sodium, magnesium, and calcium at 25°C and 0.72 ionic strength, Ph. D. Thesis, Oregon State University.

Hawley, J. and Pytkowicz, R. M., 1969, Solubility of calcium carbonate in seawater at high pressures and 2°C, *Geochim. Cosmochim. Acta* **33**, 1557–1561.

Hay, R. L., 1966, Zeolites and zeolitic reactions in sedimentary rocks, Geol. Soc. Amer. Special Paper 85.

Heath, G. R., 1974, Dissolved silica and deep-sea sediments, in *Studies in Paleo- Oceanography*, W. W. Hay, Ed. Society of Economic Paleontologists Mineralogists, Special Publication 20, pp. 73–93.

Hedberg, H. D., 1936, Gravitational compaction of clays and shales, *Am. J. Sci., 5th Ser.* **31**, 241–287.

Hemley, J. J., Meyer, C., and Richter, D. H., 1961, Some alteration reactions in the system $Na_2O–Al_2O_3–SiO_2–H_2O$, U. S. Geological Survey Professional Paper 424-D, pp. 338–340.

Hey, M. H., 1954, A new review of the chlorites, *Mineral. Mag.* **30**, 277–292.

Holeman, J. N., 1968, The sediment yield of major rivers of the world, *Water Resour. Res.* **4**, 737–747.

Holland, H. D., 1965, The history of ocean water and its effect on the chemistry of the atmosphere, *Proc. Natl. Acad. Sci.* **53**, 1173–1183.

Holland, H. D., 1968, The abundance of CO_2 in the earth's atmosphere through geologic time, in *Origin and Distribution of the Elements*, L. H. Ahrens, Ed., Pergamon Press, New York, pp. 949–954.

Holland, H. D., 1972, The geologic history of sea water—an attempt to solve the problem, *Geochim. Cosmochim. Acta* **36**, 637–657.

Holland, H. D., 1973, Systematics of the isotopic composition of sulfur in the oceans during the Phanerozoic and its implications for atmospheric oxygen, *Geochim. Cosochim. Acta* **37**, 2605–2616.

Holland, H. D., 1974, Marine evaporites and the composition of seawater during the Phanerozoic, in *Studies in Paleo-Oceanography*, W. W. Hay, Ed. Society of Economic Paleontologists Mineralogists, Special Publication 20, pp. 187–192.

Holland, H. D., Kirsipu, T. V., Huebner, J. S., and Oxburgh, U. M., 1964, On some aspects of the chemical evolution of cave waters, *J. Geol.* **72**, 36–67.

Holland, H. D., Quirk, R. F., and Mottl, M. J., 1976, The nonimportance of reverse weathering reactions in the ocean, *Geol. Soc. Am. Abstr. Programs* **8**, 922.

Holser, W. T., 1966, Diagenetic polyhalite in recent salt from Baja California, *Am. Miner.* **51**, 99–109.

Holser, W. T. and Kaplan, I. R., 1966, Isotope geochemistry of sedimentary sulfates, *Chem. Geol.* **1**, 93–135.

Honjo, S., and Erez, J., 1978, Dissolution rates of calcium carbonate in the deep ocean; an in situ experiment in the North Atlantic Ocean, *Earth Planet. Sci. Letters*, to be published.

Honnorez, J., 1972, *La Palagonitisation*, Birkhäuser Verlag, Basel, Stuttgart, 131 pp.

Hostetler, P. B., Hemley, J. J., Christ, C. L., and Montoya, J. W., 1971, Talc-chrysotile equilibrium in aqueous solutions, *Geol. Soc. Am. Abstr. Programs* **3**, 605–606.

Hower, J., Eslinger, E. V., Hower, M. E., and Perry, E. A., 1976, Mechanism of burial metamorphism of argillaceous sediments. 1. Mineralogical and chemical evidence, *Bull. Geol. Soc. Am.* **87**, 725–737.

Hsu, K. J., 1963, Solubility of dolomite and composition of Florida ground waters, *J. Hydrol.* **1**, 288–310.

Humphris, S. E., and Thompson, G., 1978, Hydrothermal alteration of oceanic basalts by seawater, *Geochim. Cosmochim. Acta* **42**, 107–125.

Hurd, D. C., 1973, Interactions of biogenic opal, sediment and sea water in the Central Equatorial Pacific, *Geochim. Cosmochim. Acta* **37**, 2257–2282.

Hurley, P. M., 1966, K-Ar dating of sediments, in *Potassium-Argon Dating*, O. A. Schaeffer and J. Zähringer, Eds. Springer-Verlag, New York, pp. 134–150.

Hurley, P. M., Heezen, B. C., Pinson, W. H. and Fairbairn, H. W., 1963, K–Ar age values in pelagic sediments of the North Atlantic, *Geochim. Cosmochim. Acta* **27**, 393–399.

Iijima, A. and Utada, M., 1966, Zeolites in sedimentary rocks, with reference to the depositional environments and zonal distribution, *Sedimentology* **7**, 327–357.

Ingle, S. E., 1975, Solubility of calcite in the ocean, *Marine Chem.* **3**, 301–319.

Isphording, W. C., 1972, Primary marine attapulgite clays of the Yucatan Platform and Southeastern United States, 21st Clay Minerals Conference, Woods Hole, Massachussetts.

Jehl, V., 1975, Le métamorphism et les fluides associés des roches océaniques de l'Atlantique Nord, Thesis, University of Nancy, France.

Junge, C. E., 1958, The distribution of ammonia and nitrate in rain water over the United States, *Trans. Am. Geophys. Union* **39**, 241–248.

Kaplan, I. R., Emery, K. O., and Rittenberg, S. C., 1963, The distribution and isotopic abundance of sulphur in recent marine sediments off southern California, *Geochim. Cosmochim. Acta* **27**, 297–332.

Kastner, M., and Gieskes, J. M., 1976, Interstitial water profiles and sites of diagenetic reactions, Leg 35, DSDP, Bellinghausen abyssal plain, *Earth Planet. Sci. Letters* **33**, 11–20.

Kazakov, A. V., 1950, Ftorapatitovaya sistema ravnovesii v usloviyakh obrazovaniya osadochnykh porod, *Akad. Nauk SSSR Inst. Geol. Nauk Tr.* **114**, Geological Series No. 40, 1–21.

Keller, W. D., 1963, Diagenesis in clay minerals—A reveiw, in *Clays and Clay Minerals, Proceedings of the Eleventh National Clay Conference*, Vol. 13, Pergamon Press, New York, pp. 136–157.

Kennedy, V. C., 1965, Mineralogy and cation-exchange capacity of sediments from selected streams, U. S. Geological Survey Professional Paper 433-D.

Kester, D. R. and Pytkowicz, R. M., 1967, Determination of the apparent dissociation constants of phosphoric acid in sea water, *Limnol. Oceanogr.* **12**, 243–252.

Kester, D. R. and Pytkowicz, R. M., 1975, Theoretical model for the formation of ion-pairs in sea water, *Marine Chem.* **3**, 365–374.

Kinsman, D. J. J., 1964, Recent carbonate sedimentation near Abu Dhabi, Trucial Coast, Persian Gulf, Ph. D. Thesis, London University.

Kinsman, D. J. J., 1966, Gypsum and anhydrite of recent age, Trucial Coast, Persian Gulf, in *Second Symposium on Salt*, Vol. 1, J. L. Rau, Ed, The Northern Ohio Geological Society, pp. 302–326.

Kinsman, D. J. J., 1967, Huntite from carbonate-evaporite environment, *Am. Miner.* 52, 1332–1340.

Kinsman, D. J. J., 1974a, Calcium sulfate minerals of evaporite deposits: Their primary mineralogy, in *Fourth Symposium on Salt*, Vol. 1, A. H. Coogan, Ed. Northern Ohio Geological Sociey, pp. 343–348.

Kinsman, D. J. J., 1974b, Evaporite deposits of continental margins, in *Fourth Symposium on Salt*, Vol. 1, A. H. Coogan, Ed. Northern Ohio Geological Society, pp. 255–259.

Kinsman, D. J. J., 1975a, Salt floors to geosynclines, *Nature* 255, 375–378.

Kinsman, D. J. J., 1975b, Rift valley basins and sedimentary history of trailing continental margins, in *Petroleum and Global Tectonics*, A. G. Fischer and S. Judson, Eds. Princeton University Press, Princeton, N. J.

Kinsman, D. J. J., 1976, Evaporites: relative humidity control of primary mineral facies, *J. Sediment Pet.* 46, 273–279.

Koblentz-Mishke, O. J., Volkovinsky, V. V., and Kabanova, J. G., 1970, Plankton primary production of the world ocean, in *Scientific Exploration of the South Pacific*, W. S. Wooster, Ed., N. A. S., Washington, pp. 183–193.

Kozary, M. T., Dunlap, J. C., and Humphrey, W. E., 1968, Incidence of saline deposits in geologic time, in *Saline Deposits, A Symposium Based on Papers from the International Conference on Saline Deposits, Houston, Texas, 1962*, Geological Society of America, Special Paper 88, pp. 43–57.

Land, L. S. and Epstein, S., 1970, Late Pleistocene diagenesis and dolomitization, North Jamaica, *Sedimentology* 14, 187–200.

Langmuir, D., 1964, Stability of carbonates in the system $CaO–MgO–CO_2–H_2O$, Ph. D. Thesis, Harvard University.

Lasaga, A. C. and Holland, H. D., 1976, Mathematical aspects of non-steady-state diagenesis, *Geochim. Cosmochim. Acta* 40, 257–266.

Lerman, A., Mackenzie. F. T., and Garrels, R. M., 1975, Modeling of geochemical cycles: phosphorus as an example in *Quantitative Studies in the Geological Sciences*, E. H. T., Whitten, Ed., Geological Society of America, Memoir, 142, pp. 205–218.

Li, Y-H., Takahashi, T., and Broecker, W. S., 1969, Degree of saturation of $CaCO_3$ in the oceans, *J. Geophys. Res.* 74, 5507–5525.

Lisitzin, A. P., 1972, *Sedimentation in the World Ocean*, Society of Economic Paleontologists and Mineralogists Special Publication No. 17, 218 pp.

Lisitsyn, A. P. and Bogdanov, Yu. A., 1968, Suspended amorphous silica in the Pacific Ocean, *Oceanol. Res.* 18, 5–45.

Lister, C. R. B., 1973, Hydrothermal convection at sea-floor spreading centers: source of power or geophysical nightmare? *Geol. Soc. Am. Abstr. Programs* 5, 74.

Livingstone, D. A., 1963, Chemical composition of rivers and lakes, in *Data of Geochemistry*, 6th ed., M. Fleischer, Ed., U.S. Geological Survey Professional Paper 440-G.

Lotze, F., 1957, *Steinsalz und Kalisalze*, 2nd rev. ed., Gebrüder Bornträger, Berlin, 465 pp.

Lyakhin, Yu. I., 1968, Calcium carbonate saturation of Pacific water, *Oceanology* (Transl.) 8, 44–58.

Mackenzie, F. T. and Garrels, R. M., 1965, Silicates: reactivity with seawater, *Science* **150**, 57–58.

Mackenzie, F. T. and Garrels, R. M., 1966a, Silica-bicarbonate balance in the ocean and early diagenesis, *J. Sedimen. Pet.* **36**, 1075–1084.

Mackenzie, F. T. and Garrels, R. M., 1966b, Chemical mass balance between rivers and oceans, *Am. J. Sci.* **264**, 507–525.

Mackenzie, F. T., Garrels, R. M., Bricker, O. P., and Bickley, F., 1967, Silica in sea-water: control by silica minerals, *Science* **155**, 1404–1405.

Manheim, F. T. and Sayles, F. L., 1970, Brines and interstitial brackish water in drill cores from the deep Gulf of Mexico, *Science* **170**, 57–61.

Matthews, D. H., 1962, Altered lavas from the floor of the Eastern Atlantic Ocean, *Nature* **194**, 368–369.

Maynard, J. B., 1976, The long-term buffering of the oceans, *Geochim. Cosmochim. Acta* **40**, 1523–1532.

McDuff, R. E., and Gieskes, J. M., 1976, Calcium and magnesium profiles in DSDP interstitial waters: diffusion or reaction?, *Earth Planet. Sci. Letters* **33**, 1–10.

McMurtry, G. M., 1975, Geochemical investigations of sediments across the Nazca Plate at 12°, Masters Thesis, University of Hawaii.

Melson, W. G. and van Andel, T. H., 1966, Metamorphism in the Mid-Atlantic Ridge, 22°N latitude, *Marine Geol.* **4**, 165–186.

Melson, W. G., Thompson, G., and van Andel, T. H., 1968, Volcanism and metamorphism in the Mid-Atlantic Ridge, 22°N latitude, *J. Geophys. Res.* **73**, 5925–5941.

Millot, G., 1964, *Géologie des Argiles*, Masson, Paris.

Millot, G., 1967, Signification des études recentes sur les roches argileuses dans l'interprétation des faciès sédimentaires (y compris les séries rouges), *Sedimentology* **8**, 259–280.

Miyashiro, A., Shido, F., and Ewing, M., 1971, Metamorphism in the Mid-Atlantic Ridge near 24° and 30°N, *Philos. Trans. R. Soc. Lond. A* **268**, 589–603.

Moberly, R., Jr., 1963, Amorphous marine muds from tropically weathered basalts, *Am. J. Sci.* **261**, 767–772.

Moberly, R., Jr., 1968, Personal communication.

Mottl, M. J., 1976, Chemical exchange between sea water and basalt during hydrothermal alteration of the oceanic crust, Ph.D. Thesis, Harvard University.

Mottl, M. J., Corr, R. F., and Holland, H. D., 1974, Chemical exchange between sea water and mid-ocean ridge basalt during hydrothermal alterations: an experimental study, *Geol. Soc. Am. Abstr. Programs* **6**, 879–880.

Mottl, M. J. and Holland, H. D., 1978, Chemical Exchange during hydrothermal alteration of basalt by seawater. I. Experimental results for major elements, *Geochim. Cosmochim. Acta*, in press.

Murray, J. and Renard, A. F., 1891, *Deep Sea Deposits, Reports Scientific Results Exploration Voyage HMS Challenger*, 1873–1876, Her Majesty's Stationery Office, 525 pp.

Nafe, J. E. and Drake, C. L., 1957, Variations with depth in shallow and deep water marine sediments of porosity, density, and the velocities of compressional and shear waves, *Geophysics* **22**, 523–552.

Nelson, B. W. and Roy, R., 1958, Synthesis of the chlorites and their structural and chemical constitution, *Am. Miner.* **43**, 707–725.

Newell, N. D., Rigby, J. K., Fischer, A. G., Whiteman, A. J., Hickox, J. E. and Bradley, J. S., 1953, *The Permian Reef Complex of the Guadalupe Mountains Region, Texas and New Mexico: A Study in Paleoecology,* Hafner, New York, 236 pp.

Newman, J. W., Parker, P. L. and Behrens, E. W., 1973, Organic carbon isotope ratios in Quaternary cores from the Gulf of Mexico, *Geochim. Cosmochim. Acta* **37**, 225–238.

Olausson, E., 1967, Climatological, geoeconomical, and paleooceanographical aspects of carbonate deposition, in *Progress In Oceanography*, Vol. 4, M. Sears, Ed., Pergamon Press, New York, 245–265.

Park, K., 1966, Deep-sea pH, *Science* **154**, 1540–1542.

Park, K., 1968, Sea water hydrogen-ion concentration: vertical distribution, *Science* **162**, 357–358.

Parry, W. T. and Reeves, C. C., Jr., 1968, Sepiolite from Pluvial Mound Lake, Lynn and Terry Counties, Texas, *Amer. Miner.* **53**, 984–993.

Pettijohn, F. J., Potter, P. W., and Siever, R., 1972, *Geology of Sand and Sandstone,* Springer–Verlag, New York, 600 pp.

Platford, R. F., 1965a, The activity coefficient of sodium chloride in seawater, *J. Marine Res.* **23**, 55–62.

Platford, R. F., 1965b, Activity coefficient of the sodium ion in seawater, *J. Fish. Res. Board Can.* **22**, 885–889.

Platford, R. F. and Dafoe, T., 1965, The activity coefficient of sodium sulfate in sea water, *J. Marine Res.* **23**, 63–68.

Platt, T. and Subba Rao, D. V., 1973, Fisheries Research Board of Canada, Technical Report No. 370.

Porrenga, D. H., 1965, Chamosite in recent sediments of the Niger and Orinoco Deltas, *Geol. Mijnbouw.* **44**, 400–403.

Porrenga, D. H., 1966, Clay minerals in recent sediments of the Niger Delta, in *Clays and Clay Minerals, Proceedings of the 14th National Conference of Clays and Clay Minerals,* Pergamon Press, New York, pp. 221–233.

Porrenga, D. H., 1967, Glauconite and chamosite as depth indicators in the marine environment, *Marine Geol.* **5**, 495–501.

Pytkowicz, R. M., 1965, Calcium carbonate saturation in the oceans, *Limnol. Oceanogr.* **10**, 220–225.

Pytkowicz, R. M. and Connors, D. N., 1964, The high pressure solubility of calcium carbonate in sea water, *Science* **144**, 840–841.

Pytkowicz, R. M., Atlas, E., and Culberson, C. H., 1975, Chemical equilibrium in seawater, in *Marine Chemistry in the Crustal Environment*, T. M. Church, Ed., American Chemical Society, Symposium Series 18, pp. 1–24.

Pytkowicz, R. M. and Kester, D. R., 1971, The physical chemistry of seawater, in *Annual Review Oceanographic Marine Biology* Vol. 9, Harold Barnes, Ed., Harper & Row, New York, pp. 11–60.

Redfield, A. C., Ketchum, B. H., and Richards, F. A., 1963, The influence of organisms on the composition of sea water, in *The Sea*, Vol. 2, M. N. Hill, Ed., Interscience, New York, pp. 26–77.

Rex, R. W., 1967, Authigenic silicates formed from basaltic glass by more than 60 million years' contact with sea water, Sylvania Guyot, Marshall Islands, in *Clays and Clay Minerals, Proceedings of the National Conference on Clay and Clay Minerals*, Vol. 15, Pergamon Press, New York, pp. 195–203.

Richards, F. A. and Redfield, A. C., 1954, A correlation between the oxygen content of seawater and the organic content of marine sediments, *Deep Sea Res.* **1**, 279–281.

Richter-Bernburg, G., 1955, Stratigraphische Gliederung des deutschen Zechsteins, *Z. Dtsch. Geol. Ges.* **105**, 843–854.

Rieke, H. H., III, and Chilingarian, G. V., 1974, *Compaction of Argillaceous Sediments*, Elsevier, New York, 424 pp.

Rivers, M. L., 1976, The chemical effects of low-temperature alteration of sea floor basalt, Bachelor's Thesis, Harvard University.

Ronov, A. B., 1958, Organic carbon in sedimentary rocks (in relation to the presence of petroleum), *Geokhimiya* (Transl.) 510–536.

Rosenberg, P. E. and Holland, H. D., 1964, Calcite-dolomite-magnesite stability relations in solutions at elevated temperatures, *Science* **145**, 700–701.

Russell, K. L., 1970, Geochemistry and halmyrolysis of clay minerals Rio Ameca, Mexico, *Geochim. Cosmochim. Acta* **34**, 893–907.

Sackett, W. M. and Thompson, R. R., 1963, Isotopic organic carbon composition of recent continental derived clastic sediments of eastern Gulf Coast, Gulf of Mexico, *Bull. Am. Assoc. Pet. Geol.* **47**, 525–531.

Sayles, F. L. and Fyfe, W. S., 1973, The crystallization of magnesite from aqueous solutions, *Geochim. Cosmochim. Acta* **37**, 87–99.

Sayles, F. L. and Mangelsdorf, P. C., Jr., 1976, The equilibration of clay minerals with seawater: exchange reactions, unpublished manuscript.

Schink, D. R., Guinasso, N. L., Jr., and Fanning, K. A., 1975, Processes affecting the concentration of silica at the sediment–water interface of the Atlantic Ocean, *J. Geophys. Res.* **80**, 3013–3031.

Scott, R. B., Rona, P. A., McGregor, B. A., and Scott, M. R., 1974, The TAG hydrothermal field, *Nature* **251**, 301–302.

Scott, R. B., Malpas, J., Rona, P. A., and Udintsev, G., 1976, Duration of hydrothermal activity at an oceanic spreading center, Mid-Atlantic Ridge (lat 26°N), *Geology* **4**, 233–236.

Shaw, D. B. and Weaver, C. E., 1965, The mineralogical composition of shales, *J. Sedimen. Pet.* **35**, 213–222.

Sheppard, R. A., Gude, A. J., III, and Griffin, J. J., 1970, Chemical composition and physical properties of phillipsite from the Pacific and Indian Oceans, *Am. Miner.* **55**, 2053–2062.

Sholkovitz, E., 1973, Interstitial water chemistry of Santa Barbara Basin sediments, *Geochim. Cosmochim. Acta* **37**, 2043–2073.

Shultz, D. J. and Calder, J. A., 1976, Organic carbon $^{13}C/^{12}C$ variations in estuarine sediments, *Geochim. Cosmochim. Acta* **40**, 381–385.

Siever, R., 1957, The silica budget in the sedimentary cycle, *Am. Miner.* **42**, 821–841.

Siever, R., 1968, Establishment of equilibrium between clays and seawater, *Earth Planet. Sci. Lett.* **5**, 106–110.

Siever, R., Beck, K. C., and Berner, R. A., 1965, Composition of interstitial waters of modern sediments, *J. Geol.* **73**, 39–73.

Siever, R. and Kastner, M., 1967, Mineralogy and petrology of some Mid-Atlantic Ridge sediments, *J. Marine Res.* **25**, 263–278.

Siever, R. and Woodford, N., 1973, Sorption of silica by clay minerals, *Geochim. Cosmochim. Acta* **37**, 1851–1880.

Silker, W. B., 1964, Variations in elemental concentrations in the Columbia River, *Limnol. Oceanogr.* **9**, 540–545.

Sillén, L. G., 1961, The physical chemistry of seawater, in *Oceanography*, Mary Sears, Ed., AAAS, Washington, D.C., pp. 549–581.

Slowey, J. F. and Hood, D. W., 1971, Copper, manganese, and zinc concentrations in Gulf of Mexico waters, *Geochim. Cosmochim. Acta* **35**, 121–138.

Spencer, C. P., 1975, The micronutrient elements, in *Chemical Oceanography*, 2nd ed., Vol. 2, Chap. 11, J. P. Riley and G. Skirrow, Ed., Academic Press, New York.

Spooner, E. T. C. and Fyfe, W. S., 1973, Sub-sea-floor metamorphism, heat and mass transfer, *Contrib. Miner. Petrol.* **42**, 287–304.

Stewart, F. H., 1963, Marine evaporites, in *Data of Geochemistry*, 6th ed., Chap. Y, U.S. Geological Survey Professional Paper 440-Y.

Strakhov, N. M., 1958, Facts and hypotheses concerning the genesis of dolomite rocks, *Izv. Akad. Nauk SSSR Ser. Geol.* No. 6, 3–22.

Strickland, J. D. H., 1965, Production of organic matter in the primary stages of the marine food chain, in *Chemical Oceanography*, Vol. 1, Chap. 12, J. P. Riley and G. Skirrow, Eds. Academic Press, New York.

Styrt, M., 1977, Variations in lithium content between igneous and sedimentary rocks, Bachelor's Thesis, Harvard University.

Sureau, J-F., 1974, Étude expérimentale de la dolomitisation de la calcite a 150°C dans le système: $CaCO_3$–$MgCl_2$–H_2O–pCO_2, Ph. D. Thesis, University of Paris.

Swindale, L. D. and Fan, P-F., 1967, Transformation of gibbsite to chlorite in ocean bottom sediments, *Science* **157**, 799–800.

Tarrant, R. F., Lu, K. C., Chen, C. S., and Bollen, W. B., 1968, Nitrogen content of precipitation in a coastal Oregon forest opening, *Tellus* **20**, 554–556.

Thompson, G., 1973, A geochemical study of the low-termperature interaction of sea water and oceanic igneous rocks, *Am. Geophys. Union Trans.* **54**, 1015–1019.

Thompson, M. E., 1966, Magnesium in sea water: an electrode measurement, *Science* **153**, 866–867.

Timm, B. C. and Maricelli, J. J., 1953, Formation waters in southwest Louisiana, *Bull. Am. Assoc. Pet. Geol.* **37**, 394–409.

Tómasson, J. and Kristmannsdóttir, H., 1972, High temperature alteration minerals and thermal brines, Reykjanes, Iceland, *Contrib. Miner. Pet.* **36**, 123–134.

Trask, P. D. and Patnode, H. W., 1942, *Source Beds of Petroleum*, The American Association of Petroleum Geologists, Tulsa, Oklahoma, 566 pp.

Turekian, K. K., 1964, The marine geochemistry of strontium, *Geochim. Cosmochim. Acta* **28**, 1479–1496.

Turekian, K. K., 1965, Some aspects of the geochemistry of marine sediments, in *Chemical Oceanography*, Vol. 2, Chap. 16, J. P. Riley and G. Skirrow, Eds., Academic Press, New York.

Turekian, K. K., 1969, The ocean, streams, and the atmosphere, in *Handbook of Geochemistry*, Vol. 1, Chap. 10, K. H. Wedepohl, Ed. Springer-Verlag, New York.

Vaccaro, R. F., 1965, Inorganic nitrogen in sea water, in *Chemical Oceanography*, Vol. 1, Chap. 9, J. P. Riley and G. Skirrow, Eds., Academic Press, New York.

van Andel, T. H., 1964, Recent marine sediments of Gulf of California, in *Marine Geology of the Gulf of California*, T. H. van Andel and G. G. Shor, Jr., Eds., Memoir 3, The American Association of Petroleum Geologists, Tulsa, Oklahoma, pp. 216–310.

Von Gärtner, H-R. and Schellmann, W., 1965, Rezente Sedimente im Küstenbereich der Halbinsel Kaloum, Guinea, *Miner. Pet. Mitt.* **10**, 349–367.

Weyl, P. K., 1967, The solution behavior of carbonate materials in seawater, in *Studies in Tropical Oceanography*, University of Miami, Miami, pp. 178–228.

Whitfield, M., 1975, The extension of chemical models for seawater to include trace components at 25°C and 1 atm pressure, *Geochim. Cosmochim. Acta* **39**, 1545–1557.

Williams, D. L. and von Herzen, R. P., 1974, Heat loss from the earth: new estimate, *Geology* **2**, 327–328.

Williams, P. M. 1968, Organic and inorganic constituents of the Amazon River, *Nature* **218**, 937–938.

Wolery, T. J. and Sleep, N. H., 1976, Hydrothermal circulation and geochemical flux at mid-ocean ridges, *J. Geol.* **84**, 249–275.

Wolgemuth, K., 1970, Barium analyses from the first Geosecs test cruise, *J. Geophys. Res.* **75**, 7686–7687.

Wolgemuth, K. and Broecker, W. S., 1970, Barium in sea water, *Earth Planet. Sci. Lett.* **8**, 372–278.

Wollast, R., 1974, The silica problem, in *The Sea*, Vol. 5, Chap. 11, E. D. Goldberg, Ed., Wiley-Interscience, New York.

Wollast, R., Mackenzie, F. T., and Bricker, O. P. 1968, Experimental precipitation and genesis of sepiolite at earth surface conditions, *Am. Miner.*, **53**, 1645–1662.

CHAPTER VI

THE CHEMISTRY OF THE ATMOSPHERE

The atmosphere envelops the earth and is a major participant in surface and near-surface processes. The hydrologic cycle, chemical and mechanical weathering, photosynthesis and decay on land and in the oceans, waves, currents, and much of marine chemistry depend on the atmosphere. References to the chemistry of the atmosphere have, therefore, been legion in earlier chapters. This chapter focuses on the atmosphere per se, and serves as a vehicle for discussions of the geochemical cycles of carbon, nitrogen, and oxygen, which depend in a major way on the composition of the atmosphere and on its communication with the solid earth, the oceans, and interplanetary space.

As a reservoir the atmosphere is similar to the oceans. Both reservoirs are reasonably homogeneous, and the residence time of most of their constituents is much less than the age of the earth; but there are some important differences. Whereas the oceans are reasonably homogeneous throughout, heterogeneities in the atmosphere become major at heights in excess of 100 km. Although virtually the entire mass of the atmosphere lies below 100 km, processes above 100 km are important for the chemistry of the atmosphere as a whole since they are critical for the escape of hydrogen and helium. In the oceans chemical disequilibrium is largely due to the sluggishness of many chemical reactions between 0 and 30°C. In the atmosphere this source of chemical disequilibrium is joined by disturbances due to solar radiation. Especially in the upper atmosphere, the presence of free radicals and the formation of chemically unstable but biologically important trace gases such as ozone owe their presence largely to the flux of energetic photons from the sun. Living organisms, which are perhaps the most spectacular non-equilibrium products of solar radiation, continue to have a profound effect on the chemistry of the atmosphere, and are themselves vulnerable to relatively minor changes in atmospheric chemistry.

Table 6-1
The Composition of the Atmosphere at Ground Level[a]

Component	Concentration at Ground Level (ppm by volume)[b]	Residence Time
Noble Gases		
Helium	5.24 ± 0.04 (1)	2×10^6 yr for escape
Neon	18.18 ± 0.04 (1)	largely accumulating
Argon	9340 ± 10 (1)	largely accumulating
Krypton	1.14 ± 0.01 (1)	largely accumulating
Xenon	0.087 ± 0.001 (1)	largely accumulating
Carbon Compounds		
Carbon dioxide, CO_2	320 (2)	ca. 10 yr for cycling through the biosphere
Carbon monoxide, CO	0.06–0.2 (10)	0.5 yr (10)
Methane, CH_4	1.4 (6)	2.6–8 yr (6)
Formaldehyde, CH_2O	0–0.01 (3)	
$CFCl_3$	130×10^{-6} (12, 13, 16)	45–68 yr (11)
CF_2Cl_2	230×10^{-6} (12, 13, 16)	45–68 yr (11)
CCl_4	$100 - 250 \times 10^{-6}$ (12, 13, 16)	
CH_3Cl	500×10^{-6} (16)	
Oxygen, Hydrogen, and their Compounds		
Oxygen	$(20.946 \pm 0.002) \times 10^4$ (1)	6000 yr for cycling through the biosphere
Hydrogen	0.55 (9)	4–7 yr (9)
Water	40–40,000 (3)	
Ozone	0.01–0.03 (3)	
Nitrogen and its Componds		
Nitrogen, N_2	$(78.084 \pm 0.004) \times 10^4$ (1)	4×10^8 yr for cycling through sediments
Nitrous Oxide, N_2O	0.33 ± 0.01 (14)	5–50 yr (7, 15)
Nitric Oxide, NO / Nitrogen Dioxide, NO_2	0.001 (2)	(<1 month)
Ammonia, NH_3	0.006–0.020	(∼1 day)
Sulfur Compounds		
Sulfur Dioxide, SO_2	0.001–0.004	hours to weeks (8)
Hydrogen Sulfide, H_2S	⩽ 0.0002 (8)	⩽ 1 day (8)
Dimethyl sulfide $(CH_3)_2S$		

[a] Sources: (1) Nicolet (1960); (2) Robinson and Robbins (1968); (3) Junge (1963); (4) Junge et al. (1971); (5) Cadle (1973); (6) Ehhalt (1973); (7) McElroy et al. (1977); Junge (1974); (8) Hill (1973); (9) Schmidt (1974); (10) Seiler (1974); (11) Wofsy et al. (1975); (12) Lovelock (1974); (13) Wilkniss et al. (1975); (14) Rasmussen (1977a); (15) Hahn and Junge (1977); (16) Rasmussen (1977b).

[b] Atmospheric background, unaffected by local pollution.

THE CHEMICAL COMPOSITION AND STRUCTURE
OF THE ATMOSPHERE

The chemical composition of the atmosphere near ground level is summarized in Table 6-1. The order follows the sequence in which the abundance of the various gases is discussed in this chapter. With the exception of water vapor, soluble gases, and particles, the troposhere is quite well mixed. Variations in the concentration of the major components are slight, and the history of the search for significant variations in the ratio of their concentration makes fascinating reading (see, for instance, Mirtov, 1964). On the other hand the distribution of some minor components,

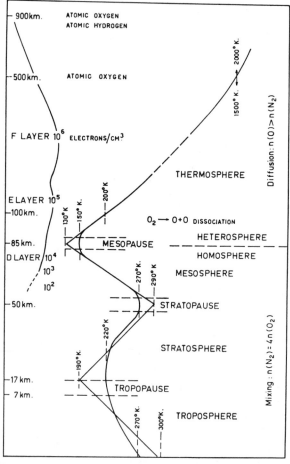

Figure 6-1. The distribution of temperature, electron density, and chemical components in the atmosphere (Nicolet, 1960).

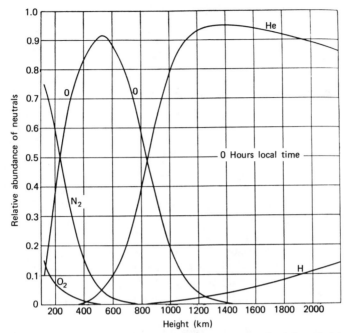

Figure 6-2. The relative abundance of neutral particles as a function of altitude at 0 hr local time (MacDonald, 1963). Copyrighted by American Geophysical Union.

particularly those due entirely or in large part to human activities, is quite heterogeneous.

Large deviations from the near-surface composition are confined to heights above 100 km. Above the mesopause in Figure 6-1 atomic oxygen rapidly becomes the most important neutral particle in the atmosphere. At still higher elevations neutral oxygen is replaced by helium, as shown in Figure 6-2, and finally by hydrogen in the region where the earth's atmosphere merges with the interplanetary gas (Chapman, 1960).

THE NOBLE GASES

With the exception of argon, the noble gases are all quite minor constituents of the earth's atmosphere. They are chemically rather dull, and this, in part, accounts for their important place in the theory of terrestrial evolution.

The dominance of argon among the noble gases in the atmosphere is due to the presence of a large quantity of ^{40}Ar. In 1937 von Weizsäcker suggested that most of the atmospheric ^{40}Ar was produced by the decay of ^{40}K within the earth, and has been subsequently released to the atmosphere. This view is now completely accepted, and has served as the starting point for a number of models for the degassing history of the earth (see, for instance, Damon and Kulp, 1958; Turekian; 1964, and Tolstikhin, 1975). There are major shortcomings in these models due to the uncertainties in the potassium content of the earth, and in the relationship between the argon content of the solid earth and the rate of argon degassing. Nevertheless, some useful and provocative results have been obtained.

The mean potassium concentration of the earth is almost certainly less than 880 ppm, the best estimate for the concentration of potassium in chondritic meteorites. Ganapathy and Anders (1974) suggest that the mean terrestrial potassium abundance is 170 ppm. The ^{40}K associated with this quantity of terrestrial potassium would have generated 14×10^{19} g of ^{40}Ar in 4.5 billion yr. The atmosphere now contains 6.8×10^{19} g of Ar, nearly all of which is ^{40}Ar (see Table 6.2). If Ganapathy and Anders' (1974) figure is correct, approximately half of the earth's ^{40}Ar is now in the atmosphere. If the potassium content of the earth is only 85 ppm, essentially all of the ^{40}Ar produced during earth history would now be in the atmosphere. Attempts to define the degree of Ar degassing more precisely by compiling analyses of crustal and mantle rocks have been largely unsuccessful, because analytical data, especially for the argon content of the mantle, are completely inadequate.

Until recently there was no compelling evidence for the presence of any residual primordial gas in the interior of the earth. The discovery of abnormal rare gases in volcanic gases by Mamyrin, et al. (1972), in volcanic xenoliths by Tolstikhin et al. (1972), in volcanic glasses by Lupton and Craig (1975), and in intermediate and deep-sea water especially near the East Pacific Rise by Clarke et al. (1969), Craig et al. (1975), and Jenkins et al. (1978) has demonstrated beyond reasonable doubt that primordial helium, and presumably other primordial gases as well, are still being discharged from the earth's interior into the atmosphere (see Table 6-3). The $^3He/^4He$ ratio of $(1.3-1.7) \times 10^{-5}$ in basaltic glasses is approximately 10 times greater than the atmospheric ratio of 0.14×10^{-5}. Since the $^3He/^4He$ ratio in natural gas and other subsurface gases is generally much smaller than in the atmosphere, a recycling origin for ^3He in basalt involving subduction and remelting of sediments is very unlikely.

Neon has also been found in basaltic glasses (Lupton and Craig, 1975). Its concentration appears to be rather variable. If the Ne concentration in mantle rocks as a whole is equal to the mean of the Ne concentration of the few currently analyzed samples, roughly equal quantities of Ne are now present in the atmosphere and in the mantle. However, it seems likely that basaltic glasses contain more Ne than average mantle material; if so, then considerably more than 50% of the earth's Ne is now in the atmosphere.

Wasserburg et al. (1963) found that the atom ratio of ^4He to radiogenic ^{40}Ar is between 0.2 and 5 in many samples of natural gas. This range agrees

Table 6-2

The Isotopic Composition of the Noble
Gases in the Atmosphere[a]

Gas	*Abundance* (%)
Helium	
^{3}He	1.4×10^{-4} [b]
^{4}He	~ 100
Neon	
^{20}Ne	90.5 ± 0.07 (2)
^{21}Ne	0.268 ± 0.002 (2)
^{22}Ne	9.23 ± 0.07 (2)
Argon	
^{36}Ar	0.337
^{38}Ar	0.063
^{40}Ar	99.600
Krypton	
^{78}Kr	0.354 ± 0.002
^{80}Kr	2.27 ± 0.01
^{82}Kr	11.56 ± 0.02
^{83}Kr	11.55 ± 0.02
^{84}Kr	56.90 ± 0.1
^{86}Kr	17.37 ± 0.2
Xenon	
^{124}Xe	0.096
^{126}Xe	0.090
^{128}Xe	1.919
^{129}Xe	26.44
^{130}Xe	4.08
^{131}Xe	21.18
^{132}Xe	26.89
^{134}Xe	10.44
^{136}Xe	8.87

[a]After Strominger et al. (1958); Eberhardt et
al. (1965).

[b]Jenkins, pers. comm. 1978.

Table 6-3
The Distribution of the Noble Gases

Gas	Atmosphere[a] (g)	Oceans[b] (g)	Continental Crust[c] (g)	Mantle (g)
Helium	3.8×10^{15}	0.01×10^{15}	(170×10^{15})	$(300 \times 10^{15})^{d}$
Neon	6.6×10^{16}	0.02×10^{16}	$(<0.2 \times 10^{16})$	$\leqslant 7 \times 10^{16e}$
Argon	6.8×10^{19}	0.07×10^{19}	(0.06×10^{19})	$\begin{cases} (<0.3 \times 10^{19})^{d} \\ (7 \times 10^{19})^{f} \end{cases}$
Krypton	1.7×10^{16}	0.04×10^{16}	$(<0.8 \times 10^{16})$	—
Xenon	2.1×10^{15}	—	$(<9. \times 10^{15})$	—

[a] Based on data in Table 6-1.

[b] Based on data of Bieri et al. (1966, 1968).

[c] Based on data of Canalas et al. (1968); Turekian and Wedepohl (1961), see text.

[d] Based on helium and argon content of ultramafic inclusions in basaltic rocks (J. J. Naughton, personal communication, 1969).

[e] Based on Ne content of basaltic glasses (Lupton and Craig, 1975).

[f] Based on an average potassium content of 170 ppm for the earth as a whole; see text.

well with calculated values based on the generation rate of ^{4}He and ^{40}Ar from the ^{238}U, ^{235}U and ^{232}Th decay series, and from ^{40}K in average crustal rocks. It is likely that the mean value of the ^{4}He/^{40}Ar ratio in gases entering the atmosphere is on the order of unity; yet the atomic ^{4}He/^{40}Ar ratio in the atmosphere is only $1/1800$. ^{4}He must, therefore, be escaping from the atmosphere on a geologically short time scale. A rough value for the residence time of ^{4}He in the atmosphere can be calculated if it is assumed that the ^{4}He/^{40}Ar ratio in gases entering the atmosphere has always been roughly unity and that both gases have entered the atmosphere at a more or less constant rate. The mean annual rate of input of ^{40}Ar, and hence of ^{4}He, has then been approximately

$$\frac{6.8 \times 10^{19} g \ ^{40}Ar}{40 g \ ^{40}Ar/mol} \times \frac{6.02 \times 10^{23} atoms/mol}{4.5 \times 10^{9} yr} = 2.3 \times 10^{32} atoms/yr$$
$$= 1.4 \times 10^{6} \ atoms/cm^{2} \ sec$$

The number of ^{4}He atoms in the atmosphere is

$$\frac{3.8 \times 10^{15} g \ ^{4}He}{4.0 g \ ^{4}He/mol} \times 6.02 \times 10^{23} atoms/mol = 5.7 \times 10^{38} atoms$$

The residence time $\tau_{^{4}He}$ is, therefore, approximately

$$\tau_{^{4}He} = \frac{0.693}{\lambda_{^{4}He}} \approx \frac{0.693 \times 5.7 \times 10^{38} \ atoms}{2.3 \times 10^{32} \ atoms/yr} \approx 1.7 \times 10^{6} \ yr$$

A variety of mechanisms for helium escape at the requisite rate have been proposed. Thermal escape at normal exospheric temperatures is too slow (MacDonald, 1963), but could be sufficiently fast during the relatively brief high-temperature episodes during which temperatures as high as 2000°K are probably reached. Axford (1968) has suggested that ion flow in the polar regions, where the magnetic field lines are open, is the dominant escape mechanism. Sheldon and Kern (1972) suggest that He is swept away by the solar wind during reversals of the geomagnetic field. Hunten (1973) has pointed out that considerable excess heating is observed at high altitudes in connection with magnetic storms, and that high polar temperatures during 2% of time could be responsible for the helium loss since He can be drawn laterally at great heights into the polar regions.

Although the escape of He from the atmosphere is still not completely understood, a 10^6 yr residence time for escape seems quite reasonable. The bothersome problem of the $^3He/^4He$ ratio in the atmosphere seems to have been solved by the discovery of ^3He leakage from the solid earth at a rate considerably in excess of the rate of ^3He production in the atmosphere by the decay of cosmic ray produced ^3H.

The terrestrial inventories of krypton and xenon are even less well known than are the inventories of helium, neon, and argon. Canalas et al. (1968) have measured noble gas concentrations in some shales, and Turekian and Wedepohl (1961) have summarized the rather scant noble gas data for igneous rocks (see Table 6-4). It is probable that at least half of the total quantity of terrestrial krypton is now in the atmosphere. Somewhat surprisingly, the xenon content of shales is sufficiently large that there may be more xenon stored in sedimentary rocks than is present in the atmosphere. If so, the total terrestrial inventory of xenon may well be 5 to 10 times the xenon content of the atmosphere.

Table 6-4

Abundance of Noble Gases in Shales and in Igneous Rocks[a]

	Shales[b]		
Gas	*Fig Tree*	*Mt. Pleasant*	*Igneous Rocks*[c]
Helium	203–4250	—	6000
Neon	8.7–27	—	7.7
Argon	—	—	2200
Krypton	5.4–21	15–16	0.42
Xenon	5.6–16	8.8–11	0.034

[a]cc at STP per 10^8 g of rock.

[b]Data from Canalas et al., 1968.

[c]Data from Turekian and Wedepohl, 1961.

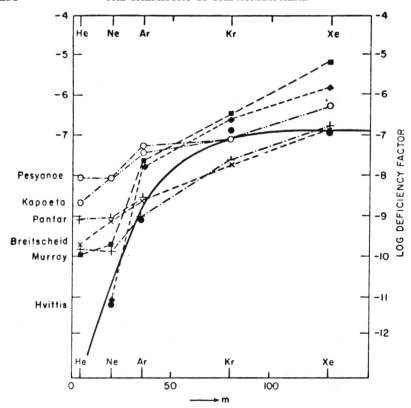

Figure 6-3. Deficiency factors of rare gases in meteorites and in the atmosphere (solid line) plotted against the atomic weight; Suess (1966).

The discovery that rather large quantities of xenon are trapped in shales removes a final difficulty in understanding the relationship between the noble gas content of the earth and of meteorites. Signer and Suess (1963) defined the deficiency factor of an element as the ratio of its abundance in a meteorite or planet to its cosmic abundance. Figure 6-3 shows that the deficiency factor of the rare gases in meteorites increases with increasing atomic number. The deficiency factor of helium in meteorites is about 10^{-8} to 10^{-10}, the deficiency factor of xenon is about 10^{-5} to 10^{-7}. The deficiency factor of these elements in the earth has been calculated by assuming that all of the rare gases are in the atmosphere, but neglecting the ^{40}Ar content of the atmosphere. The similarity of the terrestrial pattern to the meteorite pattern is rather striking. The relative deficiency of terrestrial helium is largely due to helium escape from the atmosphere; the relative deficiency of xenon is probably due to the trapping of xenon in sedimentary rocks. If the terrestrial pattern in Figure 6-3 were corrected for gases that are still retained in the mantle and for retrapped rare gases, it would probably agree remarkably well with the meteoritic abundance pattern of the noble gases. Helium would be the lone exception. This suggests that the earth and meteorites have a similar origin, and either that a primordial atmosphere rich in rare gases never surrounded the earth or that such a primordial atmosphere was somehow removed early in the evolution of the earth.

CARBON COMPOUNDS

Three carbon compounds dominate the atmospheric chemistry of carbon (see Table 6-1). CO_2 is by far the most abundant of these. Its mean concentration is currently close to 320 ppm by volume, and is expected to reach 390 ppm by the end of the century. The concentration of CH_4 varies rather little in the troposphere from its mean value of ca. 1.4 ppm by volume. In the stratosphere the ratio of methane to the sum of the other atmospheric components (mixing ratio) decreases steadily upward (see Figure 6-4) to values of about 0.25 ppm by volume at a height of 50 km. The concentration of CO in the lower troposphere is highly variable. Its concentration in the background troposphere is generally between 0.05 and 0.2 ppm; in smoggy city air CO concentrations of 40 ppm are not uncommon (Cadle, 1973). The volume fraction of CO decreases in the stratosphere, but a good deal less rapidly than the mixing ratio of methane. Ehhalt et al. (1975) report a value of 0.05 ± 0.01 ppmV in an air sample taken between an altitude of 40 and 50 km.

Among the carbon compounds present in concentrations below 100 ppb those containing halogens have recently received particular attention, because chlorine is an efficient catalyst for the photochemical destruction of ozone in the stratosphere. Wilkniss et al. (1975) reported Freon-11 ($CFCl_3$) concentrations between 82 and 148 ppt (parts in 10^{12}) and CCl_4 in concentrations between 112 and 258 ppt in air over the Greenland Sea and the Arctic Ocean in January 1974. Rasmussen (1977b) has found that Freon-11 concentrations in the troposphere are generally close to 130 ppt, Freon-12 concentrations close to 230 ppt, CH_3Cl concentrations close to 500 ppt, and CCl_4 concentrations close to 120 ppt (see Table 6-1).

The Production and Destruction of Methane

Most methane in the atmosphere is of recent biologic origin (Ehhalt, 1973, 1974). Since the ^{14}C content of methane in the atmosphere is approximately 80% that of modern wood, contributions to atmospheric CH_4 from the combustion of fossil fuels and leakage of natural gas from oil and gas wells

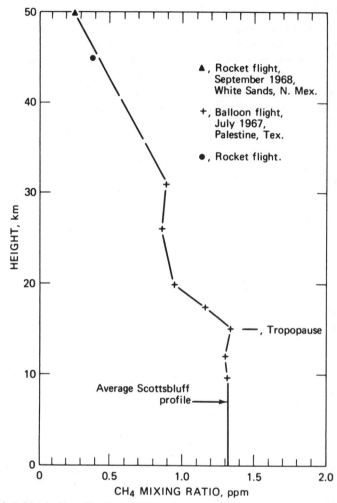

Figure 6-4. Vertical profile of the CH_4 volume mixing ratio in the stratosphere. The balloon data are supplemented by the results of a rocket flight that collected an integrated sample at altitudes between 44 and 62 km. The weighted mean altitude of this sample is 50 km; after Ehhalt (1973) and Ehhalt et al. (1975).

cannot exceed about 20% of the total production rate, and are probably considerably smaller.

The current annual production of methane is still not well known. Fermentative decay of organic matter in swamps and paddy fields and enteric fermentation in animals are apparently the major land-based sources; anoxic sediments are probably the major marine source. Ehhalt

(1973) proposes that the terrestrial production rate of methane is between 6.5 and 19×10^{14} g/yr. If correct, these figures imply that roughly 0.7 to 2% of all photosynthetically produced organic matter decays to produce atmospheric CH_4; most of the remainder decays directly to CO_2.

A large fraction of the atmospheric methane is apparently destroyed in the troposphere. Oxidation begins by reaction with OH radicals

$$CH_4 + OH \rightarrow CH_3 + H_2O \qquad (6\text{-}1)$$

passes through a rather complex reaction scheme, and ultimately yields CO_2 and H_2O. CO is an important intermediate product, so much so that the decay of methane is probably the most important single source of CO in the atmosphere.

McElroy (1976) estimates that the upward flux of CH_4 into the stratosphere is approximately 4.5×10^9 molecules/cm^2 sec. This corresponds to a worldwide flux of ca. 2.1×10^{13} g/yr, that is, a few percent of Ehhalt's (1973) estimated total rate of CH_4 production. Ehhalt (1977) has set the figure for the methane flux to the stratosphere at approximately 10% of the total annual production rate of methane. The marked decrease upward in the mixing ratio of methane in the stratosphere indicates that CH_4 is rapidly destroyed there. Although the fraction of methane that is destroyed in the stratosphere is small, the hydrogen that is released during its destruction plays an important role in determining the escape rate of hydrogen from the earth's atmosphere (see text following).

The Production and Destruction of Carbon Monoxide

The pollution problems created by the ca. 2 to 4×10^{14} g of CO produced annually during combustion processes (see McConnell et al., 1971) have tended to obscure the importance of the CO produced without benefit of man. Forest fires and the oceans probably contribute roughly 0.1 and 0.8×10^{14} g of CO/yr, respectively (Cadle, 1973), but the dominant source is almost certainly the photochemical oxidation of CH_4 (Wofsy, 1976). If all atmospheric CH_4 passes through an intermediate CO stage during its oxidation to CO_2, about 11 to 33×10^{14} g of CO are produced annually during CH_4 oxidation.

The destruction of CO in the atmosphere is quite well understood. The reaction

$$CO + OH \rightarrow CO_2 + H \qquad (6\text{-}2)$$

is of major importance (Wofsy et al., 1972), but several other reactions have been proposed, and it is possible that biological sinks and heterogeneous chemical reactions at ground level also play important roles in the removal of CO from the atmosphere.

Carbon Dioxide in the Biological Cycle

Many important biological processes are coupled to the organic and inorganic chemistry of the earth via the CO_2 content of the atmosphere. Atmospheric P_{CO_2} influences the composition of marine sediments, the global rate of photosynthesis, and the oxidation state of the atmosphere and oceans. The CO_2 pressure today and in the past is, therefore, a parameter of prime geochemical importance and a critical part of a very complicated geochemical system. Although a good deal of the data for a quantitative modeling of this system are still missing, the large body of available information does permit a rough analysis of the major features of the system.

Figure 6-5 shows in highly schematic form the involvement of atmospheric CO_2 in the biosphere. The CO_2 content of the atmosphere is well known, but the other numbers are uncertain by about a factor of 2 to 3. Despite this large uncertainty some features of the behavior of CO_2 are very clear. The

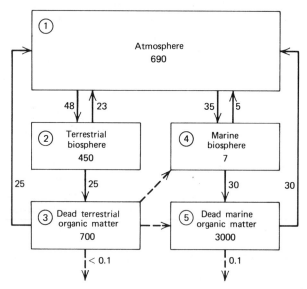

Figure 6-5. The biological cycle of carbon; carbon contents are in units of 10^{15} g of C, and transfer rates are in units of 10^{15} g of C/yr. The data are largely from Bolin (1970), Reiners (1973), Koblentz-Mishke, et al. (1970), and Whittaker and Likens (1973).

passage of CO_2 from the atmosphere through the biosphere and back is rapid. Approximately $\frac{1}{10}$ of the atmospheric CO_2 is transferred to the biosphere every year; the return of most of this CO_2 is also rapid. However, the residence time of carbon in some humus is about 100 to 1000 yr. The terrestrial biosphere is much larger than the marine biosphere, but the reservoir of dead organic matter in the oceans is larger than that on land.

The cycles in Figure 6-5 are nearly closed, and can be treated in isolation from the rest of the carbon cycle for sufficiently short periods of time. The diurnal and seasonal variations of the CO_2 content of the atmosphere depend largely on the short-term balance between the rate of photosynthesis and the rate of decay. In a forest the CO_2 content of air at ground level can rise from 320 to 400 ppm at night, when photosynthesis is shut off and respiration is active; at noon the CO_2 content at the treetop level can drop to 305 ppm (Bolin, 1970). The magnitude of seasonal fluctuations in the concentration of atmospheric CO_2 due to imbalances between the rate of photosynthesis and the rate of decay depends on latitude. Figure 6-6 shows the variation of the CO_2 content of the atmosphere at Mauna Loa, Hawaii, and at the South Pole. As might be expected, the amplitude of the annual cycle is much more pronounced in Hawaii. At both stations a similar, steady increase of the seasonally adjusted monthly averages was observed; this increase is almost certainly due to fossil fuel burning (see text following).

A first-order model for the distribution of CO_2 between the reservoirs in Figure 6-5 is easy to construct. The data in Figures 2-10 and 2-11 showed that, all other things being equal, the rate of photosynthesis of plants is related to P_{CO_2} by a relation of the form

$$\frac{dm_{1,2}}{dt} \cong a\left(1 - e^{-bP_{CO_2}}\right) - c \tag{6-3}$$

where c is the respiration rate in the absence of CO_2; the CO_2 pressure at the compensation point is

$$\left(P_{CO_2}\right)_{comp} = -\frac{1}{b}\ln\left(1 - \frac{c}{a}\right) \tag{6-4}$$

and the maximum rate of photosynthesis,

$$\left(\frac{dm_{1,2}}{dt}\right)_{max} = a - c \tag{6-5}$$

is approached asymptotically with increasing CO_2 pressure.

It is likely that the rate of CO_2 return to the atmosphere from the living biosphere, and the rate of transfer of living tissue to the reservoir of dead organic matter are roughly linear functions of the mass of the biosphere.

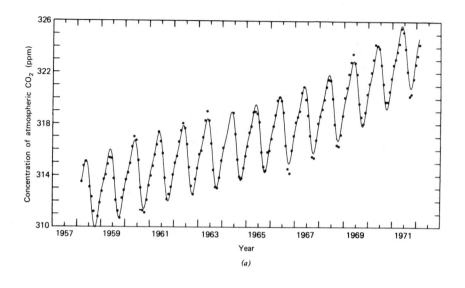

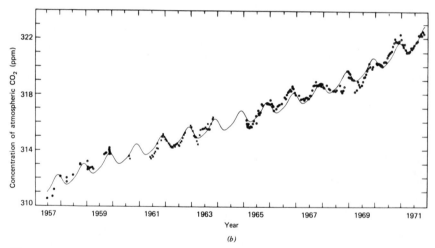

Figure 6-6. (*a*) Variation in the concentration of atmospheric CO_2 at Mauna Loa Observatory between 1958 and 1971. The circles indicate the observed monthly average concentration. The oscillatory curve is a least-squares fit to those averages of an empirical equation containing 6- and 12-month cyclic terms and a cubic-trend function, chosen to contain powers of time up to the third power; Ekdahl and Keeling (1973). (*b*) Variation in the concentration of atmospheric CO_2 at the South Pole between 1957 and 1971, daily averages based on flask analyses. x: continuous analyzer bimonthly averages. The oscillatory curve is a least-squares fit to these averages of an empirical equation having the same form as that used to derive the seasonal variation and trend for Mauna Loa; Ekdahl and Keeling (1973).

264

Similarly, the rate of return of CO_2 from the reservoir of dead organic matter to the atmosphere is probably a roughly linear function of the quantity of dead organic matter. Thus

$$\frac{dm_{2,1}}{dt} \cong em_2 \tag{6-6}$$

$$\frac{dm_{2,3}}{dt} \cong fm_2 \tag{6-7}$$

and

$$\frac{dm_{3,1}}{dt} \cong gm_3 \tag{6-8}$$

At steady state

$$\frac{dm_{1,2}}{dt} - \frac{dm_{2,1}}{dt} - \frac{dm_{2,3}}{dt} = 0 \tag{6-9}$$

$$\frac{dm_{2,3}}{dt} - \frac{dm_{3,1}}{dt} = 0 \tag{6-10}$$

and

$$\frac{dm_{3,1}}{dt} - \frac{dm_{1,2}}{dt} + \frac{dm_{2,1}}{dt} = 0 \tag{6-11}$$

It follows that

$$a\left(1 - e^{-bP_{CO_2}}\right) - c - em_2 - fm_2 = 0 \tag{6-12}$$

$$fm_2 - gm_3 = 0 \tag{6-13}$$

and

$$gm_3 - a\left(1 - e^{-bP_{CO_2}}\right) + c + em_2 = 0$$

Since P_{CO_2} is proportional to the CO_2 content of the atmosphere, $b'm_1$ can be substituted for bP_{CO_2}. Therefore

$$m_2 = \frac{a\left(1 - e^{-b'm_1}\right) - c}{e + f} \tag{6-14}$$

and

$$m_3 = \frac{f}{g} m_2 = \frac{f}{g}\left[\frac{a\left(1 - e^{-b'm_1}\right) - c}{e + f}\right] \tag{6-15}$$

If the total amount of carbon $(m_1 + m_2 + m_3)$, in the three reservoirs, is fixed, then the distribution of carbon between the three reservoirs, and hence the CO_2 pressure in the atmosphere and the rate of transfer of carbon between the reservoirs is determined as well. Similar expressions presumably control the transfer between reservoirs 1, 4, and 5 in Figure 6-5. It follows that the dependence of the rate of photosynthesis on P_{CO_2} and the very rapid biological cycling of carbon between the atmosphere and the biosphere do not determine the CO_2 content of the atmosphere per se; they only do so if the total quantity of carbon in the five reservoirs of Figure 6-5 is fixed. It is shown later that this sum is generally not fixed, but rather that the atmospheric CO_2 pressure and hence the carbon content of reservoirs 2–5 are determined largely by slow leaks in the system of Figure 6-5.

The Effect of Fossil Fuel Burning on Atmospheric Carbon Dioxide

The mean CO_2 content of the preindustrial atmosphere was approximately 290 ppm by volume, the mean CO_2 content in 1970 was 325 ppm. Figure 6-6 shows that the increase in the CO_2 content of the atmosphere has been steady, at least since 1957, and that the annual increase in P_{CO_2} at Mauna Loa has been virtually identical to that at the South Pole. This increase must be due largely to fossil fuel burning.

"Old" carbon from fossil fuels contains no ^{14}C; the addition of such carbon to the atmosphere should, therefore, be diluting the ^{14}C concentration in the atmosphere; this dilution, named the Suess Effect in honor of its discoverer, has been well established in terms of a progressive decrease in the ^{14}C content of wood in tree rings formed between 1700 A.D. and 1954 A.D. Since 1954 nuclear weapons testing has added large quantities of ^{14}C to the atmosphere, and these have obscured the effects of fossil fuel burning on the ^{14}C content of the atmosphere.

Figure 6-7 shows the growth of the input rate of industrial carbon dioxide production since 1860. The rate of CO_2 injection has been almost exactly twice the observed rate of increase in the CO_2 content of the atmosphere (Ekdahl and Keeling, 1973); some of the missing CO_2 has certainly entered the oceans and some has probably entered the biosphere. Just how the missing CO_2 is distributed between the two reservoirs is difficult to determine (Bacastow and Keeling, 1973). Sufficiently precise, direct measurements of changes in the size of the biosphere, both living and dead, are impossible to make, and models of the biosphere response are confounded by changes in agricultural practices, land clearing, urbanization, and a variety of other human perturbations of the biosphere. Woodwell and Houghton (1977) and Woodwell et al. (1978) have argued that the amount of carbon stored in the biosphere is decreasing markedly rather than increasing modestly, but Lerman et al. (1977) have pointed out the considerable

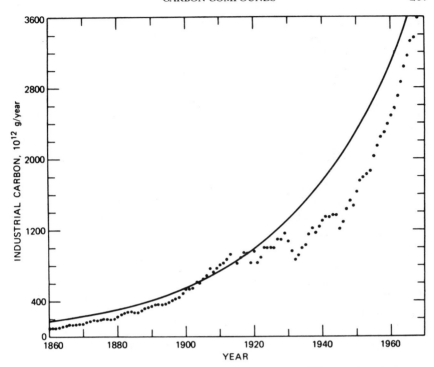

Figure 6-7. Industrial carbon production since 1860. The solid curve is an exponential function with a growth of about 3% per year; Bacastow and Keeling (1973).

uncertainties in current estimates of carbon fluxes both to and from the atmosphere. Direct measurements of changes in the CO_2 content of the oceans are also difficult to make with a sufficient degree of accuracy, but calculations of the expected CO_2 fluxes are reasonably convincing (Siegenthaler and Oeschger, 1978). The two most important reactions are the solution of CO_2 in seawater and the reaction of excess dissolved CO_2 with CO_3^{2-} to produce HCO_3^-

$$CO_3^{2-} + CO_2 + H_2O \rightarrow 2HCO_3^- \qquad (6\text{-}16)$$

The amount of CO_2 in seawater above the thermocline in North Atlantic surface water at 15°C is ca. 0.0086 mmol/kg. The depth to the thermocline is ca. 100 m. The mass of seawater above the thermocline is, therefore, ca.

$$100 \times 10^2 \text{cm} \times 3.6 \times 10^{18} \text{cm}^2 \times 1.02 \text{g/cc} \cong 3.6 \times 10^{22} \text{g}$$

and the mass of dissolved CO_2 ca.

$$3.6 \times 10^{19} \text{ kg} \times 8.6 \times 10^{-6} \frac{\text{mol}}{\text{kg}} \times 44 \frac{\text{g}}{\text{mol}} = 1.4 \times 10^{16} \text{gCO}_2$$

This is less than 1% of the CO_2 content of the atmosphere $(2.5 \times 10^{18}$ g). Since the CO_2 content of seawater is proportional to the CO_2 content of the atmosphere, less than 1% of any CO_2 added to the atmosphere enters the ocean above the thermocline as dissolved CO_2.

The use of CO_2 to neutralize CO_3^{2-} is a more important sink for atmospheric CO_2. Surface Atlantic seawater contains ca. 0.25 mmol CO_3^{2-}/kg. The quantity of CO_3^{2-} above the oceanic thermocline is, therefore,

$$3.6 \times 10^{19} \text{ kg} \times 0.25 \text{ mmol/kg} = 0.90 \times 10^{16} \text{ mol}$$

It would require 40×10^{16} g of CO_2 to convert all of this CO_3^{2-} to HCO_3^-. However, even such complete neutralization would remove only one-sixth of the present quantity of CO_2 in the atmosphere. Ocean water above the thermocline can therefore remove only minor quantities of the industrial CO_2 (Broecker et al., 1971). On the other hand the reservoir of CO_3^{2-} below the thermocline and in $CaCO_3$ sediments in shallow and deep parts of the oceans is large compared to the quantity of industrial CO_2 added to the atmosphere since the start of the Industrial Revolution. It is likely that much of the industrial CO_2 that has left the atmosphere is now in the oceans, and that the partition of industrial CO_2 between the atmosphere and the oceans has been and will continue to be determined by downward mixing into the deeper parts of the oceans.

By 1970 the input of industrial CO_2 had amounted to approximately 18% of the CO_2 content of the atmosphere (Bacastow and Keeling, 1973). The burning of all currently minable coal, oil, and natural gas would add approximately 10 times the mass of preindustrial atmospheric CO_2; projections of the present rate of increase of fossil fuel burning indicate that most of the known reserves of fossil fuels will be burned during the next 100 to 200 yr. The use of lithospheric carbon will decrease thereafter unless it becomes feasible to burn carbon present in currently uneconomical, thin coal seams, in oil shales, or in carbonaceous shales.

The presently fashionable forecast of a peak in fossil fuel burning between 2050 A.D. and 2150 A.D. followed by a more or less rapid decline in the burning rate (see, for instance, Morgan et al., 1977) implies that the CO_2 transferred to the oceans and to the biosphere prior to the period of peak CO_2 input would gradually be returned to the atmosphere during subsequent centuries, and would ultimately be removed from the atmosphere by slow, normal weathering processes until CO_2 in the atmosphere returns to pre-1900 A.D. levels.

The diagram in Figure 6-8 represents the portion of the carbon cycle that includes the reservoirs in Figure 6-5 and those additional reservoirs involved in the production and removal of industrial CO_2. It is rather interesting that

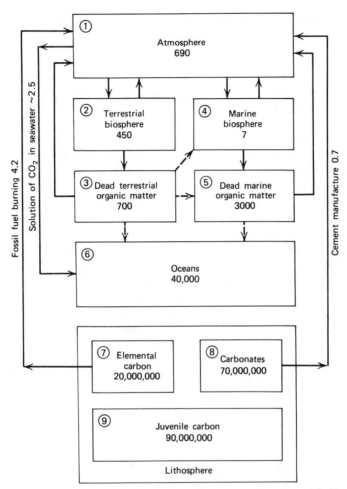

Figure 6-8. A portion of the carbon cycle emphasizing the addition of industrial CO_2 to and its removal from the atmosphere. Figures are carbon contents in units of 10^{15} g of C and transfer rates in units of 10^{15} g of C/yr.

the current rate of industrial CO_2 generation is still only ca. 7% of the rate of photosynthetic fixation of carbon and its release during the decay of organic matter. The steady increase in the CO_2 content of the atmosphere produced by fossil fuel burning suggests that the system was nearly in balance prior to 1700 A.D., and that the atmosphere is now experiencing a major transient that will decay centuries after the peak of fossil fuel burning has passed.

The Effect of Weathering and Sedimentation on Atmospheric Carbon Dioxide

During chemical weathering CO_2 is both added to and subtracted from the atmosphere. The oxidation of elemental carbon and organic compounds in rocks adds CO_2 to the atmosphere; the solution of carbonate minerals and the decomposition of Ca- and Mg-silicates during weathering remove CO_2 from the atmosphere (see Chaps. II and IV).

Table 6-5 summarizes a good deal of the available data for the distribution of elemental carbon in crustal rocks. Shales contain the largest concentration of elemental carbon, and the value chosen for the mean C^o concentration of rocks undergoing weathering depends heavily on the value chosen for the proportion of shales among sedimentary and low grade metamorphic rocks. In Table 6-5 60% has been taken as the percentage of shales in sedimentary and low grade metamorphic rocks. The proportion of these rocks represents about 75% of the total rock mass currently being eroded (Blatt and Jones, 1975), so that the mean C^o concentration in rocks undergoing weathering is approximately $0.45 \pm 0.10\%$. Apparently, most of the carbon is oxidized to CO_2 during weathering. If this were not the case, that is, if a large fraction of the elemental carbon exposed to weathering managed to remain among the detrital components, then the ^{14}C age of organic carbon in modern sediments should reflect the presence of such a "dead" component. It does not.

The addition of carbon as CO_2 due to the oxidation of elemental carbon in rocks exposed to weathering must, therefore, be approximately

$$(0.45 \pm 0.10) \times 10^{-2} \frac{\text{g C}}{\text{g rock}} \times 2 \times 10^{16} \frac{\text{g rock}}{\text{yr}} \cong (0.9 \pm 0.2) \times 10^{14} \text{g/yr}$$

During the weathering of carbonates and Ca- and Mg-silicates CO_2 reacts to form HCO_3^- (see Chap. II). In the summary Table 4-15 the Ca^{2+} and Mg^{2+} concentrations in average river water contributed by the solution of carbonates were estimated to be 0.25 and 0.065 mmol/kg, respectively. Since the solution of carbonates involves the addition of 1 mol of atmospheric CO_2 per mole of Ca^{2+} and Mg^{2+}, the total loss of CO_2 due to the dissolution of limestones and dolomites is currently approximately

$$0.31 \frac{\text{mmol}}{\text{kg}} \times 12 \times 10^{-3} \frac{\text{g C}}{\text{mmol}} \times 4.6 \times 10^{16} \frac{\text{kg}}{\text{yr}} \cong 1.7 \times 10^{14} \frac{\text{g C}}{\text{yr}}$$

The decomposition of Ca- and Mg-silicates involves the loss of 2 moles of atmospheric CO_2 per mole of cation. The concentration of Ca^{2+} and Mg^{2+} in average river water due to the decomposition of Ca- and Mg-silicates was

Table 6-5

Elemental Carbon Content of Crustal Rocks[a]

Rock Type	(%) *Elemental Carbon*
Sedimentary and Low-Grade Metamorphic rocks (ca. 75% of total rocks undergoing weathering)	
Shales (60% of sedimentary rocks)	0.9 ± 0.2 (1)
Limestones and dolomites (20%)	0.2 ± 0.1 (2)
Sandstones (20%)	< 0.1
Average sedimentary rock calculated from data above	0.6 ± 0.1
	0.5 (4)
Igneous and High Grade Metamorphic Rocks	
(25% of total rocks undergoing weathering)	< 0.1 (3)
Average rock undergoing weathering	0.45 ± 0.1

[a]Sources: (1) Clarke and Washington (1924); Trask and Patnode (1942); Nanz (1953); Gehman (1962); Ronov and Yaroshevsky (1967, 1976). (2) Gehman (1962); Ronov and Yaroshevsky (1967, 1976). (3) Clarke (1924). (4) Garrels and Perry (1974).

estimated to be 0.07 and 0.10 mmol/kg, respectively. Thus the loss of atmospheric carbon due to this process is approximately

$$0.34 \frac{\text{mmol}}{\text{kg}} \times \frac{12 \times 10^{-3} \text{g}}{\text{mmol C}} \times 4.6 \times 10^{16} \frac{\text{kg}}{\text{yr}} \cong (1.9 \pm 0.2) \times 10^{14} \frac{\text{g C}}{\text{yr}}$$

The net annual carbon loss from the atmosphere due to chemical weathering of calcium and magnesium minerals is, therefore, roughly 3.6×10^{14} g/yr; a value that is less than 0.5% of the annual use rate of CO_2 in photosynthesis, and approximately 7% of the current input rate of industrial CO_2.

During sedimentation more carbon is lost as a constituent of organic matter than was gained due to the oxidation of elemental carbon during weathering. Gehman (1962) has shown that the $C°$ content of modern pelitic sediments is similar to that of ancient shales. Trends in the $C°$ content of sediments are difficult to discern, and are probably minor. Thus the formation of modern sediments involves the addition of approximately 0.6% $C°$ to ca. 2×10^{16} g sediment/yr for a loss rate of

$$(0.6 \pm 0.1) \times 10^{-2} \frac{\text{g C}°}{\text{g sediment}} \times 2 \times 10^{16} \frac{\text{g sediment}}{\text{yr}} = (1.2 \pm 0.2) \times 10^{14} \text{ g C}°/\text{yr}$$

The net loss to the atmosphere due to the elemental carbon cycle is one-quarter of this rate (see Table 6-6).

Table 6-6

Annual Gains and Losses of Atmospheric Carbon due to Weathering and
Sedimentation[a]

	Gains	*Losses*
Weathering		
Oxidation of elemental carbon	0.9 ± 0.2	
Dissolution of limestones and dolomites		1.6 ± 0.2
Decomposition of Ca- and Mg-silicates		1.9 ± 0.2
Decomposition of Na- and K-silicates		0.8 ± 0.2
Sedimentation		
Deposition of elemental carbon		1.2 ± 0.3
Deposition of carbonates	2.2 ± 0.4	
Deposition of Mg-silicates	0.8 ± 0.2	
Deposition of Na- and K-silicates	0.8 ± 0.2	
Net Changes		
Due to C° cycle		0.3 ± 0.1
Due to carbonate cycle		0.5 ± 0.3
Total net loss		0.8 ± 0.4

[a] In units of 10^{14} g/yr.

The carbonate balance is more difficult to establish. Today very little dolomite is being formed; it was suggested in Chapter V that more than 50% of the river Mg^{2+} may be exchanged for Ca^{2+} during sea water cycling through midocean ridges, and that the total $CaCO_3$ precipitation rate of 0.40 mmol/kg is the sum of the river Ca^{2+}, the Ca^{2+} exchanged from clays in estuaries, and the Ca^{2+} extracted from basalts during seawater cycling through midocean ridges. Since the precipitation of 1 mol of $CaCO_3$ returns 1 mol of CO_2 to the atmosphere via the reaction

$$Ca^{2+} + 2HCO_3^- \rightarrow CaCO_3 \downarrow + CO_2 \uparrow + H_2O \qquad (6-17)$$

the total return is approximately

$$0.40 \frac{mmol}{kg} \times 12 \times 10^{-3} \frac{g\ C}{mmol\ CO_2} \times 4.6 \times 10^{16} \frac{kg}{yr} = 2.2 \times 10^{14} g\ C/yr$$

The fraction of the river flux of Mg^{2+} that is removed from the ocean as a constituent of silicates is not well defined (see Chap. V), but may be as large as one-half. All of the HCO_3^- that balanced this portion of the river flux of Mg^{2+} is returned to the atmosphere, that is, 2 mol of CO_2 per mol of Mg^{2+}. The return flux of CO_2 to the atmosphere is, therefore, about 0.8×10^{14} g/yr.

Table 6-6 summarizes the gains and losses of atmospheric CO_2 due to weathering and sedimentation. The net changes are understandably uncertain. Losses due to the $C°$ cycle reflect the addition of $C°$ to igneous and

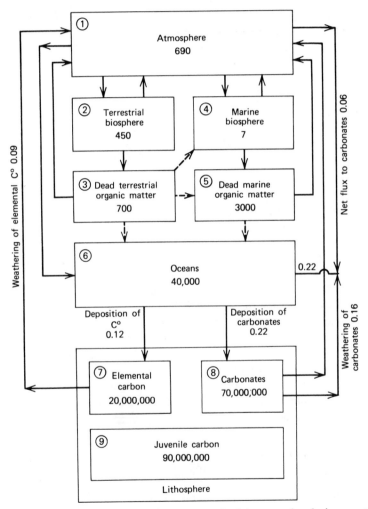

Figure 6-9. A portion of the carbon cycle emphasizing transfer during weathering and sedimentation. Figures are carbon contents in units of 10^{15} g of C and transfer rates in units of 10^{15} g of C/yr.

high grade metamorphic rocks during their conversion to average sediments; losses due to the carbonate cycle reflect the conversion of calcium silicates and some magnesium silicates to carbonates. The total net loss of $(0.8 \pm 0.4) \times 10^{14}$ g of C/yr is approximately 0.1% of the annual turnover of atmospheric CO_2 in the photosynthetic cycle. Clearly, the cycle of carbon through the biosphere is very nearly closed. However, the slow leak in the cycle is not trivial, and losses of CO_2 due to this leak can deplete the atmosphere on a geologically rapid time scale. If CO_2 in the atmosphere were not replenished, the present rate of removal would deplete the atmosphere in ca.

$$\frac{690 \times 10^{15} g}{0.8 \times 10^{14} g/yr} \sim 8600 \text{ yr}$$

and the atmosphere plus the oceans in ca.

$$\frac{40,700 \times 10^{15} g}{0.8 \times 10^{14} g} \sim 510,000 \text{ yr}$$

The carbon balance in the atmosphere–ocean system is therefore somewhat precarious. There is good evidence that the carbon content of the ocean–atmosphere system has not fluctuated wildly in the past. Thus there must always have been a reasonably close balance between the rate of removal of carbon from the ocean–atmosphere system and its replenishment from another source.

In general, these conclusions agree with those of Garrels and Perry (1974). However, their value for the sedimentation rate is only one-third the value accepted here, and their figures for the distribution of Mg between silicates and carbonates in sedimentary rocks and in marine sediments differ considerably from those developed in Chapters IV and V. They have neglected the elemental carbon demanded by the conversion of average igneous and high grade metamorphic rocks to average sedimentary rocks; for all these reasons their carbon demand for weathering and sedimentation is rather different from that in Table 6-6.

Degassing of the Earth and Controls on Atmospheric Carbon Dioxide

The only reasonable source of carbon to cover the annual deficit of the ocean–atmosphere system is the interior of the earth. Many intermediate and high grade metamorphic rocks contain calcsilicates produced by the decarbonation of limestones and dolomites; fluid inclusions in these terrains are frequently rich in CO_2 or CH_4. Much or all of the carbon present in such rocks has certainly been removed, and has probably made its way back to the surface. In addition, volcanic gases generally contain ca. 10% CO_2 by volume, and the discovery of juvenile components in at least some volcanic gases suggests that at present some of the carbon deficit of the atmosphere –ocean system is made up by juvenile CO_2 as shown in Figure 6-10 and in Table 6-7.

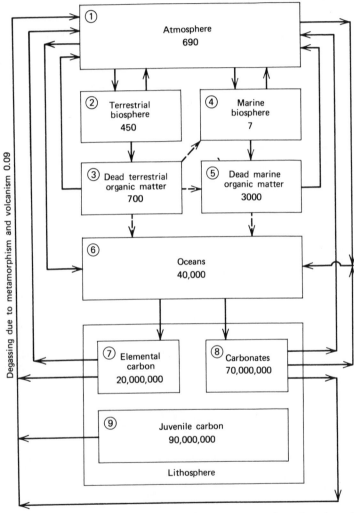

Figure 6-10. The cycle of carbon, emphasizing the transfer of carbon from the lithosphere to the atmosphere during degassing accompanying metamorphism and volcanism. Figures are carbon contents in units of 10^{15} g of C and transfer rates in units of 10^{15} g of C/yr.

Table 6-7

Data Pertaining to the Carbon Cycle

A. The Carbon Content of Important Reservoirs

	Carbon Content (in units of 10^{15} g)	Source of Data
1. Atmosphere	690	Table 6-1
2. Terrestrial biosphere	450	
3. Dead terrestrial organic matter	700	
4. Marine biosphere	7	Bolin (1970) and Reiners (1973)
5. Dead marine organic matter	3000	
6. Dissolved in seawater	40,000	Table 6-1
7. Recycled elemental carbon in the lithosphere	20,000,000	Hunt (1972)
8. Recycled carbonate carbon in the lithosphere	70,000,000	
9. Juvenile carbon	90,000,000	Assuming that the earth is half degassed

B. Transfer Rates of Carbon Between Reservoirs

Transfer Between Reservoirs				
From	To	Process	Rate (in 10^{15} g/yr)	Source of Data
1	2	Net photosynthesis on land	48	Whittaker and Likens (1973)
2	1	Rapid decay of terrestrial organic matter	23	
1	4	Net photosynthesis at sea	35	Koblentz-Mishke et al. (1970)
4	1	Rapid decay of marine organic matter	5	
2	3	Accumulation of dead terrestrial organic matter	25	Bolin (1970) and Reiners (1973)
4	5	Accumulation of dead marine organic matter	30	
3	1	Decay of dead terrestrial organic matter	25	
5	1	Decay of dead marine organic matter	30	

Table 6-7 (Continued)

| Transfer Between Reservoirs | | | Rate | |
From	To	Process	(in 10^{15} g/yr)	Source of Data
7	1	Fossil fuel burning	4.2	Rotty (1976)
8	1	Cement manufacture	0.7	
7	1	Oxidation of elemental carbon	0.09 ± 0.02	See text
6	7	Deposition of elemental carbon	0.12 ± 0.03	See text
1	6	Net flux to carbonates	0.06 ± 0.04	See text
8	6	Weathering of carbonates	0.16 ± 0.04	See text
6	8	Deposition of carbonates	0.22 ± 0.04	See text
7, 8, 9	1	Degassing due to metamorphic and igneous processes	0.09 ± 0.03	See text

The proportion of recycled and juvenile carbon in gases from the earth's interior has been a matter of considerable debate. The argument is still not settled, in part because the distribution of carbon in the earth's crust is not sufficiently well known, and in part because the history of the crust is still rather obscure. In extreme recycling models primary degassing and the formation of the continental crust are thought to have taken place early in earth history, probably in the Archean, prior to 2.5 billion yr ago. Since then the mass of volatiles in the atmosphere, oceans, and crust, and the mass of the continental crust itself are thought to have remained essentially constant. If the present erosion rates are taken to be reasonably representative, and if the rate of erosion of sedimentary rocks has typically been three times the rate of erosion of igneous and high grade metamorphic rocks, then the mean residence time of sedimentary rocks and their highly metamorphosed equivalents can be calculated. Garrels et al., (1972) have proposed a value of 3.2×10^{24} g for the mass of sedimentary rocks including volcanoclastics and recognizable metasediments. If this material is being eroded at a rate of 1.5×10^{16} g/yr today, its half-life is roughly

$$\tau_{sed} \approx \frac{3.2 \times 10^{24} \text{g}}{1.5 \times 10^{16} \text{g/yr}} = 2.1 \times 10^8 \text{yr}$$

This value is somewhat less than the mean half-mass age of 500 million yr for sedimentary rocks suggested by the age distribution of sedimentary rock masses (Garrels et al., 1972). The difference may well be due to a present-day erosion rate that is somewhat faster than the mean erosion rate.

The mass of the continental crust is roughly 20×10^{24} g, and the mass of the crust that consists of high grade metamorphic and igneous rocks is approximately $(20 - 3) \times 10^{24} = 17 \times 10^{24}$ g; if this material normally erodes at the current rate of 0.5×10^{16} g/yr, its half-life is

$$\tau_{\text{met-ign}} \approx \frac{17 \times 10^{24} \text{ g}}{0.5 \times 10^{16} \text{ g/yr}} = 3.4 \times 10^9 \text{ yr}$$

Such a long half-life is hard to reconcile with the distribution of radiometric ages in high grade metamorphic and igneous terrains. However, some of these ages reflect the effects of repeated periods of metamorphism, and little is known of the history of material near the base of the continental crust. Recycling models of this type are clearly somewhat unsatisfactory but are hard to eliminate.

At the other extreme, models of continous continental growth are similarly difficult to prove or disprove. If the oceans have accumulated gradually, and degassing of the earth's interior has continued at a more or less steady rate, the rate of evolution of water has been approximately

$$\frac{2.0 \times 10^{24} \text{ g}}{4.5 \times 10^9 \text{ yr}} \sim 4.4 \times 10^{14} \text{ g/yr}$$

The mole fraction of carbon in volcanic gases is frequently close to 0.1 (see, for instance, Giggenbach and LeGuern, 1976). Thus the mean rate of carbon evolution would be ca.

$$4.4 \times 10^{14} \text{ g/yr} \times 0.1 \times \tfrac{12}{18} \cong 0.3 \times 10^{14} \text{ g/yr}$$

that is, about the same as the degassing rate required to balance the present-day carbon deficit of the ocean–atmosphere system. Although this is reassuring, the argument is meaningful only if it can be shown that the oceans have accumulated at a more or less steady rate. A progressive increase in the mass of the continental crust is very difficult to demonstrate or refute. In part this is due to the incomplete geologic record of the Archean, in part to the absence of convincing evidence for estimating the thickness of the Archean crust.

Despite the current uncertainty in the proportion of recycled and juvenile carbon gases in the degassing products of the earth, the fundamental theorem that determines the state of the ocean–atmosphere system is clear: the system arranges itself so that the rate of carbon loss from the ocean–atmosphere system balances the rate of input of carbon compounds on a time scale of 10^5–10^6 yr. The removal of carbon from the atmosphere as a constituent of organic compounds seems to be limited by the condition that the overall oxidation state of the crust has not changed markedly during much of geologic time (see text following). The removal of carbon from the atmosphere as a constituent of carbonate minerals depends on the ratio of the rate of carbon inputs to the rate of erosion of Ca- and Mg-silicates. It was shown in Chapter IV that calcium in chemically mature sedimentary rocks is almost entirely present as a constituent of one or more carbonate minerals, whereas during most of geologic time magnesium has been distributed more or less evenly between clay minerals and carbonates. This implies that the rate of CO_2 addition to the atmosphere–ocean system has generally been sufficient to remove and precipitate as carbonates most of the calcium and part of the magnesium in the igneous and high grade metamorphic rocks that have been exposed to weathering during the course of earth history.

Atmospheric CO_2 plays a double role in the process of balancing the input and output of carbon compounds in the ocean–atmosphere system. It is an important factor both in weathering processes, which release Ca^{2+} and Mg^{2+} into solution, and in sedimentation processes that remove these cations from the ocean. The role of photosynthesis and plant decay in supplying CO_2 to weathering horizons was discussed in Chapter II. The effect of atmospheric CO_2 on chemical weathering is amplified today by the terrestrial biosphere. A decrease in atmospheric P_{CO_2} by a factor of 10 would probably produce a large decrease in the terrestrial rate of photosynthesis and hence a significant decrease in the rate of chemical weathering (see Chaps. II and IV). An increase in atmospheric P_{CO_2} by a factor of 10 would probably produce a modest increase in the terrestrial rate of photosynthesis and a minor increase in the rate of chemical weathering. A more precise statement of the functional relationship between atmospheric P_{CO_2} and the rate of chemical weathering is obviously needed, but is difficult to make convincingly with the presently available data.

The mineralogy of marine sediments is determined both by thermodynamic equilibria and by reaction rates (see Chap. V). The requirement that CO_2 be removed at a certain rate from the ocean–atmosphere system implies either that one or more carbonate minerals of calcium and magnesium are thermodynamically stable with respect to other compounds of these cations, or that the kinetics of the precipitation of carbonates of these cations are sufficiently rapid that they overcome their thermodynamic instability with respect to other calcium and magnesium compounds. Urey (1952, 1956, 1959) proposed that the low pressure of carbon dioxide in the earth's atmosphere is due to the reaction of CO_2 with silicates of magnesium, calcium, and ferrous iron to form quartz and carbonates. In particular, the reactions

$$MgSiO_3 + CO_2 \rightarrow MgCO_3 + SiO_2 \qquad (6\text{-}18)$$
$$\text{Enstatite} \qquad\qquad \text{Magnesite} \quad \text{Quartz}$$

and

$$CaSiO_3 + CO_2 \rightarrow CaCO_3 + SiO_2 \qquad (6\text{-}19)$$
$$\text{Wollastonite} \qquad\qquad \text{Calcite} \quad \text{Quartz}$$

were singled out. As shown in Figure 6-11 the CO_2 pressure at equilibrium in both of these reactions at 25°C is well below the present day CO_2 pressure in the atmosphere, a condition ascribed by Urey to the competition between CO_2 use at the earth's surface and CO_2 regeneration below the surface during metamorphic decarbonation reactions. Such an interpretation seems somewhat unlikely (Holland, 1965). Enstatite is certainly a common enough constituent of igneous rocks, but magnesite is a rare mineral in sediments. Similarly, calcite is a common constituent of sediments and sedimentary

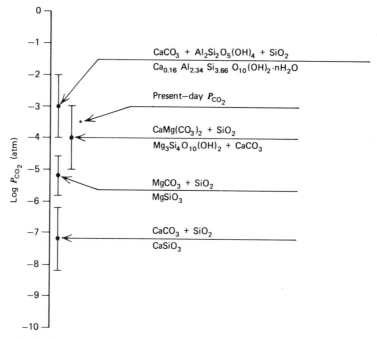

Figure 6-11. P_{CO_2} at 25°C for some carbonate–silicate equilibria of importance for the ocean–atmosphere system.

rocks, but wollastonite is not a common calcium silicate mineral. Equations 6-18 and 6-19 should, therefore, be regarded merely as shorthand notations for the removal of CO_2 from the atmosphere as a constituent of calcium and magnesium carbonates. Comparisons between atmospheric CO_2 and the equilibrium CO_2 pressure at 25°C for reactions 6-18 and 6-19 are of interest only if these are similar to the equilibrium CO_2 pressure for the reactions that are potentially or actually controlling the CO_2 content of the atmosphere. This is not the case. Montmorillonite in rivers is normally in the calcium form (see Chap. V). Rivers are normally not far from saturation with respect to calcite, tend to be somewhat supersaturated with respect to quartz (see Chap. IV), and frequently transport kaolinite together with calcium montmorillonite. It seems likely, therefore, that the CO_2 pressure for equilibrium in the reaction

$$6 \, \underset{\text{Ca-montmorillonite}}{Ca_{0.17}Al_{2.34}Si_{3.66}O_{10}(OH)_2 \cdot 2H_2O} + CO_2 \rightleftarrows \underset{\text{Calcite}}{CaCO_3}$$

$$+ 7 \, \underset{\text{Kaolinite}}{Al_2Si_2O_5(OH)_4} + 8 \, \underset{\text{Quartz}}{SiO_2} + 4 \, H_2O \qquad (6\text{-}20)$$

is not much less than the present-day atmospheric CO_2 pressure of $10^{-3.5}$ atm, and may well be somewhat greater. The instability of calcium montmorillonite in ocean water is apparently not with respect to decomposition to kaolinite, calcite, and quartz but with respect to sodium, potassium, and magnesium montmorillonites. The development of reasonably Ca-rich zeolites on the ocean floor as an alteration product of volcanic glass (see Chap. V) suggests that these minerals are also stable or nearly stable in the ocean at CO_2 pressures near $10^{-3.5}$ atm. Opinions regarding the thermodynamics of these rather complicated phases, which react very slowly at low temperatures, are, of course, in need of confirmation. However, it is likely that the present value of atmospheric P_{CO_2} is not so much a matter of overshoot beyond the very low CO_2 pressure demanded by the stability of $CaCO_3$ with respect to $CaSiO_3$ as it is a requirement for the stability of $CaCO_3$ with respect to Ca-clays and Ca-zeolites.

Equation 6-18 is suspect for similar reasons. Anhydrous magnesium silicates are unstable in the presence of water at room temperature, and alter rapidly to a variety of hydrates (see Chaps. II and V). Equilibria involving hydrated magnesium silicates are, therefore, of much greater interest than those involving anhydrous magnesium silicates. Magnesite, $MgCO_3$, is decidedly rarer than dolomite, $CaMg(CO_3)_2$, in sedimentary rocks. The reasons for this have been described in Chapter V. Equation 6-18 should, therefore, be replaced by equations relating P_{CO_2} to the stability fields of dolomite, hydrous magnesium silicates, and hydrous magnesium-aluminum silicates. In the absence of reliable thermochemical data for magnesian montmorillonites and glauconites, the reaction

$$Mg_3Si_4O_{10}(OH)_2 + 3CaCO_3 + 3CO_2 \rightarrow 3CaMg(CO_3)_2 + 4SiO_2 + H_2O$$

Talc   Calcite   Dolomite   Quartz

$$(6\text{-}21)$$

is probably of most direct interest. The appearance of talc in certain marine sediments and the relationship of its field of stability to the composition of ocean water were discussed in Chapter V. Bartholomé (1966) suggested that the CO_2 pressure in equilibrium with talc, calcite, dolomite, quartz, and water at 25°C is $10^{-2\pm1}$ atm. The redeterminations of the stability of talc by Bricker et al. (1973) and by Hemley et al. (1977) indicate that talc is slightly less stable than had been thought previously, and that Bartholomé's figure is best lowered to $10^{-4\pm1}$ atm, a range that includes the present atmospheric CO_2 pressure of $10^{-3.5}$. This may be somewhat fortuitous, because the concentration of dissolved silica in the oceans and in interstitial waters is so variable, and because the kinetics of dolomite formation are so complicated.

Whatever the quantitative aspects of the process, an increase in atmospheric CO_2 would certainly drive silicate–carbonate equilibria in the direction of greater carbonate stability. Since most river calcium is now removed from seawater as a constituent of $CaCO_3$, an increase in P_{CO_2} would do little to change the calcium output from the oceans. Dolomite would presumably become more important, but virtually nothing is known about the functional relationship between an increase in P_{CO_2} and an increase in the rate of removal of magnesium as a constituent of carbonates rather than as a constituent of silicates. A decrease in atmospheric P_{CO_2} would reduce the proportion of calcium and magnesium that leaves the ocean as a constituent of carbonates. It is likely that magnesium would be affected before calcium; but again, the functional relationships are essentially unknown.

It is reasonable that the rate of CO_2 degassing should be coupled roughly to the rate of erosion of igneous and high grade metamorphic rocks. Both depend on the rate of orogeny. A prolonged lull in tectonic activity would be accompanied by a slow erosion rate because the stand of the continents would have approached sea level, and by a slow rate of degassing because the rate of metamorphism and volcanism would be low. Conversely, a high rate of tectonic activity would lead to rapid erosion rates of all types of rocks, and would tend to be accompanied by rapid degassing. The lack of correlation between the concentration of total dissolved solids in rivers and the mean elevation of the continents that they are draining (see Chap. IV) suggests that the CO_2 budget of the atmosphere–ocean system can be balanced only if P_{CO_2} is higher during times of intense tectonic activity than during times of modest tectonic activity. However, imbalances between CO_2

supply and demand may have had a rather minor effect on P_{CO_2} (Holland, 1968). Table 4-15 shows that in world average river water the quantity of Mg^{2+} from carbonate weathering (0.13 meq/kg) is essentially equal to the quantity of Ca^{2+} from silicate weathering (0.14 meq/kg). These figures are obviously rather uncertain, but they do suggest that the quantity of "$MgCO_3$" dissolved by weathering today is nearly equal to the CO_2 demand of weathered Ca-silicates. Thus, if no CO_2 were added to the atmosphere-–ocean system today, enough CO_2 would be available from the "dolomite bank" to balance the reaction

$$\text{Ca-silicates} + \text{"}MgCO_3\text{"} \rightarrow CaCO_3 + \text{Mg-silicates} \qquad (6\text{-}22)$$

A complete cessation of CO_2 injection is, of course, rather unlikely. A partial cessation or any negative balance between the rate of CO_2 injection and the rate of CO_2 demand could be met by drawing on part of the accumulated CO_2 in dolomites.

The assessment of the effects of CO_2 imbalances is complicated by $Mg^{2+} - Ca^{2+}$ exchange during seawater cycling through midocean ridges (see Chap. V). If the relatively small proportion of dolomite in post-Triassic compared to pre-Triassic sedimentary rocks is demonstrated to be due to a more intense rate of $Mg^{2+} - Ca^{2+}$ exchange today and in the recent past, then a deficit of CO_2 injection today would produce either a reduction of P_{CO_2} and the formation of Ca-silicates in marine sediments, and/or a reduction of the Mg^{2+} concentration in seawater until the $Mg^{2+} - Ca^{2+}$ exchange in mid-ocean ridges becomes minor.

An excess of CO_2 input over CO_2 demand would have the opposite effect. If the capacity of the Mg-silicate–Mg-carbonate system were exceeded, sodium and potassium bicarbonates and carbonates would begin to act as CO_2 sinks. Although soda lakes act in this fashion, the oceans never seem to have reached such a state, at least not during the Phanerozoic Era (Holland, 1972). In fact, the geologic record suggests that the rate of CO_2 injection and the CO_2 demand have been matched surprisingly well since early pre-Cambrian time.

OXYGEN, HYDROGEN, AND OZONE

Combined oxygen is one of the most abundant constituents of all the inner planets and their satellites, but only the atmosphere of the earth contains a large amount of uncombined, molecular oxygen. The presence of O_2 in such abundance is linked to the other two terrestrial peculiarities: the large amount of free water in the oceans and the abundance of life. Hydrogen and ozone are both minor constituents of the atmosphere. They are discussed together with oxygen, because the escape of hydrogen affects the oxidation state of the atmosphere, and because ozone is an important shield against solar ultraviolet radiation at the earth's surface.

Controls on Atmospheric Oxygen

The cycle of atmospheric oxygen is linked to the cycle of all those elements that change their oxidation state at and close to the surface of the earth. Although there are more than 20 of these elements, only a few are quantitatively important. Reaction of oxygen with carbon, sulfur, and iron probably accounts for more than 90% of the oxygen flux to and from the atmosphere. The photosynthetic cycle dominates short-term oxygen fluxes. Annually about 73×10^{15} g of carbon are transferred from atmospheric CO_2 to plants (see previous section). During this transfer

$$73 \times 10^{15} \text{ g C/yr} \times \tfrac{32}{12} = 1.9 \times 10^{17} \text{ g O}_2$$

are released. Nearly all of this oxygen is subsequently consumed during respiration and the decay of organic matter; the entire inventory of atmospheric oxygen, therefore, cycles through the biosphere in approximately

$$\tau \cong \frac{1.2 \times 10^{21} \text{ g O}_2}{1.9 \times 10^{17} \text{ g O}_2/\text{yr}} \cong 6000 \text{ yr}$$

This period is much longer than the cycling time of atmospheric carbon, because the mass of O_2 is 1700 times the mass of atmospheric CO_2. For the same reason the percentage fluctuations of the oxygen pressure due to seasonal variations in the rate of photosynthesis and decay are much smaller than seasonal fluctuations in the CO_2 pressure, and the effect of fossil fuel burning on P_{O_2} is and will continue to be negligible (Broecker, 1968).

On the other hand, the small leak in the cycling of carbon between the atmosphere and the biosphere which was shown to be so important for the geochemistry of carbon, is similarly important for the geochemistry of oxygen. The burial rate of organic carbon, largely as a constituent of marine sediments, is approximately $(1.2 \pm 0.3) \times 10^{14}$ g/yr (see Table 6-7). The oxygen liberated during the production of this quantity of organic carbon is roughly $(3.2 \pm 0.8) \times 10^{14}$ g/yr. If there were no corresponding loss of oxygen from the atmosphere, atmospheric oxygen would double in

$$1.2 \times 10^{21} \text{ g O}_2/3.2 \times 10^{14} \text{ g/yr} = 4 \times 10^6 \text{ yr}$$

that is, in a geologically short period of time.

Oxygen use during weathering prevents such a rapid increase in atmospheric oxygen. The rate of oxygen use due to the oxidation of carbon in old sedimentary and metamorphic rocks undergoing weathering (see previous section) is approximately three-quarters of the rate of oxygen production via the deposition of new sediments. The oxidation of other elements, particularly S and Fe, in rocks undergoing weathering also consumes atmospheric oxygen, as does the oxidation of several of the gases that are released during volcanism and hydrothermal activity. The data in Table 6-8 indicate that ca. 0.3 wt % S^{2-} and ca. 1.9 wt. % "FeO" are available for oxidation in average surface rocks together with ca. 0.45 wt % elemental carbon (see Table 6-5). The oxygen consumption during the complete oxidation of 1 kg of average surface rock due to the oxidation of $C°$, S^{2-}, and "FeO" is, therefore, approximately

$$C + O_2 \rightarrow CO_2, \quad 12 \pm 3 \text{ g } O_2/kg$$

$$S^{2-} + 2O_2 \rightarrow SO_4^{2-}, \quad 6 \pm 2 \text{ g } O_2/kg$$

$$4\text{"FeO"} + O_2 \rightarrow 2Fe_2O_3, \quad 2 \pm 1 \text{ g } O_2/kg$$

Thus, the potential total oxygen use rate per kilogram of average weathered rock is 20 ± 6 g O_2/kg, and the total potential use rate is ca.

$$20 \pm 6 \text{ g } O_2/kg \text{ rock} \times 2 \times 10^{13} \text{ kg rock/yr} = (4 \pm 1) \times 10^{14} \text{ g } O_2/yr$$

The rough agreement between this figure and the figure for the O_2 production due to the burial of organic carbon with modern sediments is no

Table 6-8

The Sulfide Sulfur and Ferrous Iron Content of Crustal Rocks[a]

Rock Type	% Sulfide Sulfur	% FeO
Sedimentary and low-grade metamorphic rocks (ca. 75% of total rocks undergoing weathering)		
Shales (60% of sedimentary rocks)	0.27 ± 0.07 (1)	2.6 ± 1.0 (4)
Limestones and dolomites (20%)	0.13 ± 0.03 (1)	<0.5 (4)
Sandstones (20%)	0.09 ± 0.02 (1)	<0.5 (4)
Average sedimentary rock	0.4 ± 0.1 (2)	1.6 ± 0.6
Igneous and high grade metamorphic rocks (ca. 25% of total rocks undergoing weathering)	<0.05 (3)	3.5 ± 1.0 (4)
Average rock weathered	0.3 ± 0.1	1.9 ± 0.6

[a]Sources: (1) Holser and Kaplan, 1966. (2) Nikolayeva, 1968, (3) Turekian and Wedepohl, 1961. (4) Poldervaart, 1955.

accident. The elemental carbon content of modern sediments is similar to that of sedimentary rocks, and since oxygen use and oxygen consumption depend heavily on these figures, rough agreement is a foregone conclusion. On the other hand there is some chemical and isotopic evidence that the C^0/S^{2-} ratio of sedimentary rocks has varied significantly during geologic time (see Holland, 1973a). There have also been changes in the "FeO" content of sedimentary rocks during the past 2.5 billion yr, but the effects of such changes on atmospheric oxygen have probably been less important than variations in the abundance of C^0 and S^{2-}, because the quantity of O_2 consumed during the oxidation of "FeO" is small compared to the quantity of O_2 consumed during the oxidation of C^0 and S^{2-} in average crustal rocks.

The chemistry of sedimentary rocks has apparently adjusted itself during much of earth history, so that the amount of oxygen consumed during weathering has been nearly the same as the quantity generated during C^0, S^{2-} and "FeO" burial with new sediments. This is quite reasonable if the overall oxidation state of the crust and upper mantle has remained essentially constant. The amount of free oxygen in the atmosphere is small compared to the quantity of C^0, S^{2-}, and "FeO" in the crust; the oxygen content of the atmosphere has, therefore, apparently been part of a feedback system that has operated to maintain the overall oxidation state of the crust.

Qualitatively, the feedback system is fairly well understood (Holland, 1973b). Quantitatively, it is largely unexplored. The major parts of the system are shown in Figure 6-12, where the rates of oxygen production and oxygen use have been plotted as functions of the oxygen content of the atmosphere. Oxygen is produced largely by the burial of organic matter with marine sediments. It was shown in Chapter V that more than 99% of the organic matter produced in the oceans is reoxidized to CO_2 and H_2O. Most of the oxidation takes place in the water column and at the sediment–water interface; molecular O_2 dissolved in seawater is the dominant oxidizing agent. Below the sediment–water interface most of the oxidation of organic matter uses SO_4^{2-} in interstitial water as the oxygen donor, and there is little change in the oxidation state of the ocean–atmosphere system during the exchange of SO_4^{2-} for HCO_3^- in interstitial water during SO_4^{2-} reduction (see Chap. V). After SO_4^{2-} has become exhausted, fermentative decomposition of CH_2O to CH_4 and CO_2 becomes the dominant process in the destruction of organic matter. Again, no overall change in the oxidation state of the ocean–atmosphere system is associated with the reaction

$$2CH_2O \rightarrow CH_4 + CO_2 \qquad (6\text{-}23)$$

If the oxygen content of the atmosphere and the oxygen content of surface seawater were reduced by a factor of 10, then not enough O_2 would be

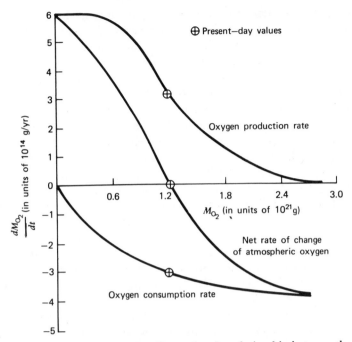

Figure 6-12. Semischematic diagram illustrating the relationship between the oxygen content of the atmosphere, the rate of oxygen production, the rate of oxygen consumption, and the net rate of change of atmospheric oxygen.

present in seawater to oxidize more than a small fraction of the organic matter produced photosynthetically in the upper parts of the oceans. Much of the ocean would, therefore, become anoxic, and a much larger quantity of organic matter would be buried, provided the present rate of photosynthesis could be maintained. Conversely, an increase in atmospheric O_2 would produce an increase in the O_2 content of seawater and hence a decrease in the burial rate of organic matter. Just how strongly the burial rate of organic matter at sea depends on P_{O_2} is not clear. Much of the organic matter in marine sediments is buried in relatively shallow parts of the oceans, where highly oxygenated water overlies carbonaceous sediments. In such settings the burial of carbon is a complex function of the sedimentation rate, the physical properties of the sediment, the rate of photosynthesis, and the oxygen content of seawater (see Chap. V). Whatever the precise nature of the functional relationship between the burial of organic matter and P_{O_2}, the rate of burial of organic matter almost certainly decreases with an increasing quantity of atmospheric oxygen, all other things being equal.

A decrease in atmospheric oxygen should produce an increase in the rate of burial of organic matter. The rate of burial of organic matter cannot, however, increase indefinitely, because the burial of organic matter removes essential nutrients from the oceans. At present roughly a quarter of the flux of marine phosphorus and half of the flux of combined nitrogen are removed as a constituent of organic matter (see Chap. V). It is unlikely, therefore, that the burial rate of organic matter can be increased by more than a factor of 2 to 4 at the present supply rate of phosphate and nitrate to the oceans for periods that are long compared to the residence time of PO_4^{3-} and NO_3^-.

At very low O_2 pressures much of the organic matter photosynthesized at sea would probably be buried. If the total burial rate cannot greatly exceed the present rate, then the total rate of photosynthesis must be much less at low P_{O_2} than today. This is simply a corollary to the observation that the very rapid rate of photosynthesis today depends on the efficient recycling of nutrients in the oceans. On a time scale greater than the residence time of carbon in the ocean–atmosphere system (ca. 1×10^5 yr) the rate of carbon burial is limited by the total rate of carbon supply to the system. Under present-day conditions this is about three times the current rate of organic carbon burial (see Table 6-7).

Weathering and the oxidation of magmatic and metamorphic fluids are the two long-term major oxygen sinks. It was pointed out earlier that oxygen use due to weathering depends on the content of C^o, S^{2-}, and "FeO" in rocks exposed to weathering, on the degree of oxidation of these constituents of eroded rocks, and on the total rate of erosion. Oxidation is reasonably complete in most rocks exposed to weathering today. Detrital sulfides are uncommon in sediments, and the ^{14}C content of carbon in modern sediments indicates that the proportion of old, detrital carbon in carbon buried at sea is small (see the discussion under Carbon Compounds). The response of the oxidation of C^o and S^{2-} to P_{O_2} is not known (see Chap. II), but it is virtually certain that a progressively smaller fraction of the C^o, S^{2-}, and "FeO" in rocks exposed to weathering will be oxidized as the O_2 content of the atmosphere is reduced. Since oxidative weathering is so nearly complete today, an increase in P_{O_2} would have relatively little effect on the rate of O_2 consumption during weathering except in areas where erosion rates are abnormally high and where mechanical disintegration dominates over chemical weathering. Some fairly indirect evidence suggests that the fraction of C^o and S^{2-} oxidized is not particularly sensitive to a decrease in P_{O_2} until P_{O_2} is reduced to values considerably less than the present value.

The other major O_2 sink is the oxidation of volcanic and metamorphic distillates. Volcanic gases all contain H_2, CO, and SO_2. The oxidation of these gases to H_2O, CO_2, and SO_3 must be essentially independent of atmospheric P_{O_2} except at P_{O_2} values that are orders of magnitude lower than 0.2 atm. The oxidation state of Hawaiian volcanic gases has been

Table 6-9
The Oxidation State of Volcanic Gases from Surtsey, Hawaii and Erta'Ale[a]

Component	Hawaii (1) Volume Percent	Hawaii (1) Ratio	Surtsey (2) Volume Percent	Surtsey (2) Ratio	Erta'Ale (3) Volume Percent	Erta'Ale (3) Ratio	Predicted Ratio (4)
H_2O	79.31	$\dfrac{P_{H_2O}}{P_{H_2}} = 137$	86.1	$\dfrac{P_{H_2O}}{P_{H_2}} > 18$	79.4	$\dfrac{P_{H_2O}}{P_{H_2}} = 53$	$\dfrac{P_{H_2O}}{P_{H_2}} = 60^{+45}_{-18}$
H_2	0.58		$\leqslant 4.7$		1.49		
CO_2	11.61	$\dfrac{P_{CO_2}}{P_{CO}} = 31$	5.7	$\dfrac{P_{CO_2}}{P_{CO}} > 15$	10.4	$\dfrac{P_{CO_2}}{P_{CO}} = 23$	$\dfrac{P_{CO_2}}{P_{CO}} = 17^{+13}_{-8}$
CO	0.37		< 0.37		0.46		
SO_2	6.48		2.6		⎫		
S_2	0.24		—		⎬ 7.36		
H_2S	—		—		⎭		

[a]Sources: (1) Eaton and Murata (1960). (2) Sigvaldason and Elisson (1968); average of samples 17, 22, and 24. (3) Giggenbach and LeGuern (1976). (4) Fudali (1965), see text.

discussed by Eaton and Murata (1960). Heald et al. (1963) have reported on a number of analyses of gases from the 1959–1960 eruption of Kilauea (Hawaii). Most of their samples were contaminated with air, but they were able to show that samples collected under favorable conditions would have been in equilibrium with a partial pressure of oxygen of 10^{-8} atm at 1200°C.

Sigvaldason and Elisson (1968) have reported a number of analyses of volcanic gases collected between 1964 and 1967 at the Icelandic volcano Surtsey. Air contamination was serious in several of their samples. All but one of the more or less uncontaminated samples were taken through a stainless steel sampling tube, and it is possible that the volcanic gases reacted with these tubes. The recorded oxidation state of their samples may well be somewhat too low.

The best analyses of volcanic gases from Hawaii, from Surtsey, and from Erta'Ale have been summarized in Table 6-9. Water and carbon dioxide are their most abundant components, but H_2 and CO are frequently present in considerable quantities. Sulfur is dominantly present as a constituent of SO_2. The ratio of water to carbon and sulfur compounds is quite variable, and seems to change during the course of the evolution of gases from a particular batch of magma. Infrared measurements of gases in a lava fountain at Kilauea (Naughton, personal communication, 1969) suggest that the proportion of H_2O to CO_2 and SO_2 in gases emitted by the fountaining lava is $95:4:1$. The higher concentration of CO_2 and SO_2 reported in Table 6-9 may reflect a progressive increase in the proportion of these gases toward the later stages of eruption, during which direct sampling becomes easier.

Fortunately, the oxidation state of volcanic gases at their source can be determined by an independent method. Kennedy (1948) first pointed out that the ratio of the concentration of ferric to ferrous iron in a silicate magma is directly related to the oxidation state of gases with which it is in equilibrium. Since then Muan and Osborn (1956), and Osborn (1959) have added a great deal of data relating the chemistry of silicate melts to the oxidation state of associated gases. Fudali (1965) has observed that the oxygen fugacity of gases in equilibrium with melts of basaltic composition at 1200°C typically lies between $10^{-8.0}$ and $10^{-8.5}$ atm. At this temperature the equilibrium constant K_{VI-24} for the reaction

$$H_2 + \tfrac{1}{2}O_2 \rightleftarrows H_2O \qquad\qquad (6\text{--}24)$$

is

$$K_{VI\text{-}24} = \frac{f_{H_2O}}{f_{H_2} \cdot \left(f_{O_2}\right)^{1/2}} = 10^{5.89}$$

At low total pressures and high temperatures fugacities are virtually equal to partial pressures; thus

$$\frac{f_{H_2O}}{f_{H_2}} \cong \frac{P_{H_2O}}{P_{H_2}} = 10^{5.89} \times 10^{-4.25 \pm 0.25} = 60^{+45}_{-18}$$

in gases equilibrated with typical basaltic magmas at 1200°C.

For the reaction

$$CO + \tfrac{1}{2}O_2 \rightleftarrows CO_2 \qquad (6\text{-}25)$$

the equilibrium constant is equal to

$$K_{VI-25} = \frac{f_{CO_2}}{f_{CO} \cdot (f_{O_2})^{1/2}} = 10^{5.47}$$

at 1200°C. It follows that the CO_2/CO ratio in volcanic gases in equilibrium with typical basaltic magmas at 1200°C is

$$\frac{f_{CO_2}}{f_{CO}} \cong \frac{P_{CO_2}}{P_{CO}} = 10^{5.47} \times 10^{-4.25 \pm 0.25} = 17^{+13}_{-8}$$

These values agree rather well with those for the H_2O/H_2 and for the CO_2/CO ratios in volcanic gases from Hawaii and Erta'Ale, and suggest that the oxidation state of these volcanic gases is rather typical of those associated with present-day basaltic volcanism. Green and Poldervaart (1955) have pointed out that the Fe_2O_3 content of basaltic lavas does not vary systematically with geologic age. This has been confirmed by Steinthórssen (1974), and it seems likely that the oxidation state of volcanic gases associated with basaltic lavas has remained more or less constant during the past 2 to 2.5 billion yr.

If all of the carbon used to convert igneous and high grade metamorphic rocks to average sedimentary rocks were to enter the ocean–atmosphere system as a constituent of volcanic gases such at those in Table 6-9, then each mole of carbon would be entering as a constituent of ca. 0.96 mol of CO_2 and 0.04 mol of CO, and would be accompanied by ca. 0.15 mol of H_2 and 0.6 mol of SO_2. At the estimated rate of carbon degassing of $(0.09 \pm 0.03) \times 10^{15}$ g/yr in Table 6-7, the rate of CO input would be

$$(0.09 \pm 0.03) \times 10^{15} \text{g C/yr} \times \frac{1}{12 \text{ g C/mol}} \times \frac{0.04 \text{ mol CO}}{\text{mol C}}$$

$$= (3 \pm 1) \times 10^{11} \text{ mol/yr}$$

the rate of H_2 input would be $(11 \pm 3) \times 10^{11}$ mol/yr and the rate of SO_2 input $(14 \pm 5) \times 10^{11}$ mol/yr. Since the oxidation of 1 mol of each of these gases requires $\frac{1}{2}$ mol of oxygen, the total annual oxygen requirement would be approximately $(30 \pm 10) \times 10^{11}$ mol, that is $(1.0 \pm 0.3) \times 10^{14}$ g of O_2/yr. This rate of O_2 use is approximately one-quarter of the rate of O_2 use in the oxidation of $C°$, S^{2-}, and "FeO" during chemical weathering and is, therefore, by no means negligible. However, volcanic gases of this type contain embarrassingly large quantities of sulfur, and it is difficult to reconcile their $\Sigma C/\Sigma S$ ratio with that of average sedimentary rocks. Other sources, therefore, seem to be more important in the carbon metabolism of the crust.

Degassing of carbon from rocks undergoing metamorphism takes place via decarbonation reactions and via the "solubilization" of graphite and graphitic carbon. The reaction of graphite with water to produce methane and CO_2 is similar to the bacterial decomposition of organic matter at low temperatures:

$$2C + 2H_2O \rightarrow CO_2 + CH_4 \qquad (6\text{-}26)$$

The reaction of graphite with Fe_3O_4, Fe_2O_3, and clay minerals to produce CO_2 and FeO-bearing aluminosilicates also liberates CO_2, and reactions that convert graphite, pyrite, magnetite, and clay minerals to CO_2, pyrrhotite, and FeO-bearing silicates are not uncommon. Gaseous products of the graphite–water reaction have a bulk oxidation state identical to that of the starting materials in equation 6-26. The gaseous products of the other reactions are more oxidized than the starting materials, but they leave behind a more reduced mineral assemblage.

Graphite loss is most pronounced from intermediate and high grade metamorphic rocks. Since metamorphic fluids released at depth in the crust react with lower grade rocks during their flow upward, the fluids that are discharged at the surface either directly or after mixing with ground water may have a composition rather different from that of fluids at greater depths. Fluids in hydrothermal ore deposits are typically less reducing and contain much less sulfur than the volcanic gases in Table 6-9 (see, for instance, Roedder, 1972). 1×10^{14} g of O_2/yr is, therefore, an upper limit for the quantity of oxygen consumed annually during the oxidation of volcanic and metamorphic fluids.

The net oxygen flux to the atmosphere is the difference between the total oxygen production rate and the total oxygen use rate. Figure 6-12 has been drawn so that the net oxygen flux is zero when the oxygen content of the atmosphere is equal to its present value. As long as the oxygen production and consumption curves remain fixed, the point $P_{O_2} = 0.2$ atm is a stable control point. If excess oxygen were suddenly introduced into the atmosphere, P_{O_2} would rise instantaneously, oxygen consumption would exceed oxygen production, and P_{O_2} would decrease gradually to its original value. Conversely, if O_2 were suddenly removed, P_{O_2} would drop instantaneously, oxygen production would exceed oxygen consumption, and P_{O_2} would gradually return to its initial value.

The rate at which P_{O_2} returns to the equilibrium value depends on the slope of the net oxygen flux curve near the equilibrium point. A straight line is a reasonable approximation to the net flux curve near $P_{O_2} = 0.2$ atm. In the vicinity of this point

$$\frac{dM_{O_2}}{dt} \cong a - bM_{O_2} \tag{6-27}$$

that is, the rate of approach of P_{O_2} to equilibrium is approximately proportional to the deviation of P_{O_2} from the steady state pressure; the variation of M_{O_2} with time, therefore, tends to follow the solution of equation 6-27

$$M_{O_2} = \frac{a}{b} + \left(M_{O_2}^0 - \frac{a}{b} \right) e^{-bt} \tag{6-28}$$

$$= 1.2 \times 10^{21} + \left[M_{O_2}^0 - (1.2 \times 10^{21}) \right] e^{-bt} \tag{6-29}$$

The slope b is no better known than the response of the O_2 production rate and the O_2 use rate to changes in P_{O_2}. As drawn in Figure 6-12, the value of b is 7×10^{-7} yr^{-1}, and the half-life of the decay of disturbances in P_{O_2} is approximately

$$\tau_{1/2} \sim \frac{0.693}{b} \sim 1 \times 10^6 \text{ yr}$$

If P_{O_2} dropped suddenly by 10%, that is, from 0.20 to 0.18 atm, the net rate of O_2 regeneration would be approximately

$$\frac{dM_{O_2}}{dt} \approx (8.4 \times 10^{14}) - (7.6 \times 10^{14}) \approx 0.8 \times 10^{14} \text{ g/yr}$$

This corresponds to an increase in the rate of deposition of organic carbon of approximately

$$\tfrac{12}{32} \times 0.8 \times 10^{14} = 0.03 \times 10^{15} \text{ g/yr.}$$

that is, an increase of ca. 25% in the current deposition rate (see Table 6-7).

Sudden changes in P_{O_2} such as those used in the example above are very unlikely. However, changes in the equilibrium value of P_{O_2} due to changes in the position of the oxygen production and oxygen use curves in Figure 6-12 are probably unavoidable. Both curves depend on the rate and location of orogenic activity, as well as on climatic and biologic factors, and it is unlikely that the effects of these variables have compensated each other so well that the equilibrium value of P_{O_2} has always been equal to 0.2 atm. However, the surprising constancy of the isotopic composition of elemental and carbonate carbon in sedimentary rocks during most of geologic time indicates that changes in the equilibrium value of atmospheric P_{O_2} have generally been slow compared to the residence time of carbon in the ocean–atmosphere system. The $\delta^{13}C$ value of elemental carbon has almost always been $-25 \pm 5\%$ PDB, the $\delta^{13}C$ value of carbonate carbon has generally been close to $0 \pm 2\%$ PDB (Eichmann and Schidlowski, 1975).

The average $\delta^{13}C$ value of volcanic and metamorphic degassing products is not well known but is probably $-7 \pm 3\%$ PDB. During weathering, carbon released from the carbonate reservoir and from the elemental carbon reservoir is mixed with degassed carbon. Carbon removal from the ocean–atmosphere system into new carbonates and new elemental organic carbon then takes place. The fractionation of carbon isotopes between oceanic HCO_3^- and $CaCO_3$ is very small; however, the fractionation of carbon isotopes between oceanic HCO_3^- and organic carbon in marine sediments is intense and amounts to ca. 25% . The constancy of the isotopic composition of carbon in the carbonate and elemental carbon reservoirs can only be explained if the proportion of carbon contributed by carbonate weathering during the residence time of carbon in the ocean–atmosphere system is nearly the same as the proportion of carbon in carbonates deposited during the same time interval. If this is not the case, large deviations in the $\delta^{13}C$ value of organic carbon and carbonate carbon can be produced.

This can be demonstrated by taking the values for the estimated rates of weathering in Table 6-7 and the mean $\delta^{13}C$ values estimated above, and calculating the effect of varying the proportion of organic and carbonate carbon deposited today on the $\delta^{13}C$ value of these two types of carbon. Table 6-10 shows the pertinent input–output data together with the estimated $\delta^{13}C$ values. The system seems to be close to isotopic steady state. If the proportion of carbon removed from the atmosphere–ocean system as a constituent of organic matter were suddenly doubled to 0.24×10^{15} g/yr, and the rate of carbonate deposition decreased to 0.10×10^{15} g/yr, the system would still be in balance as far as carbon is concerned, but would not be in isotopic equilibrium. The isotopic composition of carbon in the

Table 6-10

Inputs and Outputs of Carbon and Their Isotopic Composition[a]

	Rate $(10^{15} g/yr)$	$\delta^{13}C°/_{00}$
Inputs to Ocean–Atmosphere System		
Oxidation of organic carbon	0.09 ± 0.02	-25 ± 5
Weathering of carbonates	0.16 ± 0.04	0 ± 2
Degassing due to igneous and metamorphic processes	0.09 ± 0.03	-7 ± 3
Total inputs	0.34 ± 0.09	-8 ± 3
Outputs from the Ocean–Atmosphere System		
Deposition of organic carbon	0.12 ± 0.03	-23 ± 3
Deposition of carbonate carbon	0.22 ± 0.04	0 ± 1
Total outputs	0.34 ± 0.07	-8 ± 2

[a]Rate data from Table 6-7.

atmosphere–ocean system in new organic matter and in new carbonate sediments would gradually shift until a new isotopic steady state is reached. If the difference between the $\delta^{13}C$ value of organic matter and carbonate carbon remains $-23°/_{00}$, then the $\delta^{13}C$ value of the oceans and of carbonate sediments would approach $+8°/_{00}$ and the $\delta^{13}C$ value of organic carbon would approach $-15°/_{00}$. The rate equation for the approach to equilibrium is similar to that for the sulfur system (Holland, 1973a) and follows the pattern outlined by Junge et al. (1975). The response time is roughly 1×10^5 yr, the residence time of carbon in the ocean–atmosphere system.

Carbonate rocks containing carbon with $\delta^{13}C$ values more positive than $+3°/_{00}$ are rare (Schidlowski and Eichmann, 1977). Imbalances in the ocean–atmosphere system as large as 0.12×10^{15} g of C/yr between the input and output of elemental carbon have, therefore, also been rare, unless the present-day total carbon input and output rates are abnormally small. Generally, the net gain in atmospheric oxygen has, therefore, been a good deal less than 0.32×10^{15} g/yr, due to the operation of the carbon cycle. The operation of the sulfur and iron cycles has probably always influenced the oxidation state of the atmosphere less than has the operation of the carbon cycle (see text preceding). Thus changes of P_{O_2} in excess of 10% of the present atmospheric level have probably required at least a time interval of 10^6 yr. This implies that atmospheric oxygen is not a good buffer against imbalances due to environmental changes such as ice ages, which have time constants of about 10^4 to 10^5 yr.

Hydrogen Escape and the Oxidation State of the Atmosphere

In the previous section it was assumed that the earth is a closed system, or at least that transfer of material between the earth and interplanetary space is of negligible importance. The validity of this assumption is not obvious. The escape of ^{3}He and ^{4}He indicates that hydrogen is also capable of escaping from the atmosphere; since hydrogen escape increases the oxidation state of the atmosphere, hydrogen loss can only be neglected if its rate is trivially slow. It now looks as if this is the case. Hunten (1973) and Hunten and Strobel (1974) have shown that the rate of hydrogen escape from the upper atmosphere depends largely on the rate of upward diffusion of hydrogen from the lower atmosphere. The cold trap in the stratosphere strongly limits the upward transport of hydrogen as a constituent of water vapor, and it is likely that methane is the major source of stratospheric hydrogen today (see, however, Schmidt, 1974).

Current estimates place the escape rate of hydrogen close to 2×10^8 atoms/cm^2 sec. This corresponds to a flux of 5×10^{10} g of H/yr, and a net gain of atmospheric oxygen of 4.5×10^{11} g/yr. At this rate about 3×10^9 yr would be required to generate the present amount of atmospheric oxygen. It was shown earlier that the entire inventory of atmospheric oxygen is consumed by weathering, and is regenerated during sedimentation in a few million yr. The contribution of hydrogen escape has, therefore, been unimportant for the oxygen balance of the atmosphere, at least since P_{O_2} reached values comparable to those of the present day, unless past methane fluxes to the atmosphere have been several orders of magnitude greater than today. Hydrogen escape from an anoxic atmosphere containing sizable quantities of reducing gases is probably orders of magnitude more rapid than it is from the atmosphere today, and could have had an important influence on the chemistry of the early atmosphere of the earth.

Atmospheric Ozone

Several atmospheric constituents absorb short wavelength solar ultraviolet radiation effectively. Water vapor absorption extends up to 2000 Å, absorption by CO_2 to 1800 Å, and by O_2 to 2200 Å. Ozone is the only gas in the earth's atmosphere that absorbs solar ultraviolet effectively up to 3000 Å (Berkner and Marshall, 1965); by virtue of this property ozone is a particularly important ultraviolet radiation shield for the biosphere.

The quantity of ozone acting as such an effective radiation shield today amounts to only 0.16 to 0.4 cm STP (Junge, 1963). If ozone were distributed homogeneously in the atmosphere, its concentration would be about 0.4 ppm. It is, however, very unevenly distributed. At sea level uncontaminated air contains about 0.1 to 0.03 ppm of ozone; in highly polluted air concentrations of nearly 1 ppm have been reported (Tebbens, 1968). The concentration of ozone increases upward, passes through a maximum between 20 and 30 km (see, for instance, Breiland, 1969) and then decreases toward greater altitudes (see Fig. 6-13). The total ozone column can vary a good deal from day to day (see Figure 6-14), and monthly averages show clear evidence of seasonal variations. Annual trends in Figure 6-15 for total ozone in the northern hemisphere suggest a slow increase since 1960, the period for which good data are available.

Atmospheric ozone is formed by a series of reactions initiated by the photolysis of O_2

$$O_2 + h\nu \rightarrow O + O \qquad (6\text{-}30)$$

O atoms may react with O_2 to form O_3

$$O + O_2 + M \rightarrow O_3 + M \qquad (6\text{-}31)$$

where M denotes an appropriate third body. Ozone can be destroyed by the reaction

$$O_3 + h\nu \rightarrow O + O_2 \qquad (6\text{-}32)$$

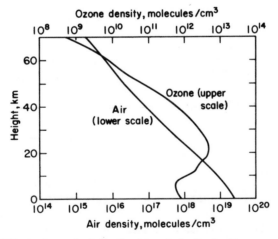

Figure 6-13. Midlatitude vertical distributions of air density (lower scale) and ozone density (upper scale) in the troposphere, stratosphere, and lower mesosphere. The air density distribution is from the U.S. Standard Atmosphere (1962); the ozone density distribution is from Krueger and Minzner (1973). Note that the air density scale is 10^6 times the ozone density scale, so that the ratio of the two curves at any height subjectively gives the ozone mixing ratio in ppm (National Research Council, 1975). Reproduced with the permission of the National Academy of Sciences.

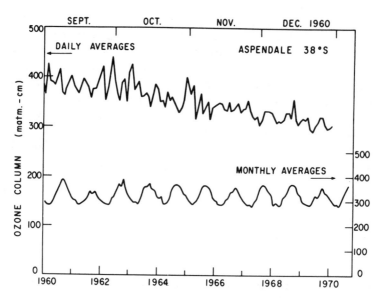

Figure 6-14. Total ozone column at Aspendale, Australia, showing daily fluctuations and seasonal variations (National Research Council, 1975). Reproduced with the permission of the National Academy of Sciences.

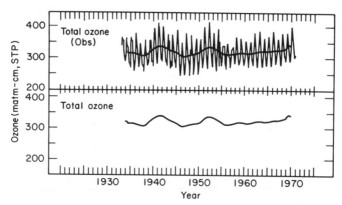

Figure 6-15. Long-term variations in total ozone in the northern hemishpere. The sharp seasonal variations (compare Figure 6-14) are shown in the top panel. An 11-yr running mean is inscribed in the top panel and repeated in the bottom panel. A possible 11-yr cycle is indicated before 1960. An increasing trend is indicated for 1960–1970, but the 1970 high value is still less than high values before 1960. It is generally recognized that the data base before 1960 is too sparse to offer firm support to these suggested long-term trends (National Research Council, 1975). Reproduced with the permission of the National Academy of Sciences.

298

and odd oxygen can be removed by the reaction

$$O + O_3 \rightarrow O_2 + O_2 \qquad (6\text{-}33)$$

Calculations of the ozone distribution based on these reactions alone predict too much ozone at low latitudes and too little ozone at high latitudes. These discrepancies are largely due to ozone transport and to the effects of compounds of H, N, and possibly Cl. Most ozone is manufactured and destroyed in the latitude band between 30°N and 30°S at altitudes above 30 km. A small fraction of the ozone leaks poleward out of this region. North of 30°N and south of 30°S the ozone concentration is determined by a balance between supply by local winds and slow chemical loss. These processes together explain the relatively high ozone content of the atmosphere at high latitudes (see Figure 6-16) and its more intense seasonal variation in the polar regions (see Figure 6-17).

At lower altitudes decomposition by nitric oxide via the reaction

$$NO + O_3 \rightarrow NO_2 + O_2 \qquad (6\text{-}34)$$

plays a major role (Crutzen, 1970; Johnston, 1971), and may account for most of the removal of odd oxygen. Cosmic rays and solar protons probably

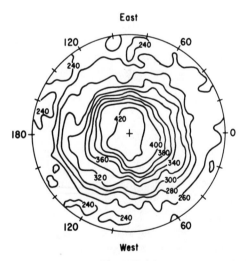

Figure 6-16. Total ozone in the northern hemisphere as measured from Nimbus IV satellite, May 1969. Contour lines are in units of milliatmosphere-centimeters STP (Dobson units) (National Research Council, 1975). Reproduced with the permission of the National Academy of Sciences.

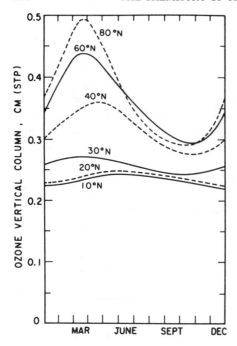

Figure 6-17. Variation of integrated vertical ozone column with season at tropical, temperate, and polar zones (National Research Council, 1975). Reproduced with the permission of the National Academy of Sciences.

exert a minor effect on both the spatial and temporal distribution of ozone via their effect on the concentration of NO_x in the stratosphere.

Figure 6-18 shows a comparison of observed and calculated O_3 profiles during late summer at a solar declination of 12° at latitude 30°N (McElroy, 1976). The agreement between the observed ozone profiles and the computed profiles A and D is satisfactory. The disagreement between the observed ozone profiles and profiles B and C is quite serious; in the computation of profile B the effects of NO_x were neglected; in computing profile C the effects of both HO_x and NO_x on ozone were neglected. The necessity for including the effect of both HO_x and NO_x is obvious. However, even when they are included, the one-dimensional profiles constructed to date give only a rough account of the distribution of stratospheric O_3.

More refined models are needed to give quantitatively meaningful answers to questions that are being raised concerning the effects of atmospheric pollutants on the distribution of stratospheric ozone (Anderson et al., 1977). Some major questions of atmospheric chemistry remain to be answered, and three-dimensional models probably have to be developed before the effects of NO_x emissions by SST flights, the increased fixation of nitrogen due to the production of fertilizers, and the injection of a variety of halogen compounds can be predicted with the necessary accuracy.

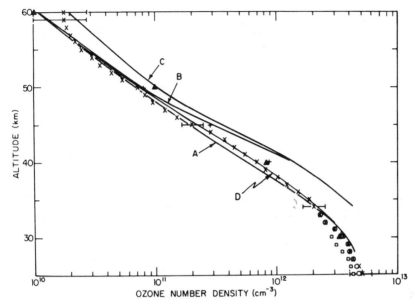

Figure 6-18. Observed and calculated ozone profiles for the normal atmosphere during late summer, +12° solar declination, 30°N latitude. Curves A and D: computed profiles; effects of NO_x and HO_x included. Curve B: computed profile; effects of NO_x omitted. Curve C: computed profile; effects of NO_x and HO_x omitted. −×− Chemiluminescent rocketsonde, and ○ Mast-Brewer electrochemical balloonsonde; Wallops Island (30°N), Sept. 16th, 1968; Hilsenrath et al. (1969). ▲ Rocketsonde solar uv absorption, White Sands, N.M. (32°N), June 14th, 1949; Johnson et al. (1952) as corrected by Evans et al. (1968). + Rocketsonde solar uv absorption from Pt. Mugu, California (34°N), June 18th, 1970; Krueger et al. (1972). □ Mast-Brewer balloonsondes from Tallahassee, Florida (30.4°N), 1966 summer average; Hering and Borden (1967).

NITROGEN AND ITS COMPOUNDS

The geochemical cycle of nitrogen is tightly linked to the geochemical cycle of carbon, because both elements are essential constituents of the biosphere. The distribution of the two elements is, however, quite different. Nearly all of the earth's carbon is contained in the crust (Table 6-7), whereas a sizable fraction, probably between 30 and 70%, of all the terrestrial nitrogen is present as molecular nitrogen in the atmosphere (see Table 6-11). The compounds of nitrogen in the atmosphere contribute little to the total store of atmospheric nitrogen (see Table 6-1). Nitrous oxide, N_2O, the most abundant of the nitrogen oxides accounts for only 0.33 ppm of the atmosphere. Nitric oxide, NO; nitrogen dioxide, NO_2; and ammonia, NH_3, are present at the parts per billion level.

Table 6-11

Data Pertaining to the Nitrogen Cycle

A. The Nitrogen Content of Important Reservoirs

		Nitrogen Content (in units of 10^{15} g)	Source of Data
1.	Atmosphere	3,950,000	Table 6-1
2.	Terrestrial biosphere	6	Table 6-7 and C/N = 79 g/g
		12	Delwiche and Likens (1977)
3.	Dead terrestrial organic matter	70	Table 6-7 and C/N = 10 g/g
		760	Delwiche (1970)
		340	Delwiche and Likens (1977)
4.	Marine biosphere	1.2	Table 6-7 and C/N = 5.7 g/g
		1.0	Delwiche and Likens (1977)
5.	Dead marine organic matter	500	Table 6-7 and C/N = 6 g/g
		900	Delwiche (1970)
		89	Delwiche and Likens (1977)
6.	Oceans		
	Dissolved organic + combined nitrogen ($NO_2^- + NO_3^-$)	920	Vaccaro (1965)
		100	Delwiche and Likens (1977)
	Dissolved N_2	20,000	Delwiche (1970)
7.	Recycled nitrogen in the lithosphere	2,000,000	Table 6-7 and assuming C/N = 10 g/g in elemental carbon reservoir
		4,000,000	Delwiche (1970)
8.	Inorganic nitrogen (probably largely in NH_4^+)	1,000,000	Based on Rayleigh's (1939) figure for igneous rocks and on NH_4^+–N data by Stevenson (1962)
9.	Juvenile nitrogen	7,000,000	Assuming that the earth is half degassed

B. Transfer Rates of Nitrogen between Reservoirs

Transfer Between Reservoirs				
From	To	Process	Rate (in units of 10^{12} g/yr)	Source of Data
(1+3)	2	Photosynthesis on land	600	Table 6-7 and C/N = 79 g/g
6	4	Marine photosynthesis	4700	Table 6-7 and C/N = 5.7 g/g
1	6	Nitrogen transfer from the atmosphere to the oceans		
		(i) Marine nitrogen fixation	30	Delwiche and Likens (1977)

Table 6-11 (Continued)

Transfer Between Reservoirs				
From	To	Process	Rate (*in units of* $10^{12}g/yr$)	Source of Data
		(ii) River flux of NO_3^-	10	See Chapter IV
		(iii) Rain		
		(a) Combustion + industrial effects	40	See text
		(b) Atmospheric N_2 fixation	5	Delwiche (1970) pro-rated between land and oceans
1	2	Nitrogen transfer from the atmosphere to the terrestrial biosphere		
		(i) Terrestrial nitrogen fixation	100	Delwiche and Likens (1977)
		(ii) Rain		
		(a) Combustion + industrial effects	58	See text and Delwiche and Likens (1977)
		(b) N_2 fixation in atmosphere	7	Delwiche and Likens (1977)
		(iii) Fertilizers	40	McElroy et al. (1976) McElroy (personal communication, 1977)
5	7	Burial of marine organic matter	20	See text
6	1	Return of NH_3 and "denitrified" nitrogen to the atmosphere	45	See text
3	7	Burial of terrestrial organic matter at sea	2	See text
3	1	Return of NH_3 and "denitrified" nitrogen to the atmosphere	$\sim$138	Delwiche and Likens (1977)
7	1	Release of nitrogen from sediments to the atmosphere		
		(i) Weathering of sedimentary rocks	15	Table 6-7 and $C/N = 6\,g/g$
		(ii) Metamorphism of sedimentary rocks	5	Table 6-7 and $C/N = 6\,g/g$
8	1	Release of nitrogen from igneous rocks to the atmosphere		
		(i) Weathering of igneous rocks	$\sim$1	—
		(ii) Metamorphism of igneous rocks	$\sim$1	—
9	1	Juvenile nitrogen degassing	$\sim$2	Mean degassing rate

Atomspheric Nitrogen and the Nitrogen Cycle

The large concentration of N_2 in the atmosphere compared to that of CO_2 is due in part to the much greater solubility of the common nitrate minerals than that of the common carbonates; but the gross thermochemical disequilibrium between N_2 and O_2 in the atmosphere, and the pH and NO_3^- content of seawater is probably a much more important factor. Numerous writers (see, for instance, Hutchinson, 1954, and Sillén, 1967) have pointed out that at equilibrium with the present-day atmosphere the oceans should contain very large concentrations of nitrate. The last row of Table 6-12 shows that a difference of some 10 orders of magnitude exists between the actual and the equilibrium nitrate concentration of present-day seawater as defined by equilibrium in the reaction

$$N_2 + \tfrac{5}{2}O_2 + H_2O \rightleftharpoons 2H^+ + 2NO_3^- \qquad (6\text{-}35)$$

$$K_{VI-35} = \frac{a_{NO_3^-}^2 \cdot a_{H^+}^2}{P_{N_2} \cdot P_{O_2}^{5/2}} \qquad (6\text{-}36)$$

This difference is a testimonial to the ability of living organisms to maintain drastic disequilibrium conditions in the chemistry of natural systems. The difference between the observed and equilibrium values of the concentration of ammonia and of the nitrogen oxides is partly due to biologic effects as well, but is also influenced strongly by photochemical reactions due to solar ultraviolet radiation (see, for instance, McEwan and Phillips, 1975).

During photosynthesis on land and in the oceans CO_2 is the major source of carbon. However, rather little of the nitrogen used in plant growth comes directly from molecular nitrogen. N_2 can be fixed by microorganisms, both "free-living" and symbiotic, on land and in the oceans (see, for instance, Delwiche, 1970), but these organisms supply only a small fraction of the

Table 6-12

Actual and Equilibrium Partial Pressures and Concentrations of Nitrogen Compounds

Species	Environment	Actual Concentration	Equilibrium Concentration
N_2O	Atmosphere	0.33 ± 0.01 ppm	$10^{-12.7}$ ppm
NO	Atmosphere	0.001 ppm	$10^{-9.6}$ ppm
NO_2	Atmosphere	0.001 ppm	$10^{-3.8}$ ppm
NH_3	Atmosphere	0.006–0.020 ppm	$10^{-51.5}$ ppm
NO_3^-	Ocean water	0 to 30×10^{-6} mol/kg	$a_{NO_3^-} = 10^{+5.7}$

nitrogen used annually in photosynthesis. The bulk of the combined nitrogen is recycled. Deevey (1973) estimated that the weight ratio of C/N in terrestrial plants is about 79; in marine plants the C/N weight ratio is ca. 5.7. These ratios and the estimated rates of carbon fixation in Table 6-7 yield values of 600×10^{12} and 4700×10^{12} g/yr, respectively, for the annual rate of nitrogen use in photosynthesis by the terrestrial and marine biosphere.

The estimates of the rate of N_2 fixation in Table 6-11 are quite imprecise, but they show clearly that recycling of combined nitrogen is very efficient. The annual supply of combined nitrogen to the oceans is ca. 85×10^{12} g of N/yr. About 30×10^{12} g of N/yr are contributed by nitrogen fixation in the oceans, ca. 10×10^{12} g of N/yr by the river flux of NO_3^- to the oceans, and roughly 45×10^{12} g of N/yr by NO_3^- and NH_4^+ in rain at sea. Of the last quantity, roughly 5×10^{12} g of N/yr are contributed by nitrogen fixed in the atmosphere, the remainder by the return of bacterially produced NH_3 and by a variety of combustion processes. Nitrogen fixed during a given year represents only a trivial part of the total amount of nitrogen used annually during photosynthesis by the marine biosphere. The cycling of marine nitrate from the oceans through the marine biosphere and back is, therefore, very nearly complete. Of the ca. 85×10^{12} g/yr of "new" combined nitrogen, roughly 20×10^{12} leave the system as a constituent of organic matter buried with marine sediments; the remainder is decomposed and presumably returns to the atmosphere, partly as N_2, partly as N_2O, and a small amount as NH_3.

The addition of fixed nitrogen to the terrestrial biosphere currently amounts to about 200×10^{12} g/yr (see Table 6-11). Of this quantity approximately 100×10^{12} g of N/yr are fixed directly, 65×10^{12} g of N/yr are added as NO_3^- and NH_4^+ in rain, and 40×10^{12} g of N/yr are currently added in the form of fertilizers. Since the total rate of nitrogen incorporation is ca. 600×10^{12} g of N/yr, ca. 400×10^{12} g of N/yr of fixed nitrogen are annually cycled from the reservoir of dead terrestrial organic matter back to the terrestrial biosphere if the system is at dynamic equilibrium.

The contribution of terrestrial organic matter to the total organic matter buried at sea is still in dispute, but seems to be small (see Chap. V). In Table 5-11 and Figure 6-19 the proportion of terrestrial organic matter has been set at 9%. Most of the nitrogen fixed annually in soils, and most of the atmospherically fixed nitrogen that is used by the terrestrial biosphere are destroyed during repeated cycling. The recycling of nitrogen on land seems to be a good deal less efficient than at sea; on land roughly 20% of the annual budget of nitrogen is probably lost; at sea the loss is apparently about 1%.

The loss of nitrogen from the ocean–atmosphere system, largely as a constituent of marine sediments, is probably balanced by gains from

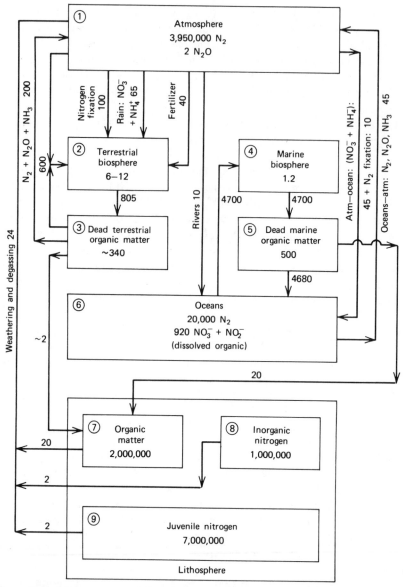

Figure 6-19. Major aspects of the nitrogen cycle; nitrogen in reservoirs in units of 10^{15} g; transfer rates in units of 10^{12} g/yr; see Table 6-11 for sources and uncertainties of data.

weathering, metamorphism, and magmatism. The nitrogen content of the atmosphere is so large, that imbalances between losses and gains can be sustained for geologically long periods of time. At the present loss rate it would take approximately

$$\frac{3,950,000 \times 10^{15} \text{ g}}{22 \times 10^{12} \text{ g/yr}} \approx 2 \times 10^{8} \text{ yr}$$

to remove all of the atmospheric N_2 as a constituent of marine sediments. Arguments for a currently balanced nitrogen budget based on the geochemistry of nitrogen alone are, therefore, unconvincing. The losses and gains of nitrogen are, however, closely tied to those of carbon, and since the budget of this element seems to have been roughly in balance for geologically long periods of time (see Carbon Compounds), the same has probably also been true for nitrogen.

Production and Loss of the Nitrogen Oxides and Ammonia

N_2O is produced during denitrification both on land and at sea. Albrecht, et al. (1970) among others, have shown that the mixing ratio of N_2O in air from soil profiles is distinctly greater than in the atmosphere; Hahn (1974, 1975) has shown that the N_2O content of North Atlantic water is supersaturated with N_2O. Yoshinari (1976) has shown that N_2O concentrations in seawater tend to be high in the oxygen minimum layer. However, estimates of the global rate of N_2O production are still very uncertain. McElroy et al. (1976) have suggested that 20 to 70% of all denitrifying events result in the emission of N_2O to the atmosphere. If this is correct, then denitrification on land produces approximately

$$400 \times 10^{12} \frac{\text{g N}}{\text{yr}} \times \frac{44 \text{ g } N_2O/\text{mol}}{28 \text{ g } N_2/\text{mol}} \times (0.2{-}0.7) = (120{-}420) \times 10^{12} \frac{\text{g } N_2O}{\text{yr}}$$

and denitrification at sea approximately

$$45 \times 10^{12} \frac{\text{g N}}{\text{yr}} \times \frac{44 \text{ g } N_2O/\text{mol}}{28 \text{ g } N_2/\text{mol}} \times (0.2{-}0.7) = (15{-}50) \times 10^{12} \frac{\text{g } N_2O}{\text{yr}}$$

Hahn (1977), on the other hand, prefers to believe that the oceans are the major source of N_2O, followed in importance by soil and fresh water sources. The total quantity of N_2O in the atmosphere is close to 2×10^{15} g. At steady state, and if there are no significant sources of N_2O other than denitrification, the residence time of N_2O in the atmosphere is approximately

$$\tau_{N_2O} \approx \frac{2 \times 10^{15} \text{ g}}{(135\text{--}470) \times 10^{12} \text{ g/yr}} \approx (5\text{-}15) \text{ yr}$$

This value is in agreement with Junge's (1974) estimate of ca. 10 yr based on other criteria, but could be low by as much as a factor of 8.

The mixing ratio of N_2O in the atmosphere decreases upward (see Figure 6-20). At a height of 40–50 km Ehhalt et al. (1975) found 3 ± 7 ppb of N_2O by volume, that is, approximately 1% of the mixing ratio of N_2O at sea level. This drastic decrease in the N_2O abundance is probably due in part to the photodissociation of N_2O

$$N_2O + h\nu \ (\lambda < 337 \text{ nm}) \rightarrow N_2 + O^1(D) \tag{6-37}$$

$$N_2O + h\nu \ (\lambda < 210 \text{ nm}) \rightarrow N_2 + O^1(S) \tag{6-38}$$

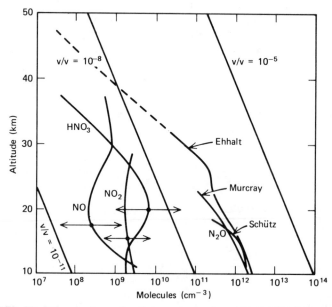

Figure 6-20. Variation in the molecular concentration of N_2O, NO, NO_2 and HNO_3 with altitude (National Research Council, 1975). Reproduced with the permission of the National Academy of Sciences.

and in part to the reactions of N_2O with $O^1(D)$ in the stratosphere (McEwan and Phillips, 1975, p. 97).

$$N_2O + O^1(D) \rightarrow 2NO \qquad (6\text{-}39)$$

$$N_2O + O^1(D) \rightarrow N_2 + O_2 \qquad (6\text{-}40)$$

Their estimated N_2O loss rate of $(2\text{--}3) \times 10^9$ molecules/cm^2 sec may be too low; it corresponds to only $(20\text{--}30) \times 10^{12}$ g N_2O/yr, that is, a rate nearly an order of magnitude less than the estimated rate of N_2O production during denitrification. At present it is not clear whether a major sink of N_2O has been overlooked, or whether the global production rate of N_2O has been considerably overestimated.

There is little doubt that human intervention in the nitrogen cycle will assume major proportions during the next 30 yr. Industrial fixation of nitrogen for agricultural purposes has grown by a factor of 10 during the past 25 yr. Estimates of the rate of nitrogen fixation accompanying fertilizer production during the year 2000 A.D. range from 100×10^{12} g/yr to 200×10^{12} g/yr (Hardy and Havelka, 1975). Nitrogen fixation by other industrial processes may amount to an additional 100×10^{12} g/yr (McElroy et al., 1976). The rate of industrial nitrogen fixation by the year 2000 A.D. may, therefore, exceed the estimated rate of "natural" nitrogen fixation by as much as a factor of 6. The time scale of the mixing and denitrification of this fixed nitrogen is not known, but the N_2O content of the atmosphere will certainly increase during the next century.

The rate of NO production in the stratosphere is, therefore, also apt to rise, and the rate of ozone destruction via the reaction

$$NO + O_3 \rightarrow NO_2 + O_2 \qquad (6\text{-}41)$$

will increase. There is no indication that the ozone concentration in the stratosphere has decreased due to the rising use of fertilizer and halocarbon compounds during the past decade (see Figure 6-14), but the relationships between nitrogen fixation and the chemistry of the atmosphere obviously deserve careful analysis and continuing observation.

NO can participate in a large number of reactions in the upper atmosphere (see Table 6-13). Some of these lead to its decomposition, others to its oxidation to NO_2, NO_3, N_2O_5, and HNO_3. Since HNO_3 is very soluble in water, its removal together with ammonia in rain water is an important loss mechanism of fixed nitrogen from the atmosphere.

Some 6 to 20 ppb ammonia are present in unpolluted air. Bacterial decomposition of organic matter and the burning of coal and fuel oil are thought to be its main sources. Removal of ammonia from the atmosphere

Table 6-13

Photochemical Reactions in the Stratosphere Classified in
Various Ways

a.	Classic Chapman, pure air reactions	
	(1)	$O_2 + h\nu$ (below 242 nm)$\rightarrow O + O$
	(2)	$O + O_2 + M \rightarrow O_3 + M$
	(3)	$O_3 + h\nu$ (uv and visible)$\rightarrow O_2 + O$
	(4)	$O_3 + O \rightarrow O_2 + O_2$
b.	Nitrogen oxide (NO_x) catalyzed destruction of ozone	
	(5)	$O_3 + NO \rightarrow O_2 + NO_2$
	(6)	$NO_2 + O \rightarrow NO + O_2$
	(7)	$NO_2 + h\nu$ (below 400 nm)$\rightarrow NO + O$
c.	Pure air reactions, considering excited oxygen atoms	
	(3a)	$O_3 + h\nu$ (below 310 nm)$\rightarrow O_2 + O(^1D)$
	(3b)	$O_3 + h\nu$ (above 310 nm)$\rightarrow O_2 + O(^3P)$
	(4a)	$O_3 + O(^1D) \rightarrow O_2 + O_2$
	(4b)	$O_3 + O(^3P) \rightarrow O_2 + O_2$
	(8)	$O(^1D) + M \rightarrow O(^3P) + M$
d.	Destruction of ozone by free radicals (HO_x) based on water	
	(9)	$HO + O_3 \rightarrow HOO + O_2$
	(10)	$HOO + O_3 \rightarrow HO + O_2 + O_2$
	(11)	$HO + O \rightarrow H + O_2$
	(12)	$HOO + O \rightarrow HO + O_2$
	(13)	$H + O_3 \rightarrow HO + O_2$
	(14)	$H + O_2 + M \rightarrow HOO + M$
e.	Sources and sinks of NO_x	
	(15a)	$O(^1D) + N_2O \rightarrow 2NO$
	(15b)	$O(^1D) + N_2O \rightarrow N_2 + O_2$
	(16)	$N_2O + h\nu \rightarrow N_2 + O(^1D)$
	(17)	$NO + h\nu$ (below 192 nm)$\rightarrow N + O$
	(18)	$N + NO \rightarrow N_2 + O$
	(19)	$N + O_2 \rightarrow NO + O$
	(20)	$N + O_3 \rightarrow NO + O_2$
	(21)	$N_2 + \text{cosmic rays} \rightarrow N + N$
	(22)	$HO + NO_2 \xrightarrow{M} HNO_3$
	(23)	$HNO_3 + h\nu \rightarrow HO + NO_2$
	(24)	$HO + HNO_3 \rightarrow H_2O + NO_3$
f.	Sources and sinks of HO_x radicals	
	(25a)	$O(^1D) + H_2O \rightarrow 2HO$
	(25b)	$O(^1D) + H_2 \rightarrow H + HO$
	(25c)	$O(^1D) + CH_4 \rightarrow HO + CH_3$
	(26)	$HO + CH_4 \rightarrow H_2O + CH_3$

310

Table 6-13   (*Continued*)

(27)	$CH_3 + O_2 \rightarrow CH_3OO + M$
(28)	$CH_3OO + NO \rightarrow CH_3O + NO_2$
(29)	$CH_3O + O_2 \rightarrow H_2CO + HOO$
(30)	$HO + HOO \rightarrow H_2O + O_2$
(31)	$HOO + HOO \rightarrow H_2O_2 + O_2$
(32)	$H_2CO + h\nu \rightarrow H + HCO$
(33)	$HCO + O_2 \rightarrow CO + HOO$
(34)	$H_2O_2 + h\nu \rightarrow HO + HO$
(35)	$HO + H_2O_2 \rightarrow H_2O + HOO$
(36)	$O + H_2O_2 \rightarrow HO + HOO$

g. Coupling of NO_x and HO_x systems

(37)	$HOO + NO \rightarrow HO + NO_2$
(22)	$HO + NO_2 \overset{M}{\rightarrow} HNO_3$
(23)	$HNO_3 + h\nu \rightarrow HO + NO_2$
(24)	$HO + HNO_3 \rightarrow H_2O + NO_3$
(38a)	$NO_3 + NO \rightarrow 2NO_2$
(38b)	$NO_3 + h\nu \ (<580 \text{ nm}) \rightarrow NO_2 + O$

h. NO_x reactions during the night

(39)	$NO + O_3 \rightarrow NO_2 + O_2$
(40)	$NO_2 + O_3 \rightarrow NO_3 + O_2$
(41)	$NO_3 + NO_2 \overset{M}{\rightarrow} N_2O_5$
(42)	(N_2O_5 photolyzed during the day)
(43)	$NO_3 + NO_2 \rightarrow NO_2 + O_2 + NO$

i. Reaction of chlorine species

(44)	$Cl_2 + h\nu \rightarrow Cl + Cl$
(45)	$HO + HCl \rightarrow H_2O + Cl$
(46)	$O(^1D) + CF_2Cl_2 \rightarrow ClO + CF_2Cl$
(47)	$CF_2Cl_2 + h\nu \ (\text{below } 210 \text{ nm}) \rightarrow CF_2Cl + Cl$
(48)	$Cl + CH_4 \rightarrow HCl + CH_3$
(49)	$Cl + O_3 \rightarrow ClO + O_2$
(50)	$ClO + O \rightarrow Cl + O_2$
(51)	$ClO + NO \rightarrow Cl + NO_2$

j. Ionic reactions

takes place largely by its neutralization and condensation in droplets as a constituent of $(NH_4)_2SO_4$ and NH_4NO_3. These reactions are rapid, and the residence time of ammonia in the atmosphere is probably on the order of a day. Robinson and Robbins (1968) have estimated that the world-wide urban rate of emission of ammonia is currently 3.8×10^{12} g/yr. Roughly three-quarters of this quantity is a by-product of the burning of coal; most of the remainder accompanies the burning of fuel oil. It is interesting to compare the figures for industrial input with the rate of removal of nitrogen with rain as ammonia and as nitrate. Eriksson (1952) has summarized a large body of the older data for ammonia and nitrate in rain. A continental world mean of (0.6–0.8) g N/m^2 yr is suggested by these results.* Ångström and Högberg (1952) studied the distribution of NH_4^+–nitrogen in rain in Sweden, and found an annual mean infall of about (0.25 ± 0.05) g of NH_4^+/m^2 yr. The quantity of nitrogen transferred to the ground in precipitation in the form of NO_3^- nitrogen was found to be half the quantity transferred as NH_4^+; the total nitrogen deposition rate was, therefore, approximately 0.3 g N/m^2 yr. Junge (1958) has reported on a rather extensive study of ammonia and nitrate in precipitation within the United States. His results show very clearly that there are large regional and seasonal variations in the ammonia and nitrate content of rainfall. These variations confirm the short residence time of NH_4^+ and NO_3^- in the atmosphere. Spring maxima—probably due to biologic activity—were observed in unpolluted areas; winter maxima—presumably due to fossil fuel burning—were observed in heavily populated areas.

The NH_4^+ concentration in rain varies from less than 0.02 to roughly 2 mg/kg. The most frequent values over the continents are in the range of 0.1 to 0.2 mg/kg (Junge, 1963). As the mean annual precipitation (see Table 3-1) is 100 cm, the mean fall-out of NH_4^+ on the continents is about 0.1 to 0.2 g/m^2 yr. The range of concentration of nitrate in precipitation is about 1.0 to 2.5 mg/kg in Europe and between 0.3 and 2.0 mg/kg in other areas (Junge, 1963). The total nitrogen accompanying rain over continents is, therefore, about

$$(0.1-0.2) \times \tfrac{14}{18} + (1.0-1.8) \times \tfrac{14}{62} \cong (0.4 \pm 0.2) \ g/m^2 \ yr$$

This corresponds to a total continental outfall of

$$(0.4 \pm 0.2) \ g \ N/m^2 \times 1.6 \times 10^{14} \ m^2 \cong (65 \pm 30) \times 10^{12} \ g \ N/yr$$

Rain over the oceans contains on the average less ammonia and nitrate than over continental areas; the available data are quite scant (see Junge, 1963). A mean value roughly one-third of that for the continents is probably a reasonable estimate. The total outfall of combined nitrogen with rain to the oceans is, therefore, roughly

$$(0.13 \pm 0.07) \ g \ N/m^2 \times 3.6 \times 10^{14} \ m^2 \cong (45 \pm 25) \times 10^{12} \ g \ N/yr$$

---

*Robinson and Robbins (1968) seem to have erred by a factor of ten in their conversion from kg/hectares to g/m^2. This error invalidates much of their discussion of the quantitative aspects of the nitrogen cycle.

These figures, uncertain though they are, were used in Table 6-11 and Figure 6-19 to estimate the input of fixed nitrogen to oceans and to the terrestrial biosphere.

COMPOUNDS OF SULFUR

The atmosphere is a very minor reservoir for compounds of sulfur (see Fig. 6-21). SO_2, the most abundant sulfur compound, is present at the parts per billion level in unpolluted air. H_2S is normally present at concentrations below instrumental detection limits. Organic sulfur compounds, such as dimethyl sulfide (DMS) and dimethyl disulfide (DMDS) are probably present (Lovelock, et al., 1972; Rasmussen, 1974) in the atmosphere but at concentrations well below those of SO_2.

The low concentration of SO_2 and H_2S in the atmosphere is due to the rapid oxidation and wash-out of both gases. The input of sulfur dioxide due to fossil fuel burning and industrial smelting and petroleum refining is

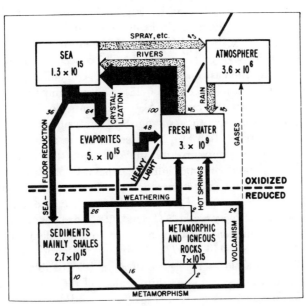

Figure 6-21. Holser and Kaplan's (1966) representation of the geochemical cycle of sulfur. Masses are in metric tons. Most material above the dashed line is oxidized to sulfate, most below this line is reduced to sulfide; above the solid line heavy sulfur predominates ($\delta > +5\%$). Long-term (dark) and short-term fluxes of sulfur between reservoirs are indicated on the basis of 100 for the long-term component of fresh water sulfate flowing to the sea.

currently about 65×10^{12} g/yr (see Table 4-4). The gas is probably oxidized to sulfate via a number of mechanisms; these include photochemical reactions in air and catalytic reactions on particles and in solution in water droplets. Considerable uncertainty remains concerning the relative importance of these various potential reactions of SO_2 in the atmosphere (Hill, 1973).

The residence time of SO_2 is apparently highly dependent on the presence of other pollutants. In urban areas the residence time of SO_2 may be measured in hours, whereas in rural regions it may be as long as a week. The mean value is probably a few days (Weber, 1970; Rodhe, 1972). The short residence time of SO_2 ensures that a large fraction of the atmospheric SO_2 pollution is removed within about 1000 km of its source, and that the SO_2 content of air is normally much greater in highly industrialized than in rural areas (see, for instance, Meetham, 1959; Fischer et al., 1969). The oxidation of SO_2 to SO_3 is usually followed by the removal of H_2SO_4 and of other sulfates in rain water. In areas of intense SO_2 pollution rain tends to be acid, and rain with a pH of less than 3 has been reported in heavily polluted areas (Likens et al., 1972). Some SO_2 is apparently taken up by vegetation (Hill, 1973; Friend, 1973), but the quantity is probably small compared to that removed from the atmosphere as a constituent of rain water.

The distribution of SO_4^{2-} in rain on land is consistent with a terrestrial origin for most of the SO_4^{2-}. Contours of the Cl^- concentration in rain tends to follow shorelines, whereas contours of the SO_4^{2-} content of rain are nearly independent of shorelines (see Figure 6-22). The very imperfect correlation in Figure 6-22 between the average SO_4^{2-} concentration of rain and the location of industrial centers suggests that SO_4^{2-} in rain on land is not only due to human pollution. High SO_4^{2-} concentrations in arid areas may well be due to wind-injected gypsum and anhydrite from the land surface. In swampy areas H_2S and organic sulfur compounds produced during the bacterial decay of organic matter may add significant quantities of sulfur to the atmosphere. However, all of the estimates that have been proposed for the production rate of biogenic H_2S (Friend, 1973) are based on differences between uncertain numbers, because there are no direct data for the rate of biogenic H_2S emission (Hill, 1973). Many of the proposed figures fall between 50×10^{12} and 110×10^{12} g/yr (Friend, 1973, Table 4); these estimates are probably excessive. The sulfur content of the biosphere is roughly 0.1% (see, for instance, Deevey, 1973, Brooks and Kaplan, 1972). The annual rate of incorporation and release of sulfur from the biosphere is, therefore, approximately

$$73 \times 10^{15} \text{ g C/yr} \times \frac{1 \text{ g S}}{550 \text{ g C}} \cong 130 \times 10^{12} \text{ g S/yr}$$

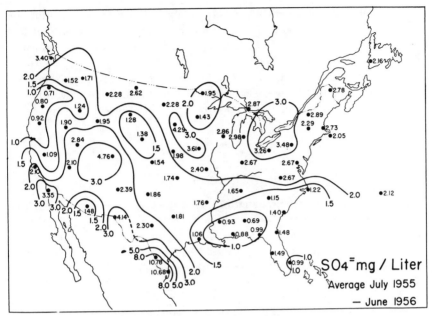

Figure 6-22. Average SO_4^{2-} concentration in precipitation over the United States (Junge, 1960). Copyrighted by American Geophysical Union.

Since photochemical oxidation of H_2S to H_2SO_4 in contact with free O_2 is very rapid, sulfur additions to the atmosphere from the decay of organic matter probably amount to no more than a few percent of 130×10^{12} g/yr, and are probably not of major importance for the sulfur balance of the atmosphere.

The major additions of sulfur to the atmosphere appear to be SO_4^{2-} in sea spray at sea and in coastal areas, SO_4^{2-} in terrestrial aerosols and in solution, from sulfates in soils on land, and anthropogenic sulfur, consisting largely of SO_2, from fossil fuel burning. Biogenic effects and volcanic inputs are apparently of secondary importance. The increasing use of coal as a source of energy during the next 25 to 50 yr implies a progressively greater rate of SO_2 injection into the atmosphere unless sulfur is either removed more efficiently prior to burning, or SO_2 is removed more efficiently prior to release of the stack gases to the atmosphere. Most of the damage due to excessive SO_2 emission will be relatively local; excellent examples of extreme SO_2 damage to vegetation in the vicinity of old smelting centers such as Ducktown, Tennessee, and Sudbury, Ontario, are readily available for predicting the consequences of intense SO_2 emission. Fortunately, the effects of SO_2 pollution are reversible. Several major cities have shown that the problem of SO_2 pollution can be alleviated by controlling the type of fuel that is burned and the kind of smoke emitted from domestic and industrial heating plants. An outstanding example of the effectiveness of such control is London, "where a gloomy mixture of smoke and fog, redolent of Sherlock Holmes and Jack the Ripper, is no longer the most characteristic feature of a winter's evening" (McEwan and Phillips, 1975, pp. 230–231).

SUMMARY

Most of the rate data accumulated in this chapter are summarized in Table 6-14. Unfortunately, many of the figures are still very uncertain, and should be regarded as little more than order-of-magnitude estimates. There are numerous sources and sinks for atmospheric gases. The process by which they are transferred among the several reservoirs can be divided into three major categories:

1. Processes that depend on the internal energy of the earth.
2. Processes that depend on solar energy inputs.
3. Anthropogenic processes.

Processes in the first category are typically much slower than those in the second category; in large part this is because the rate of energy input from the sun is about 10^4 times greater than the energy released by radioactive decay within the earth. Thus, the atmospheric flux of CO_2 due to photosynthesis and decay is about 10^3 times greater than the rate of CO_2 degassing, and the biologically related flux of nitrogen and its compounds is at least 10 times greater than the corresponding rate of degassing. Terrestrial degassing dominates the total flux of only the rare gases.

The earth has acted as a nearly closed system since early in its history. Helium and hydrogen are the only two gases that are escaping from the earth at a significant rate today. The balance between helium gain by degassing and helium loss by escape into interplanetary space determines the helium content of the atmosphere. Hydrogen loss is currently of minor importance for the oxidation state of the atmosphere, but could have been significant during the early history of the earth.

The partial pressure of most atmospheric gases is determined by the condition that the rate of output is equal to the rate of input. Ne, Ar, Kr, and possibly Xe are the only important exceptions. The relationship between abundance and flux intensities is quite different for the several atmospheric constituents. The total inventory of carbon compounds in the atmosphere is less than one-thousandth of the inventory of nitrogen and its compounds in spite of the greater absolute flux intensity of carbon compounds. This difference is due in part to the chemical inertness of N_2 compared to that of CO_2, in part to the low CO_2 pressure required to stabilize $CaCO_3$ and $CaMg(CO_3)_2$ with respect to silicates of calcium and magnesium that are stable in the presence of water between 0 and 50°C.

The CO_2 pressure of the atmosphere is apparently close to the equilibrium value for carbonate and silicate minerals in the oceans. On the other hand atmospheric N_2 and O_2 are very far out of equilibrium with respect to the NO_3^- content and the pH of seawater. The large oxygen content of the atmosphere is apparently required to oxidize all but a few tenths of a percent of the organic matter produced annually, and is maintained by a negative feedback system involving oxygen production via photosynthesis and oxygen loss via oxidation reactions.

Ozone owes its existence in the main to photochemical reactions of oxygen in the atmosphere. Its production in the stratosphere is largely balanced by photochemical destruction both at higher and lower altitudes. Anthropogenic inputs of halogen compounds and nitrogen oxides are approaching levels that can probably produce a nonnegligible reduction in the ozone content of the atmosphere.

Compounds of sulfur are very much underrepresented in the atmosphere. Sulfur is strongly involved in biological processes, but SO_2 and H_2S are removed so rapidly from the atmosphere that their concentration is extremely low except in the immediate vicinity of strong sources.

The chemistry of the atmosphere is closely linked to that of the oceans and the biosphere. Atmospheric CO_2 is linked directly to the marine carbonate cycle, atmospheric O_2 is linked via its solubility in seawater to the decay of organic matter in the oceans, and the production of minor gases such as N_2O and CH_4 in the oceans influences the composition and behavior of the atmosphere. Such close links between the atmosphere, the oceans, and the biosphere do not, however, imply that the biosphere as a whole is an adaptive control system capable of maintaining the earth in homoeostasis, as proposed by Lovelock and Margulis (1974). The ocean–atmosphere system has responded to biologic inventions such as photosynthesis by adjusting its composition so that the bulk chemistry of the crust is maintained. The biosphere has responded to these adjustments by optimizing the use of available free energy. The long-term stability of the system is due to the continuing interplay between the numerous inorganic and biologic processes that have shaped the surface of the earth.

Table 6-14
The Major Sources and Sinks of Atmospheric Gases[a]

	Rare Gases					Carbon Compounds				
	He	Ne	Ar	Kr	Xe	CO_2	CH_4	CO	$CFCl_{3}+$ CCl_2F_2	$CHCl_{3}+$ CH_3Cl
Sources										
1. Interior of the Earth (volcanic emissions, etc.)	$(\sim 2 \times 10^9)$	$(\sim 1 \times 10^7)$	$(\sim 1 \times 10^{10})$	$(\sim 4 \times 10^6)$	$(\sim 1 \times 10^6)$	0.3×10^{15}	m	0.1×10^{14}	—	—
2. Surface reactions										
Rock weathering	?	?	?	?	?	0.9×10^{15}	—	—	—	—
Other reactions	—	—	—	—	—	—	—	—	—	—
3. Biologic processes										
Photosynthesis	—	—	—	—	—	—	—	—	—	—
Respiration and decay of organic matter	—	—	—	—	—	300×10^{15}	1×10^{15}	1.6×10^{14}	—	—
Forest fires	—	—	—	—	—	m?	—	0.6×10^{14}	—	—
4. Anthropogenic effects										
Combustion processes	—	—	—	—	—	15×10^{15}	$<.2 \times 10^{15}$	3×10^{14}	—	—
Industrial syntheses, etc.	—	—	—	—	—	2.6×10^{15}	—	—	6×10^{11}	$(0.3-6) \times 10^{12}$
5. Reactions within the atmosphere	—	—	—	—	—	m	—	30×10^{14}	—	—

Sinks

1. Planetary escape   (~2×10^9)

2. Reactions within the atmosphere 1×10^{15} 33×10^{14} M   M

3. Anthropogenic effects
   Combustion processes
   Industrial syntheses

4. Biologic processes
   Photosynthesis 300×10^{15} $\sim2\times10^{14}$
   Respiration and decay of organic matter

5. Removal with rain   tr

6. Surface reactions
   Rock weathering 0.6×10^{15} m   m(?)
   Other reactions 0.5×10^{15} tr?

Table 6-14 (Continued)

Sources	Oxygen, Hydrogen, and Ozone			Nitrogen and Its Compounds				Compounds of Sulfur		
	O_2	H_2	O_3	N_2	N_2O	$NO+NO_2$ $+N_2O_5$	NH_3	SO_2	H_2S	$DMS+DMDS$
1. Interior of the Earth (volcanic emissions, etc.)	—	2×10^{12}	—	2×10^{12}	—	—	—	10×10^{12}	tr	—
2. Surface reactions										
Rock weathering	—	—	—	20×10^{12}	—	—	tr(?)	—	—	—
Other reactions	—	—	—	—	—	—	—	—	—	—
3. Biologic processes										
Photosynthesis	190×10^{15}	—	—	—	—	—	—	—	—	—
Respiration and decay of organic matter	—	M(?)	—	180×10^{12}	200×10^{12}	tr(?)	M	m(?)	$<50\times10^{12}$	M
Forest fires	—	—	—	m(?)	—	—?	—?	m(?)	—	—
4. Anthropogenic effects										
Combustion processes	—	m(?)	m(?)	m(?)	—	M	4×10^{12}	65×10^{12}	M(?)	—
Industrial syntheses, etc.	—	—	—	—	—	—	m(?)	—	—	—
5. Reactions within the atmosphere	m(?)	M(?)	M	tr	—	M	—	M	—	—

320

Sinks									
1. Planetary escape	—	0.05×10^{12}	—	—	—	—	—	—	—
2. Reactions within the atmosphere	m	M	M	12×10^{12}	200×10^{12}	M	65×10^{12}	$<50 \times 10^{12}$	M(?)
3. Anthropogenic effects									
Combustion processes	11×10^{15}	—	—	58×10^{12}	—	—	—	—	—
Industrial syntheses	—	—	—	40×10^{12}	—	—	—	—	—
4. Biologic processes									
Photosynthesis	—	—	—	—	tr(?)	—	tr(?)	—	—
Respiration and decay of organic matter	190×10^{15}	m(?)	—	130×10^{12}	—	—	—	—	—
5. Removal with rain	—	—	—	—	m	M	100×10^{12}	M	?
6. Surface reactions									
Rock weathering	0.3×10^{15}	tr(?)	—	—	—	—	tr(?)	—	—
Other reactions	—	—	—	—	—	—	m(?)	m(?)	m(?)

^{a}M = major, m = minor, tr = trace process; units of rates in g/yr; see text for sources and uncertainties of data.

REFERENCES

Albrecht, B., Junge, C., and Zakosek, H., 1970, Der N_2O-Gehalt der Bodenluft in drei Bodenprofilen, *Z. Pflanz. Bodenk.* **125**, 205–211.

Anderson, J. G., Margitan, J. J., and Stedman, D. H., 1977, Atomic chlorine and the chlorine monoxide radical in the stratosphere: three in situ observations, *Science* **198**, 501–503.

Ångström, A. and Högberg, L., 1952, On the content of nitrogen in atmospheric precipitation in Sweden, II, *Tellus* **4**, 271–279.

Axford, W. I., 1968, The polar wind and the terrestrial helium budget, *J. Geophys. Res.* **73**, 6855–6859.

Bacastow, R. and Keeling, C. D., 1973, Atmospheric carbon dioxide and radio-carbon in the natural carbon cycle, II, Changes from A.D. 1700 to 2070 as deduced from a geochemical model, in *Carbon and the Biosphere*, G. M. Woodwell and E. V. Pecan, Ed., U.S. Atomic Energy Commission, Technical Information Center, pp. 86–135.

Bartholomé, P., 1966, The abundance of dolomite and sepiolite in the sedimentary record, *Chem. Geol.* **1**, 33–48.

Berkner, L. V. and Marshall, L. C., 1965, On the origin and rise of oxygen concentration in the earth's atmosphere, *J. Atmos. Sci.* **22**, 225–261.

Bieri, R. H., Koide, M., and Goldberg, E. D., 1966, The noble gas contents of Pacific seawaters, *J. Geophys. Res.* **71**, 5243–5265.

Bieri, R. H., Koide, M., and Goldberg, E. D., 1968, Noble gas contents of marine waters, *Earth Planet. Sci. Lett.* **4**, 329–340.

Blatt, H. and Jones, R. L., 1975, Proportions of exposed igneous, metamorphic, and sedimentary rocks, *Bull. Geol. Soc. Am.* **86**, 1085–1088.

Bolin, B., 1970, The carbon cycle, *Sci. Am.* **223**, Sept., 124–132.

Breiland, J. G., 1969, Variations in the vertical distribution of atmospheric ozone during the passage of a short wave in the Westerlies, *J. Geophys. Res.* **74**, 4501–4510.

Bricker, O. P., Nesbitt, H. W., and Gunter, W. D., 1973, The stability of talc, *Am. Miner.* **58**, 64–72.

Broecker, W. S., 1968, Man's oxygen reserves, *Science* **168**, 1537–1538.

Broecker, W. S., Li, Y-H., and Peng, T-H., 1971, Carbon dioxide—Man's unseen artifact, in *Impingement of Man on the Oceans*, Chap. 11, D. W. Hood, Ed., Wiley-Interscience, New York.

Brooks, R. R. and Kaplan, I. R., 1972, Biogeochemistry, in *The Encyclopedia of Geochemistry and Environmental Sciences*, Vol. IVA, R. W. Fairbridge, Ed., Van Nostrand Reinhold Co., New York, pp. 74–82.

Cadle, R. D., 1973, Atmospheric carbon monoxide, in *Carbon and the Biosphere*, G. M. Woodwell and E. V. Pecan, Eds., U.S. Atomic Energy Commission, Technical Information Center, pp. 136–143.

Canalas, R. A., Alexander, E. C., Jr., and Manuel, O. K., 1968, Terrestrial abundance of noble gases, *J. Geophys. Res.* **73**, 3331–3334.

Chapman, S., 1960, The thermosphere-the earth's outermost atmosphere, in *Physics of the Upper Atmosphere*, Chap. 1, J. A. Ratcliffe, Ed., Academic Press, New York.

Clarke, F. W., 1924, *The Data of Geochemistry*, 5th ed., U.S. Geological Survey Bulletin 770.

Clarke, F. W. and Washington, H. S., 1924, The composition of the earth's crust, U.S. Geological Survey, Professional Paper 127.

Clarke, W. B., Beg, M. A., and Craig, H., 1969, Excess ^{3}He in the sea: evidence for terrestial primordial helium, *Earth Planet. Sci. Lett.* **6**, 213–220.

Craig, H., Clarke, W. B., and Beg, M. A., 1975, Excess ^{3}He in deep water on the East Pacific Rise, *Earth Planet. Sci. Lett.* **26**, 125–132.

Crutzen, P. J., 1970, The influence of nitrogen oxides on the atmosphere ozone content, *Q. J. R. Meteorological Soc.* **96**, 320–325.

Damon, P. E. and Kulp, J. L., 1958, Inert gases and the evolution of the atmosphere, *Geochim. Cosmochim. Acta* **13**, 280–292.

Deevey, E. S., Jr., 1973, Sulfur, nitrogen, and carbon in the biosphere, in *Carbon and the Biosphere*, G. M. Woodwell and E. V. Pecan, Eds., U.S. Atomic Energy Commission, Technical Information Center, pp. 182–190.

Delwiche, C. C., 1970, The nitrogen cycle, *Sci. Am.* **223**, Sept., 136–146.

Delwiche, C. C. and Likens, G. E., 1977, Biological response to fossil fuel products, *Global Chemical Cycles and Their Alterations by Man*, Dahlem Konferenzen.

Eaton, J. P., and Murata, K. J., 1960. How volcanoes grow, *Science* **132**, 925–938.

Eberhardt, P., Eugster, O., and Marti, K., 1965, A redetermination of the isotopic composition of atmospheric neon, *Z. Naturforsch.* **209**, 623–624.

Ehhalt, D. H., 1973, Methane in the atmosphere, in *Carbon and the Biosphere*, G. M. Woodwell and E. V. Pecan, Ed., U.S. Atomic Energy Commission, Technical Information Center, pp. 144–158.

Ehhalt, D. H., 1974, The atmospheric cycle of methane, *Tellus* **26**, 58–70.

Ehhalt, D. H., 1977, The atmospheric cycle of methane, International Symposium on the Influence of the Biosphere upon the Atmosphere, Mainz, Germany.

Ehhalt, D. H., Heidt, L. E., Lueb, R. H., and Martell, E. A., 1975, Concentrations of CH_4, CO, CO_2, H_2, H_2O, and N_2O in the upper stratosphere, *J. Atmos. Sci.* **32**, 163–169.

Eichmann, R. and Schidlowski, M., 1975, Isotopic fractionation between coexisting organic carbon–carbonate pairs in Precambrian sediments, *Geochim. Cosmochim. Acta* **39**, 585–595.

Ekdahl, C. A. and Keeling, C. D., 1973, Atmospheric carbon dioxide and radio-carbon in the natural carbon cycle. I. Quantitative deductions from records at Mauna Loa Observatory and at the South Pole, in *Carbon and the Biosphere*, G. M. Woodwell and E. V. Pecan, Eds., U.S. Atomic Energy Commission, Technical Information Center, pp. 51–85.

Eriksson, E., 1952, Composition of atmospheric precipitation; I. Nitrogen compounds, *Tellus* **4**, 215–232.

Evans, W. F. J., Hunten, D. M., Llewellyn, E. J., and Jones, A. V., 1968, Altitude profile of the infrared atmosphere system of oxygen in the dayglow, *J. Geophys. Res.* **73**, 2885–2896.

Fischer, W. H., Lodge, J. P., Jr., Pate, J. B., and Cadle, R. D., 1969, Antarctic atmospheric chemistry: Preliminary exploration, *Science* **164**, 66–67.

Friend, J. P., 1973, The global sulfur cycle, in *Chemistry of the Lower Atmosphere*, Chap. 4, S. I. Rasool, Ed., Plenum Press, New York.

Fudali, R. F., 1965, Oxygen fugacities of basaltic and andesitic magmas, *Geochim. Cosmochim. Acta* **29**, 1063–1075.

Ganapathy, R. and Anders, E., 1974, Bulk compositions of the moon and earth estimated from meteorites, in Lunar Science V, Abstracts, Lunar Science Institute, Houston, Texas, pp. 254–256.

Garrels, R. M., Mackenzie, F. T., and Siever, R., 1972, Sedimentary cycling in relation to the history of the continents and oceans, in *The Nature of the Solid Earth*, Chap. 5, E. C. Robertson, Ed., McGraw-Hill, New York.

Garrels, R. M. and Perry, E. A., Jr., 1974, Cycling of carbon, sulfur and oxygen through geologic time, *The Sea*, Vol. 5, Chap. 9, E. D. Goldberg, Ed., Wiley-Interscience, New York.

Gehman, H. M., Jr., 1962, Organic matter in limestones, *Geochim. Cosmochim. Acta* **26**, 885–897.

Giggenbach, W. F. and LeGuern, F., 1976, The chemistry of magmatic gases from Erta'Ale, Ethiopia, *Geochim. Cosmochim. Acta* **40**, 25–30.

Green, J. and Poldervaart, A., 1955, Some basaltic provinces, *Geochim. Cosmochim. Acta* **7**, 177–188.

Hahn, J., 1974, The North Atlantic Ocean as a source of atmospheric N_2O, *Tellus* **26**, 160–168.

Hahn, J., 1975, N_2O measurements in the Northeast Atlantic Ocean, *Meteor. Forsch. Ergebnisse Reihe A* **16**, 1.

Hahn, J., 1977, The cycle of atmospheric nitrous oxide, International Symposium on the Influence of the Biosphere upon the Atmosphere, Mainz, Germany.

Hahn, J. and Junge, C. E., 1977 Atmospheric nitrous oxide: a critical review, *Z. Naturforsch A.* **32**, 190–214.

Hardy, R. W. F. and Havelka, O. D., 1975, Nitrogen fixation research: a key to world food?, *Science* **188**, 633–643.

Heald, E. F., Naughton, J. J., and Barnes, I. L. Jr., 1963, The chemistry of volcanic gases. 2. Use of equilibrium calculations in the interpretation of volcanic gas samples, *J. Geophys. Res.* **68**, 545–557.

Hemley, J. J., Montoya, J. W., Christ, C. L., and Hostetler, P. B., 1977, Mineral equilibria in the MgO–SiO_2–H_2O system: I. Talc–chrysotile–forsterite–brucite stability relations, *Am. J. Sci.* **277**, 322–351.

Hering, W. S. and Borden, T. R., 1967, Ozonesonde observations over North

America, Vol. 4, Air Force Cambridge Research Laboratories, Environmental Research Papers.

Hill, F. B., 1973, Atmospheric sulfur and its links to the biota, in *Carbon and the Biosphere*, G. M. Woodwell and E. V. Pecan, Ed., U.S. Atomic Energy Commission, Technical Information Center, pp. 159–181.

Hilsenrath, E., Seiden, L., and Goodman, P., 1969, An ozone measurement in the mesosphere and stratosphere by means of a rocketsonde, *J. Geophys. Res.* **74**, 6873–6880.

Holland, H. D., 1965, The history of ocean water and its effect on the chemistry of the atmosphere, *Proc. Natl. Acad. Sci.* **53**, 1173–1183.

Holland, H. D., 1968, The abundance of CO_2 in the earth's atmosphere through geologic time, in *Origin and Distribution of the Elements*, L. H. Ahrens, Ed., Pergamon Press, New York, pp. 949–954.

Holland, H. D., 1972, The geologic history of seawater—an attempt to solve the problem, *Geochim. Cosmochim. Acta* **36**, 637–657.

Holland, H. D., 1973a, Systematics of the isotopic composition of sulfur in the oceans during the Phanerozoic and its implications for atmospheric oxygen, *Geochim. Cosmochim. Acta* **37**, 2605–2616.

Holland, H. D., 1973b, Ocean water, nutrients, and atmospheric oxygen, in *Hydrogeochemistry*, of *Proceedings of Symposium of Hydrogeochemistry and Biogeochemistry*, Vol. 1, E. Ingerson, Ed., The Clark Co., Washington, D.C., pp. 68–81.

Holser, W. T. and Kaplan, I. R., 1966, Isotope geochemistry of sedimentary sulfates, *Chem. Geol.* **1**, 93–135.

Hunt, J. M., 1972, Distribution of carbon in crust of earth, *Bull. Am. Assoc. Pet. Geol.* **56**, 2273–2277.

Hunten, D. M., 1973, The escape of light gases from planetary atmospheres, *J. Atmos. Sci.* **30**, 1481–1494.

Hunten, D. M. and Strobel, D. F., 1974, Production and escape of terrestrial hydrogen, *J. Atmos. Sci.* **31**, 305–317.

Hutchinson, G. E., 1954, The biochemistry of the terrestrial atmosphere, in *The Earth as a Planet*, Vol. 2, Chap. 8, G. P. Kuiper, Ed., University of Chicago Press, Chicago.

Jenkins, W. J., 1978, Personal communication.

Jenkins, W. J., Edmond, J. M., and Corliss, J. B., 1978, Excess ^{3}He and ^{4}He in Galapagos submarine hydrothermal waters, submitted to *Nature*.

Johnson, F. S., Purcell, J. D., Tousey, R., and Watanabe, K., 1952, Direct measurements of the vertical distribution of atmospheric ozone to 70 km altitude, *J. Geophys. Res.* **57**, 157–176.

Johnston, H., 1971, Reduction of stratospheric ozone by nitrogen oxide catalysts from exhaust, *Science* **173**, 517–522.

Junge, C. E., 1958, The distribution of ammonia and nitrate in rain water over the United States, *Trans. Am. Geophys. Union* **39**, 241–248.

Junge, C. E., 1960, Sulfur in the atmosphere, *J. Geophys. Res.* **65**, 227–237.

Junge, C. E., 1963, *Air Chemistry and Radioactivity*, Academic Press, New York, 382 pp.

Junge, C. E., 1974, Residence time and variability of tropospheric trace gases, *Tellus* **26**, 477–488.

Junge, C. E., Seiler, W., and Warneck, P., 1971, The atmospheric ^{12}CO and ^{14}CO budget, *J. Geophys. Res.* **76**, 2866–2879.

Junge, C. E., Schidlowski, M., Eichmann, R., and Pietrek, H., 1975, Model calculations for the terrestrial carbon cycle: carbon isotope geochemistry and evolution of photosynthetic oxygen, *J. Geophys. Res.* **80**, 4542–4552.

Kennedy, G. C., 1948, Equilibrium between volatiles and iron oxides in igneous rocks, *Am. J. Sci.* **246**, 529–549.

Koblentz-Mishke, O. J., Volkovinsky, V. V., and Kabanova, J. G., 1970, Plankton primary production of the world ocean, in *Scientific Exploration of the South Pacific*, W. S. Wooster, Ed., N.A.S., Washington, D.C., pp. 183–193.

Krueger, A. J., Heath, D. F., and Mateer, C. L., 1972, Variations in the stratospheric ozone field inferred from Nimbus satellite observations, NASA TM-X-66108, NTIS-N73-12379.

Krueger, A. J. and Minzner, A., 1973, A proposed mid-latitude ozone model for the U.S. Standard Atmosphere (summary).

Lerman, A., et al., 1977, Fossil fuel burning: its effects on the biosphere and biogeochemical cycles; group report, in *Global Chemical Cycles and Their Alteration by Man*, W. Stumm, Ed., Dahlem Konferenzen, Berlin, pp. 275–289.

Likens, G. E., Bormann, F. H., and Johnson, N. M., 1972, Acid rain, *Environment* **14**, March, 33–40.

Lovelock, J. E., 1974, Atmospheric halocarbons and stratospheric ozone, *Nature*, **252**, 292–294.

Lovelock, J. E., Maggs, R. J., and Rasmussen, R. A., 1972, Atmospheric dimethyl sulfide and the natural sulphur cycle, *Nature* **237**, 452–453.

Lovelock, J. E. and Margulis, L., 1974, Atmospheric homeostasis by and for the biosphere: the gaia hypothesis, *Tellus* **26**, 2–10.

Lupton, J. E. and Craig, H., 1975, Excess ^3He in oceanic basalts; evidence for terrestrial primordial helium, *Earth Planet. Sci. Lett.* **26**, 133–139.

MacDonald, G. J. F., 1963, The escape of helium from the earth's atmosphere, *Rev. Geophys.* **1**, 305–349.

Mamyrin, B. A., Tolstikhin, I. N., Anufriev, G. S. and Kamensky, I. L., 1972, Isotopic composition of helium in thermal springs of Iceland, *Geokhimiya*, 1396–1397.

McConnell, J. C., McElroy, M. B., and Wofsy, S. C., 1971, Natural sources of atmospheric CO, *Nature* **233**, 187–188.

McElroy, M. B., 1976, Chemical processes in the solar system: a kinetic perspective, in *Chemical Kinetics*, Vol. 9, Physical Chemistry Series Two, Chap. 5, D. R. Herschbach, Ed., Butterworths, Boston.

McElroy, M. B., Wofsy, S. C., and Yung, Y. L., 1977, The nitrogen cycle: perturbations due to man and their impact on atmospheric N_2O and O_3, *R. Soc. Lond., Philo. Trans., Ser. B.* **277**, 159–181.

McEwan, M. J. and Phillips, L. F., 1975, *Chemistry of the Atmosphere*, John Wiley & Sons, New York.

Meetham, A. R., 1959, The behavior of sulphur dioxide in the atmosphere, in *Atmospheric Chemistry of Chloride and Sulfur Compounds*, J. P. Lodge, Jr., Ed., Geophysical Monograph No. 3, American Geophysical Union of the National Academy of Sciences—National Research Council, pp. 115–121.

Mirtov, B. A., 1964, *Gaseous Composition of the Atmosphere and its Analysis*, Izdatel Akad. Nauk, U.S.S.R., Moscow, 1961, Translated from the Russian, Israel Program for Scientific Translations, Jerusalem.

Morgan, J. J. et al., 1977, Source functions: fossil fuel combustion products, radio-nuclides, trace metals, and heat; Group report, in *Global Chemical Cycles and Their Alterations by Man*, W. Stumm, Ed., Dahlem Konferenzen, Berlin, pp. 291–311.

Muan, A. and Osborn, E. F., 1956, Phase equilibria at liquidus temperatures in the system $MgO-FeO-Fe_2O_3-SiO_2$, *J. Am. Ceram. Soc.* **39**, 121–140.

Nanz, R. H., Jr., 1953, Chemical composition of Precambrian slates with notes on the geochemical evolution of lutites, *J. Geol.* **61**, 51–64.

National Research Council, 1975, *Environmental Impact of Stratospheric Flight: Biological and Climatic Effects of Aircraft Emissions in the Stratosphere*, Climatic Impact Committee, National Research Council, National Academy of Sciences, Washington, D.C., 348 pp.

Naughton, J. J., 1969, personal communication.

Nicolet, M., 1960, The properties and constitution of the upper atmosphere, in *Physics of the Upper Atmosphere*, Chap. 2, J. A. Ratcliffe, Ed., Academic Press, New York.

Nikolayeva, O. V., 1968, Background contents of various forms of sulfur in sedimentary rocks and marine sediments, *Geochem. Int.* **5**, 1037–1041.

Osborn, E. F., 1959, Role of oxygen pressure in the crystallization and differentiation of basaltic magma, *Am. J. Sci.* **257**, 609–647.

Poldervaart, A., 1955, Chemistry of the earth's crust, in *Crust of the Earth*, A. Poldervaart, Ed., Geological Society of America Special Paper 62, pp. 119–144.

Rasmussen, R. A., 1974, Emission of biogenic hydrogen sulfide, *Tellus* **26**, 254–260.

Rasmussen, R. A., 1977a, A comparison of global and regional N_2O measurements, International Symposium on the Influence of the Biosphere upon the Atmosphere, Mainz, Germany.

Rasmussen, R. A., 1977b, Measurement techniques and observed data of CFM, International Symposium on the Influence of the Biosphere upon the Atmosphere, Mainz, Germany.

Rayleigh, Lord (R. J. Strutt), 1939, Nitrogen, argon and neon in the earth's crust with applications to cosmology, *Proc. R. Soc. London Ser. A*, **170**, 451–464.

Reiners, W. A., 1973, A summary of the world carbon cycle and recommendations for critical research, in *Carbon and the Biosphere*, G. M. Woodwell and E. V. Pecan, Eds., U.S. Atomic Energy Commission, Technical Information Center, pp. 368–382.

Robinson, E. and Robbins, R. C., 1968, Sources, abundances and fate of gaseous atmospheric pollutants, Final Report, S.R.I. Project PR-6755, Stanford Research Institute.

Rodhe, H., 1972, A study of the sulfur budget for the atmosphere over northern Europe, *Tellus* **24**, 128–138.

Roedder, E., 1972, Composition of fluid inclusions, in *Data of Geochemistry*, 6th ed., Chapter JJ, M. Fleischer, Ed., U.S. Geological Survey Professional Paper 440-JJ.

Ronov, A. B. and Yaroshevsky, A. A., 1967, Chemical structure of the earth's crust, *Geokhimiya*, 1285–1309.

Ronov, A. B. and Yaroshevsky, A. A., 1976, A new model for the chemical structure of the earth's crust, *Geokhimiya*, 1761–1795.

Rotty, R. M., 1976, Global carbon dioxide production from fossil fuels and cement, A.D. 1950–A.D. 2000, ONR Conference on the Fate of Fossil Fuel Carbon, Honolulu, Hawaii.

Schidlowski, M. and Eichmann, R., 1977, Evolution of the terrestrial oxygen budget, in *Chemical Evolution of the Precambian*, C. Ponnamperuma, Ed., Academic Press, New York.

Schmidt, U., 1974, Molecular hydrogen in the atmosphere, *Tellus* **26**, 78–90.

Seiler, W., 1974, The cycle of atmospheric CO, *Tellus* **26**, 116–135.

Sheldon, W. R. and Kern, J. W., 1972, Atmospheric helium and geomagnetic field reversals, *J. Geophys. Res.* **77**, 6194–6201.

Siegenthaler, U. and Oeschger, H., 1978, Predicting future atmospheric carbon dioxide levels, *Science* **199**, 388–395.

Signer, P. and Suess, H. E., 1963, Rare gases in the sun, in the atmosphere and in meteorites, in *Earth Science and Meteoritics*, Chap. 13, J. Geiss and E. D. Goldberg, Eds., Interscience, New York.

Sigvaldason, G. E. and Elisson, G., 1968, Collection and analysis of volcanic gases at Surtsey, Iceland, *Geochim. Cosmochim. Acta* **32**, 797–805.

Sillén, L. G., 1967, The ocean as a chemical system, *Science* **156**, 1189–1197.

Steinthórssen, S., 1974, The oxide mineralogy, initial oxidation state, and deuteric alteration in some Precambrian diabase dike swarms in Canada, Ph.D. Thesis, Princeton University.

Stevenson, F. J., 1962, Chemical state of the nitrogen in rocks, *Geochim. Cosmochim. Acta* **26**, 797–809.

Strominger, D., Hollander, J. M., and Seaborg, G. T., 1958, Table of isotopes, *Rev. Mod. Phys.* **30**, 585–904.

Suess, H. E., 1966, Some chemical aspects of the evolution of the terrestial atmosphere, *Tellus* **18**, 207–211.

Tebbens, B. D., 1968, Gaseous pollutants in the air, in *Air Pollution*, 2nd ed., Vol. I, Chap. 2, A. C. Stern, Ed., Academic Press, New York.

Tolstikhin, I. N., 1975, Helium isotopes in the earth's interior and in the atmosphere: a degassing model of the earth, *Earth Planet. Sci. Lett.* **26**, 88–96.

Tolstikhin, I. N., Mamyrin, B. A., and Khabarin, L. V., 1972, Anomalous isotopic composition of helium in some xenolithes, *Geokhimiya*, 629–631.

Trask, P. D. and Patnode, H. W., 1942, *Source Beds of Petroleum*, The American Association of Petroleum Geologists, Tulsa, Oklahoma, 566 pp.

Turekian, K. K., 1964, Degassing of argon and helium from the earth, in *The Origin and Evolution of Atmospheres and Oceans*, P. J. Brancazio and A. G. W. Cameron, Eds., Wiley, New York, pp. 74–82.

Turekian, K. K. and Wedepohl, K. H., 1961, Distribution of the elements in some major units of the earth's crust, *Bull. Geol. Soc. Am.* **72**, 175–192.

Urey, H. C., 1952, *The Planets, Their Origin and Development*, Yale University Press, New Haven, 245 pp.

Urey, H. C., 1956, Regarding the early history of the earth's atmosphere, *Bull. Geol. Soc. Am.* **67**, 1125–1128.

Urey, H. C., 1959, The atmospheres of the Planets, in *Handbuch der Physik*, Vol. 52, S. Függe, Ed., Springer-Verlag, Berlin, pp. 363–418.

Vaccaro, R. F., 1965, Inorganic nitrogen in sea water, in *Chemical Oceanography*, Vol. I, Chap. 9, J. P. Riley and G. Skirrow, Ed., Academic Press, New York.

Von Weizsäcker, C. F., 1937, The possibility of a dual β-decomposition in potassium, *Physik. Z.* **38**, 623–24.

Wasserburg, G. J., Mazor, E., and Zartman, R. E., 1963, Isotopic and chemical composition of some terrestrial natural gases, in *Earth Science and Meteoritics*, Chap. 12, J. Geiss and E. D. Goldberg, Eds., Interscience, New York.

Weber, E., 1970, Contribution to the residence time of sulfur dioxide in a polluted atmosphere, *J. Geophys. Res.* **75**, 2909–2914.

Whittaker, R. H. and Likens, G. E., 1973, Carbon in the biota, in *Carbon and the Biosphere*, G. M. Woodwell and E. V. Pecan, Eds., U.S. Atomic Energy Commission, Technical Information Center, pp. 281–302.

Wilkniss, P. E., Swinnerton, J. W., Bressan, D. J., Lamontagne, R. A. and Larson, R. E., 1975, CO, CCl_4, Freon-11, CH_4 and Rn-222 concentrations at low altitude over the Arctic Ocean in January 1974, *J. Atmos. Sci.* **32**, 158–162.

Wofsy, S. C., 1976, Interactions of CH_4 and CO in the earth's atmosphere, *Annual Review of Earth and Planetary Sciences*, Vol. 4, F. A. Donath, F. G. Stehli, G. W. Wetherill, Ed., Palo Alto, California, pp. 441–469.

Wofsy, S. C., McConnell, J. C., and McElroy, M. B., 1972, Atmospheric CH_4, CO, and CO_2, *J. Geophys. Res.* **77**, 4477–4493.

Wofsy, S. C., McElroy, M. B., and Sze, N. D., 1975, Freon consumption: implications for atmospheric ozone, *Science* **187**, 535–537.

Woodwell, G. W. and Houghton, R. A., 1977, in *Dahleem Workshop on Global Chemical Cycles and Their Alteration by Man*, W. Stumm, Ed., (Abakon, Berlin), pp. 61–72.

Woodwell G. M., Whittaker, R. H., Reiners, W. A., Likens G. E., Delwiche, C. C., and Botkin D. B., 1978, The biota and the world carbon budget, *Science* **199**, 141–146.

Yoshinari, T., 1976, Nitrous oxide in the sea, *Marine Chem.* **4**, 189–202.

NAME INDEX

331

SUBJECT INDEX